5

Plant Cell Monographs

Series Editor: David G. Robinson

AF212618

Available online at
SpringerLink.com

Plant Cell Monographs

Recently Published Titles

The Expanding Cell
Volume Editors: Verbelen, J.-P., Vissenberg, K.
Vol. 5, 2007

The Plant Endoplasmic Reticulum
Volume Editor: Robinson, D. G.
Vol. 4, 2006

The Pollen Tube
A Cellular and Molecular Perspective
Volume Editor: Malhó, R.
Vol. 3, 2006

Somatic Embryogenesis
Volume Editors: Mujib, A., Šamaj, J.
Vol. 2, 2006

Plant Endocytosis
Volume Editors:
Šamaj, J., Baluška, F., Menzel, D.
Vol. 1, 2005

The Expanding Cell

Volume Editors: Jean-Pierre Verbelen, Kris Vissenberg

With 53 Figures and 6 Tables

Springer

Volume Editors:

Professor Dr. Jean-Pierre Verbelen
Dr. Kris Vissenberg

Department of Biology
University of Antwerp CGB
Groenenborgerlan 171
B-2020 Antwerpen
Belgium

Series Editor:

Professor Dr. David G. Robinson
Ruprecht-Karls-University of Heidelberg
Heidelberger Institute for Plant Sciences (HIP)
Department Cell Biology
Im Neuenheimer Feld 230
D-69120 Heidelberg
Germany

Cover picture is courtesy of Mr. J. Van Orden

ISSN 1861-1370
ISBN 978-3-642-07255-0
e-ISBN 978-3-540-39116-6
DOI 10.1007/11536338

This work is subject to copyright. All rights are reserved, whether the whole or part of the material is concerned, specifically the rights of translation, reprinting, reuse of illustrations, recitation, broadcasting, reproduction on microfilm or in any other way, and storage in data banks. Duplication of this publication or parts thereof is permitted only under the provisions of the German Copyright Law of September 9, 1965, in its current version, and permission for use must always be obtained from Springer. Violations are liable for prosecution under the German Copyright Law.

Springer is a part of Springer Science+Business Media

springer.com

© Springer-Verlag Berlin Heidelberg 2010

The use of registered names, trademarks, etc. in this publication does not imply, even in the absence of a specific statement, that such names are exempt from the relevant protective laws and regulations and therefore free for general use.

Editor: Dr. Christina Eckey, Heidelberg
Desk Editor: Anette Lindqvist, Heidelberg
Cover design: WMXDesign GmbH, Heidelberg

Printed on acid-free paper 149/3100 YL – 5 4 3 2 1 0

Editors

Jean-Pierre Verbelen studied botany (M.Sc., 1969) at the University of Gent and obtained his Ph.D. in 1976 at the University of Antwerp, where he is currently a professor in the Department of Biology. His major research interests are plant development, cell expansion and cytoskeleton involvement.

Kris Vissenberg studied biology (M.Sc., 1997) and obtained his Ph.D. at the University of Antwerp in 2001. Currently he is a post-doc at the Fund for Scientific Research – Flanders (FWO-Vlaanderen). His major research interests are cell expansion, cell wall metabolism and general plant development.

Preface

Plant cells go through distinct phases of development. They are generally formed in meristems and undergo expansion before differentiating and acquiring their final functions. In this sequence, cell expansion is the process that mainly contributes to the cell's size and hence to the plant's size and morphology. Furthermore, it allows the plant to rapidly adapt to changes in the environment and to respond to several hormone signals.

Cell expansion primarily occurs by the uptake of water in the cytoplasm and vacuole of the plant cell. This process is driven by osmotic forces generated by accumulation of solutes by several classes of (transporter) proteins. This causes the vacuole to expand and to exert a pressure against the cell wall. In order to enlarge, the cell wall has to yield to the stress imposed by the turgor pressure. Several families of proteins and enzymes, as well as the composition and architecture of the cell wall itself, render the wall stiff and tough but at the same time modifiable for a drastic increase in surface area. The direction of cell expansion is hereby governed by cytoskeletal elements in the cytoplasm as well as by load-bearing elements in the cell wall. The process of cell expansion is thus a complex process brought about by activities at different levels in both the symplast and apoplast.

This book addresses the involvement of the different actors in plant cell expansion and its control by integrating the up-to-date views of cell biologists, biochemists, physiologists, molecular biologists, biophysicists and microscopists. The combination of these different views, resulting from different experimental techniques and methodologies (explained in distinct boxes), gives a timely summary on what is currently known and believed to occur during the cell expansion process.

August 2006 Jean-Pierre Verbelen & Kris Vissenberg

Contents

Plant Cell Monogr (5)
J.-P. Verbelen · K. Vissenberg: The Expanding Cell
DOI 10.1007/7089_2006_068/Published online: 20 October 2006
© Springer-Verlag Berlin Heidelberg 2006

Cell Expansion: Past, Present and Perspectives

Jean-Pierre Verbelen (✉) · Kris Vissenberg

Biology Dept., Plant Physiology and Morphology, University of Antwerp,
Groenenborgerlaan 171, 2020 Antwerp, Belgium
jean-pierre.verbelen@ua.ac.be

Plant size and organ size are dependent both on cell division and cell expansion (Lyndon 1990). Cell division is the process whereby one cell divides into two daughter cells; expansion is the growth in volume beyond the size of the mother cell before mitosis.

Both cell division and cell expansion were correctly defined in the 19th century on the basis of careful microscopic observations. Wilhelm Hofmeister (1867) demonstrated that the nucleus of a mother cell divides and that one half of the contents of the mother cell collects around each of the two daughter nuclei when a new cell wall forms between the daughter nuclei. Julius Sachs (1882) on the other hand clearly depicted the changes in appearance of parenchyma cells during cell expansion in a growing root, with reference to the volume increase of the central vacuole. He further emphasized cell turgor and water uptake as instrumental in causing expansion. He also pointed to the fact that during expansion the existing cell wall was stretched and thinned, but that new material was added keeping wall thickness rather constant (Fig. 1).

In plant organs the peak activities of both events are separated in time or space, a fact also known since the 19th century, as elegantly described and depicted by Sachs (1874). His figures of growing seedling roots gained an immediate popularity and were copied in Strasburger's famous handbook (Strasburger et al. 1894). They remained there as reference illustrations at least up to the 30th edition, published in 1971 (Strasburger et al. 1971).

Since that period of fundamental discoveries the process of mitosis and cytokinesis has been explored intensively and, during the last decades especially, the picture of both aspects has become extremely detailed. It has turned out that the mitotic machinery and its control do resemble that of animal systems but that they are plant-specific and very elaborate (for reviews see Dewitte et al. 2003; De Veylder et al. 2003). The formation of the cell plate, the new cell wall separating the newly formed daughter cells, turns out to be a highly complex cellular activity implying a precise orchestration of cytoskeleton activity, and synthesis and transport of wall components (Otegui and Staehelin 2000a,b; Otegui et al. 2001, and references therein).

Cell expansion has no equivalent in animal systems and progress in the understanding of the process was slow. As mentioned above, from the be-

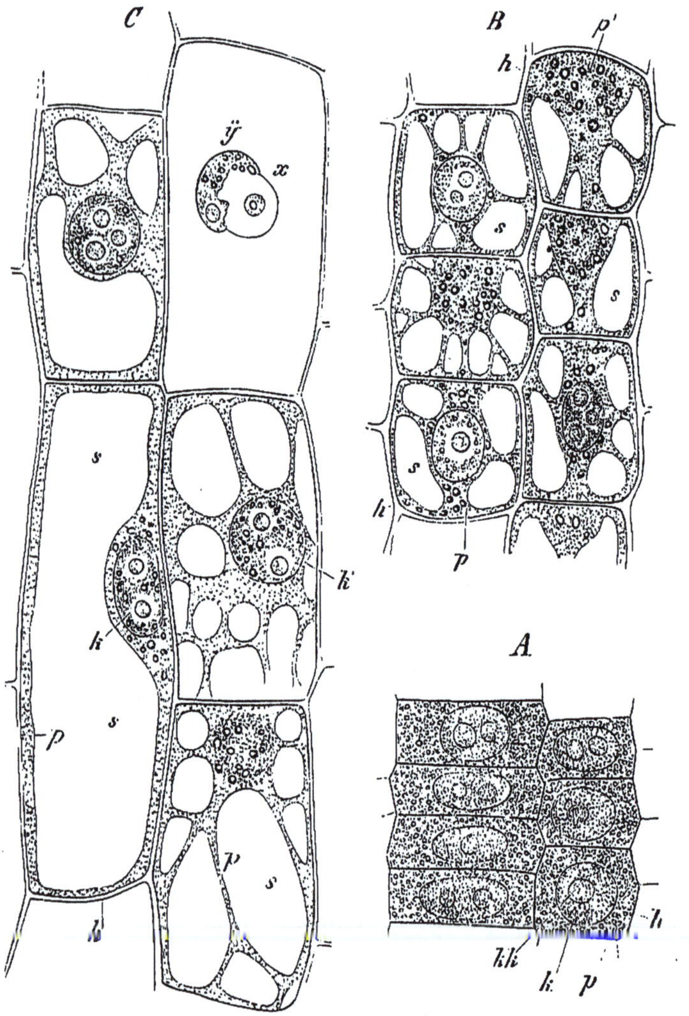

Fig. 352. Parenchymzellen aus der mittleren Schicht der Wurzelrinde von Fritillaria imperialis; Längsschnitte, nach 550maliger Vergrößerung. *A* dicht über der Wurzelspitze liegende, sehr junge Zellen, noch ohne Zellsaft; *B* die gleichnamigen Zellen etwa 2 Millimeter über der Wurzelspitze, der Zellsaft *s* bildet im Protoplasma *p* einzelne Tropfen, zwischen denen Protoplasmawände liegen; *C* die gleichnamigen Zellen etwa 7—8 Millimeter über der Wurzelspitze; die beiden Zellen rechts unten sind von der Vorderfläche gesehen, die große Zelle links unten im optischen Durchschnitt gesehen; die Zelle rechts oben durch den Schnitt geöffnet; der Zellkern lässt unter dem Einfluss des eindringenden Wassers eine eigenthümliche Quellungserscheinung wahrnehmen (*x y*).

Fig. 1 Parenchyma cells from the cortex of the root of *Fritillaria imperialis* in a longitudinal section of fresh material. *A* cells immediately above the root tip without vacuoles. *B* cells about 2 mm above the root tip with small developing vacuoles. *C* cells 7–8 mm away from the root tip with large vacuoles

ginning botanists knew that during cell expansion it was mainly the vacuole that grew considerably in volume and also that the existing cell wall became thinner as it was stretched but "reinforced" by addition of new wall material. A crucial step for the understanding of the physiology behind expansion was made by the discovery that auxin affects elongation and its control (Went and Thimann 1937). Most of the research, however, only refined the existing descriptive knowledge (Avery and Burkholder 1936; Erickson and Sax 1956). Interest within the scientific community was indeed very moderate, as witnessed by the limited attention to cell expansion in notorious handbooks (Esau 1960; Clowes and Juniper 1968; Wareing and Philips 1973; Fahn 1974; Bidwell 1979).

A reliable view on the state of the art in the early 1960s can be found in the *Encyclopedia of Plant Physiology, vol XIV* on growth and growth substances (Ruhland 1961). It clearly depicts the nascent interest in the process of cell expansion. Cell expansion receives little attention in the anatomy chapter (one sentence) but is treated in detail in the chapters "Cell expansion and metabolism (Ziegler H)", "Physics of cell elongation (Burström H)" and "The growth of the cell wall (Preston RD)". These chapters contain detailed information on in vitro extensibility of cell wall preparations and on changes in cell wall composition (cellulose, hemicelluloses, pectin and proteins) in elongating coleoptiles and hypocotyls.

Around that time, the attempts to understand cell expansion shifted into a new gear. On the theoretical side, Lockhart (1965) summarized a lot of experimental data on wall extensibility in a formula that was readily comprehensible for the whole scientific community and that continued life as the "Lockhart equation":

$$r = \Phi(P - Y)$$

where r is growth rate, Φ is extensibility of the cell wall, P is turgor pressure (i.e. the source of cell wall stress) and Y is yield threshold (i.e. the minimum pressure required for growth).

This simple equation clearly states that the rate of cell expansion is a product of the imbalance between turgor pressure and the mechanical properties of the cell wall, emphasizing that the principal players are thus to be found in the symplast as well as in the apoplast.

Since then, detailed data were gathered on the composition and the interaction of the primary cell wall and its then-known components: cellulose, hemicelluloses, pectins and proteins. Cellulose was found to be synthesized by cellulose synthases (Arioli et al. 1998) that are organized in cellulose synthase complexes (Kimura et al. 1999). Fluorescent labelling of these rosettes pointed to the role of the cytoskeleton in the orientation of the cellulose microfibrils in the wall (Paredez et al. 2006). The acid growth theory was substantiated by the discovery of expansins (McQueen-Mason et al. 1992), while many other proteins and processes with putative roles in cell wall loosening were described (Cosgrove 2005). Mechanisms emerged that counteract the loosening of the cell wall and so arrest cell expansion (Cooper and Varner 1984). Aquaporins were described as universal facilitators of water transport through vacuolar membranes (Crispeels and Maurel 1994). The mode of action of auxins and of the other plant growth regulators became much clearer (e.g. Weijers and Jurgens 2004). These are the scene and the actors that make the content of this volume. Most of the recently published reviews focus on or are limited to the cell wall. As stated above, the Lockhart equation indicates that both apoplastic and symplastic players are involved in cell elongation. This volume therefore combines actual state-of-the-art papers on the different aspects of the cell's biology involved in expansion and its control. Nuclear ploidy is often related to cell expansion (Nagl 1979). As this is only the case in about half of the plant species, endoreduplication does not seem fundamental for expansion. It will therefore not be treated. It also needs to be mentioned that cell expansion includes diffuse expansion (in most cells) and tip growth (in certain specific cells). The latter method of cell growth will not be treated as it has been covered by Rui Malhó in another volume of this series (Malhó 2006).

During cell expansion, the cell wall clearly is a centre of activity. Up to now, however, an adequate model of the cell wall structure and how this structure permits both an increase in surface and the incorporation of new wall material still remains elusive. Using wall microscopy, selective extraction of components followed by structural analysis and in situ spectroscopic approaches, several artificial models have been proposed. Cosgrove (2000) mentions and discusses three models that differ only in the types of interaction and spacing of the different components. These specific associations and locations of the components need to be further elaborated to fully understand the mechanism of cell wall enlargement.

At the onset, during, and at the end of cell expansion, undoubtedly different sets of genes and proteins are expressed and active/inactive. Several of these crucial genes and proteins are starting to emerge, but the complete picture is far from clear. The combination of the knowledge on the three-dimensional architecture (e.g. of the *Arabidopsis* root, which is well-described) and cell type-specific expression profiling as performed by Birnbaum et al. (2005) could eventually provide the complete transcriptome of single cells in the root apex. This information could then provide all of the changes in gene expression that occur when a cell switches from a meristematic to an expanding cell or when a cell responds to environmental and hormonal stimuli. Similar proteomic approaches could give complementary information on protein involvement in the cell's crucial developmental processes and switches.

References

Arioli T, Peng L, Betzner AS, Burn J, Wittke W, Herth W, Camilleri C, Höfte H, Plazinski J, Birch R, Cork A, Glover J, Redmond J, Williamson RE (1998) Molecular analysis of cellulose biosynthesis in *Arabidopsis*. Science 279:717–720

Avery GS, Burkholder PR (1936) Polarized growth and cell studies on the Avena coleoptile, phytohormone test object. Bull Torrey Bot Club 63:1–15

Bidwell RGS (1979) Plant physiology. 2nd edn. MacMillan, New York

Birnbaum K, Jung JW, Wang JY, Lambert GM, Hirst JA, Galbraith DW, Benfey PN (2005) Cell type-specific expression profiling in plants via cell sorting of protoplasts from fluorescent reporter lines. Nat Methods 2:615–619

Clowes FAL, Juniper BE (1968) In: Burnett JH (ed) Plant cells. Botanical Monographs. Blackwell, Oxford

Cooper JB, Varner JE (1984) Cross-linking of soluble extensin in isolated cell walls. Plant Physiol 76:414–417

Cosgrove DJ (2000) Expansive growth of plant cell walls. Plant Physiol Biochem 38:109–124

Cosgrove DJ (2005) Growth of the plant cell wall. Nature Rev Mol Cell Biol 6:850–861

Crispeels MJ, Maurel C (1994) Aquaporins : the molecular basis of facilitated water movement through living plant cells. Plant Physiol 105:9–13

De Veylder L, Joubes J, Inzé D (2003) Plant cell cycle transitions. Curr Opin Plant Biol 6:536–543

Dewitte W, Murray JAH (2003) The plant cell cycle. Ann Rev Plant Biol 54:235–264

Erickson RO, Sax KB (1956) Elemental growth rate of the primary root of Zea mays. Proc Amer Philos Soc 100:487–498

Esau K (1960) Plant anatomy. 2nd edn. Wiley, New York

Fahn A (1974) Plant anatomy. 2nd edn. Pergamon, Oxford

Fry SC (2000) The growing plant cell wall. Blackburn, UK

Hofmeister W (1867) Die Lehre von der Pflanzenzelle. W Engelman, Leipzig

Kimura S, Laosinchai W, Itoh T, Cui X, Linder CR, Brown RM Jr (1999) Immunogold labeling of rosette terminal cellulose-synthesizing complexes in the vascular plant *Vigna angularis*. Plant Cell 11:2075–2085

Lockhart JA (1965) An analysis of irreversible plant cell elongation. J Theoret Biol 8:264–275

Lyndon RF (1990) In: Black M, Chapman J (eds) Plant development: the cellular basis. Topics in plant physiology, vol 3. Unwin Hyman, UK

Malhó R (ed) (2006) The pollen tube: a cellular and molecular perspective. Plant Cell Monographs, vol 3. Springer, Berlin Heidelberg, New York

McQueen-Mason S, Durachko DM, Cosgrove DJ (1992) Two endogenous proteins that induce cell wall expansion in plants. Plant Cell 4:1425–1433

Nagl W (1979) Differential DNA replication in plants: a critical review. Zeitschr Pflanzenphys 85:283–314

Otegui M, Staehelin LA (2000a) Cytokinesis in flowering plants: more than one way to divide a cell. Curr Opin Plant Biol 3:493–502

Otegui M, Staehelin LA (2000b) Syncytial-type cell plates: a novel kind of cell plate involved in endosperm cellularization of Arabidopsis. Plant Cell 12:933–947

Otegui MS, Mastronarde DN, Kang B-H, Bednarek SY, Staehelin LA (2001) Three-dimensional analysis of syncytial-type cell plates during endosperm cellularization visualized by high resolution electron tomography. Plant Cell 13:2033–2052

Paredez AR, Somerville CR, Ehrhardt DW (2006) Visualization of cellulose synthase demonstrates functional association with microtubules. Science 312:1491–1495

Ruhland W (1961) Growth and growth substances. Encyclopedia of plant physiology, vol 14. Springer, Berlin Heidelberg New York
Sachs J (1874) Ueber das Wachstum der Haupt- und Nebenwurzeln. Arb D Bot Inst Wurzb 1:385–474
Sachs J (1882) Vorlesungen uber Pflanzen – Physiologie: XXXIII Vorlesung. Engelman W, Leipzig
Strasburger E, Noll F, Schenck H, Schimper AFW (1894) Lehrbuch der Botanik für Hochschulen. 1st edn. Gustav Fischer, Jena
Strasburger E, Noll F, Schenck H, Schimper AFW (1971) Lehrbuch der Botanik für Hochschulen, 30th edn. Gustav Fischer, Stuttgart
Wareing PF, Philips IDJ (1973) The control of growth and differentiation in plants. Pergamon, Oxford
Weijers D, Jurgens G (2004) Funneling auxin action: specificity in signal transduction. Curr Opin Plant Biol 7:687–693
Went FW, Thimann KV (1937) Phytohormones. Macmillan, New-York

Plant Cell Monogr (5)
J.-P. Verbelen · K. Vissenberg: The Expanding Cell
DOI 10.1007/7089_2006_069/Published online: 7 October 2006
© Springer-Verlag Berlin Heidelberg 2006

Solute and Water Relations of Growing Plant Cells

Wieland Fricke[1,2] · François Chaumont[3] (✉)

[1]Division of Biological Sciences, University of Paisley, Paisley PA1 2BE, UK

[2]School of Biology and Evironmental Science, Science Centre West,
University College Dublin (UCD), Belfield, Dublin 4, Ireland

[3]Unité de Biochimie Physiologique, Institut des Sciences de la Vie,
Université catholique de Louvain, Croix du Sud 5-15, B-1348 Louvain-la-Neuve, Belgium
chaumont@fysa.ucl.ac.be

Abstract Cell expansion requires the continuous uptake of water into cells, which in turn is driven through osmotic forces generated by accumulation of solutes. Herein, we assess the significance of water and solute transport across cell membranes as a rate-limiting step during cell expansion. Two membranes are considered, the tonoplast, which separates the largest intracellular storage compartment (vacuole) from the portion of the protoplasts where most enzymatic reactions take place (cytoplasm), and the plasma membrane, which constitutes the site of exchange between protoplasts and apoplast (cell wall). Most of the solutes that generate the bulk of osmolality are heterogeneously distributed between cells, tissues and cell compartments, and this heterogeneity must be taken into consideration in studies on growth. Because of differences in transmembrane potential at the plasma membrane (significantly negative) and tonoplast (close to zero), ion channels and transporters are likely to make different contributions to solute transport across these two membranes. The osmotic permeability of the tonoplast exceeds that of the plasma membrane by a factor of 100. This aids cell-internal osmotic equilibration and renders the plasma membrane rate-limiting for water uptake into cells or transcellular water transport. Candidate aquaporins, ion channels and transporters which could mediate solute and water transport specifically into growing cells are reviewed in this work.

1 Introduction

1.1 Growth

Growth requires the co-ordination of many processes and has to be adjusted to environmental constraints. Since growth is a prerequisite for any organism to reach its full potential, it is intricately linked to development. The definition of growth varies, depending on which variables are used as the reference system, yet there are some aspects of growth which do not vary: (i) growth is irreversible; (ii) growth of multicellular organisms is due to expansion of (some of) its individual units—cells; and (iii) as cells grow, they pass through well-defined developmental stages. In addition, in multi-cellular organisms, growth is often restricted to specialized regions, "growth zones". Although growth requires the production of new cells and expansion of these cells, it is the latter that is responsible for the bulk of size increase, particularly in plants, which have, on average, larger cells compared to animals.

The above has several implications for the study of growth. A particular process must be unidirectional, when integrated over the entire growth period. For example, water moves in and out of a growing cell, but at the end of the growth period, cell water content and wall extension have increased irreversibly. Growth must also be studied at the level of the individual cell, since turgor pressure, the mechanical force driving wall extension, is defined at cell level. A range of tissues has to be analyzed since organ expansion requires the coordinated expansion of cells of different tissues. A cell goes through various developmental stages as it passes through the growth zone and it is likely that the molecular cause of growth limitation changes with development. For example, as cells elongate and mature, the cellular ratio of vacuole:cytosol changes. The two compartments differ in solute relations and this impacts on solute requirements of cells. Since cells are growing in specialized regions in which they act as sinks for energy, water, carbon and solutes, their demands have to be met—possibly in competition with other growing regions—by those regions which provide these resources. The question is not so much "what limits growth?" but "which factor limits growth at a particular developmental stage of a cell?", "which particular tissue or which cell type within a tissue limits expansion of the organ (Peters and Tomos 1996)?" and "where does the limitation originate, within the plant or externally?" If we are to manipulate the growth of cells and yield of plants, we need to identify molecular targets.

1.2
Walls, Water and Solutes

From the biophysical point of view, the three main factors potentially limiting cell expansion are cell wall, water and solutes (Fricke 2002a). Wall mass per cell increases during expansion and cell wall polymers must give in—"yield"—to cell turgor pressure. Water must flow into cells to increase volume and maintain turgor, while solutes are needed to generate the osmotic force driving water uptake into cells. Since water cannot be pumped actively into cells, the only way to increase cell water content is by generating a downhill gradient in chemical potential of water, i.e. the water potential, through osmotic forces across membranes.

The idea that cell wall properties differ between growing cells (yielding walls) and non-growing cells (non-yielding walls) or that environmental stressors affect growth through alteration of wall properties has been supported by numerous studies (e.g. Cramer 1992; for review, see Cosgrove 1993;

Hsiao and Xu 2000). More recently, these changes have been related to specific proteins (expansins, McQueen-Mason et al. 2006, in this volume), enzymes (xyloglucan endotransglycosylase/hydrolase (XTH, Nishitani and Vissenberg 2006, in this volume); peroxidase) and wall components (Cosgrove et al. 2002; de Souza and MacAdam 1998; Fry 1998; Huang et al. 2000; Reidy et al. 2001; Ruan et al. 2001; Thompson et al. 1997; Schunmann et al. 1997; Palmer and Davies 1996; Rose et al. 2002; Yokoyama et al. 2004). It is not known how, at the molecular level, changes in wall properties are modulated or wall-modifying proteins are regulated (De Cnodder et al. 2007, in this volume). In contrast, although water and solutes have received less attention, more is known about the regulation at the molecular level of candidate transporters and channels (Chaumont et al. 2005; Cherel 2004; Luu and Maurel 2005; Tornroth-Horsefield et al. 2006; Tournaire-Roux et al. 2003; Very and Sentenac 2002). This provides an ideal basis for identifying molecular mechanisms through which cell expansion is controlled.

Aquaporins are channels facilitating the movement of water and/or small neutral solutes across biological membranes. They have been found in a wide range of organisms and account for a considerable portion of membrane protein (Chaumont et al. 2005; Johansson et al. 2000; Maurel et al. 2002; Schaffner 1998). Classification of aquaporins occurs according to their subcellular location (TIPs, tonoplast intrinsic proteins, PIPs, plasma membrane intrinsic proteins), their organ of discovery (NIPs, nodule, or NOD26-like intrinsic proteins) or their molecular size (SIPs, small basic intrinsic proteins). Most aquaporins transport primarily water and increase the osmotic water permeability coefficient of the membrane (P_f) several-fold when expressed in *Xenopus* oocytes. Other aquaporins are less specific and also transport solutes such as glycerol, urea, boron, hydrogen peroxide or, as recently suggested, ammonia, carbon dioxide and silicon (Biela et al. 1999; Dordas et al. 2000; Gerbeau et al. 1999; Hanba et al. 2004; Henzler and Steudle 2000; Loque et al. 2005; Ma et al. 2006; Uehlein et al. 2003). It has also been suggested that aquaporins fulfil less of a transport role but can function either as osmo- and turgor sensors (Hill et al. 2004) or as markers for targeting vesicles to the central vacuole (Ma et al. 2004).

Before the discovery of water channels, it was assumed that cells have little control over the regulation of diffusional flow of water through membranes. Since then, several studies have shown that facilitated movement of water through aquaporins accounts for most of the transmembrane flow of water in plant cells. Whereas membrane water diffusion is characterized by equal osmotic water permeability (P_f) and diffusional water permeability (P_d) coefficients, and a high Arrhenius activation energy (E_a), facilitated water transport through pores has a P_f higher than P_d, a low E_a and, in addition, is blocked by mercurial compounds indicative of the proteinaceous nature of the pore (Maurel 1997). According to the composite transport model of water flow through tissues, water moves along three major pathways: along the apoplast (wall space), along the symplast (through plasmodesmata) or through membranes (including passage through aquaporins) (Steudle and Peterson 1998). Notably, the driving force for water movement differs between apoplast (hydrostatic gradients) and transmembrane flow (osmotic gradients); the driving force for water movement through plasmodesmata may be either. The hydraulic conductivity of the apoplastic path is at least one order of magnitude higher than the hydraulic conductivity of the transmembrane path (for review, see Steudle and Peterson 1998).

The likelihood that hydraulic properties of tissues limit growth and that aquaporins are involved very much depends on the main paths along which water moves from a plant internal source (xylem, phloem) to peripheral tissues (e.g. mesophyll, epidermis). For example, in maize roots, the growing tip region is supplied with water mostly from phloem via the symplasmic path, whereas more distal regions are supplied via the transmembrane path (Hukin et al. 2002). Alteration of aquaporin activity in the tip region should have little effect on growth. In contrast, in growing hypocotyl tissue of soybean, a large number of small-volume xylem parenchyma cells exists, through which water has to pass as it moves from (inner-lying) xylem to peripheral tissues. Water has to cross many membranes per distance travel as it passes through xylem parenchyma. This creates a hydraulic bottleneck and significant growth-induced water potential gradients (Boyer 1985; Nonami et al. 1997). In the growing grass leaf, xylem parenchyma or mestome and parenchymatous bundle sheath cells may fulfil similar functions (Boyer and Silk 2004; Fricke 2002a). Over-expression or increase in activity of aquaporins in these tissues should overcome some of the hydraulic limitation in growth—provided growth is limited hydraulically in the first place!

What surprises most about existing work on biophysical limitation of growth is the scarcity of studies on the solute aspect, in particular on solute transport properties specific to growing tissues (for review, see Van Volkenburgh 1999).

In the following sections, we will look in more detail at the potential roles which water and solute transport play during cell elongation. We will focus on two membranes, the tonoplast, separating vacuole from cytoplasm, and the plasma membrane, forming the interface between protoplasm and wall space. Since solute requirements of vacuole and cytosol and of different leaf tissues and cell types differ, it is possible that one particular solute, for example, Ca, limits growth in one cell type but not in another. Therefore, we will briefly review the distribution of solutes between cell compartments, cells and tissues, particularly for leaves. For water, the situation is easier since it is present in the same chemical form in each compartment. However, the number of different aquaporins per plant species is considerable (Chaumont et al. 2001; Johanson et al. 2001; Sakurai et al. 2005), probably reflecting tissue- and cell-specific regulation of water transport through membranes.

2
Solutes

2.1
Solutes—Continuous Deposition During Growth

Theory predicts that as cells expand and cellular contents are diluted, solutes must be deposited (Silk and Erickson 1979) continuously and at high rates to maintain the osmotic force driving water uptake into cells. If solute accumulation did not occur in parallel to cell expansion, cell osmolality and turgor would continue to decrease until close to zero. For example, a grass leaf epidermal cell can elongate to 50 times or more its original size as it passes through the elongation zone (Schnyder et al. 1990). Without solute uptake, an initial osmolality of 300–400 mosmol kg^{-1} (Fricke 2004a) and turgor of around 0.5 MPa (Fricke 2002b) would be "diluted" half-way through the elongation zone to 12–16 mosmol kg^{-1} and 0.02 MPa, respectively. This would be insufficient to expand a wall and drive water uptake in an apoplastic environment which has either a significant tension or solute potential.

There exist few studies on grass leaves, and some on roots, where osmolality in the elongation zone has actually been determined at the level of the cell (see Technical Box) (Pritchard 1996; Fricke 1997, 2002a; Fricke and Peters 2002; Martre et al. 1999). These studies show that cell osmolality changes little along the elongation zone. In maize roots, cells expand in volume by as much as 50% h^{-1} (Pritchard 1994) and this means that cell solute contents must also increase by about 50% h^{-1}—a considerable task for a cell which has a total solute concentration of 250–350 mM. In grass leaves, cells elongate at relative rates as high as 12–20% h^{-1} and have total solute concentrations in the range 300–400 mosmol kg^{-1}. Estimated solute flux rates per cell surface are in the upper region of values for plant cells (although considerably lower than rates for guard cells which require fast movement of solutes across membranes for functioning; reviewed in Fricke and Flowers 1998).

In barley leaves, turgor increases after cells have exited the elongation zone, while osmolality stays the same (Fricke 1997). This suggests that net uptake of solutes by cells is linked to elongation, regardless of whether solute uptake regulates growth or vice versa, whereas turgor remains at a certain level during growth due to continuous yielding of the wall and rises, once wall properties are modified (stiffened).

2.2
Two Major Cellular Solute Compartments: Vacuole and Cytosol

Most solutes entering a growing plant cell are destined to the large central vacuole. When a new cell is produced through division and commences elongation, it contains a number of smaller-sized vacuoles ("vacuon"). As a cell elongates to reach its mature size, a large central vacuole forms by fusion of vacuon in an autophagic process (for review, see Marty 1997). This central vacuole can account for almost 100% of the cell's volume (epidermis; ca. 99%) or "only" about 60% (mesophyll). The other major cellular compartment to which solutes are destined is the cytosol/cytoplasm (for simplicity, we do not distinguish here between these two terms but use "cytosol" throughout), which comprises around 40% of the cell's volume in fully-expanded mesophyll cells but only about 1% of the cell's volume in the epidermis. Cytosol and vacuole are separated by one membrane, the tonoplast. Membranes cannot expand by more than 3% in surface area (Wolfe and Steponkus 1981) and, therefore, vacuole and cytosol must be iso-osmotic, their total osmotically active solute concentrations must match each other (if not, hydrostatic pressure differences would develop which would distort or even rupture the tonoplast). Fortunately—to a plant cell, not to those studying it! —requirements for particular solutes differ between vacuole and cytosol. Cytosolic solute concentrations have to be in tune with specific metabolic demands and maintained within a narrow range, whereas vacuolar solutes fulfil less specific functions and concentrations vary more. Leigh once termed this pointedly "the selective cytosol and the promiscuous vacuole" (Leigh and Wyn Jones 1986).

The best-studied solutes are K, in particular, and those ions (Na, Cl, heavy metals) that are linked to specific environmental stresses (e.g. salinity). Cytosolic K is crucial for many metabolic processes such as protein biosynthesis and enzyme activation. As a result cytoplasmic K concentrations are maintained at 60–80 mM whereas vacuolar concentrations can exceed 300 mM, particularly in the epidermis (Cuin et al. 2003; Fricke et al. 1996; Walker et al. 1996), or decrease close to zero (Fricke et al. 1996). The metabolic function of K in the cytosol cannot be replaced by any other solute, but its function as a major vacuolar osmolyte can be replaced by cations such as Na and Ca (Box and Schachtman 2000).

In leaf epidermal cells of barley, vacuolar Ca concentration exceeds cytosolic Ca concentration by a factor 10^5 to 10^6 (Fricke et al. 1995) and in salinized plants, vacuolar Na and Cl can exceed 500 mM, while cytosolic concentrations are below 100 mM. On the one hand, the vacuole serves as a buffer and exchange site of solutes for the cytosol; on the other hand, the vacuole represents the potentially largest hazard to the cytosol.

Cell elongation affects the cellular ratio of cytosol:vacuole. Vacuolar volume increases manifold, whereas cytosolic volume increases little or may not increase at all (root hairs). Leaf epidermal cells of grasses elongate to 50 times or more their original volume. A cell which commences elongation and has a cytosol : vacuole ratio of 1 : 1, may only increase vacuolar volume and finish with a ratio of 1 : 49 – 99—a ratio observed in mature epidermal cells (1–2% cytosol; 98–99% vacuole). A mature mesophyll cell contains about 40% cytosol. However, mesophyll cells are by factor 10 – 100 smaller than epidermal cells (grasses), and the increase in total amount of cytosol per cell during elongation will be small. If it was not for the conflicting demands on solutes of vacuole and cytosol, a growing cell may not have to take up any solute for the cytosol. The real challenge for maintaining solute homeostasis in the cytosol during cell expansion is the vacuole. Being an infinitely larger osmotic sink which requires and stores solutes as it expands, the vacuole drains solutes from the cytosol and threatens to flood the cytosol with those solutes that are stored at much higher concentrations.

2.3 Vacuolar Solutes: Few and Heterogeneous

Plant cells accumulate a range of solutes in the vacuole to generate osmolality. Different tissues and different cell types within one tissue accumulate different solutes (Fricke et al. 1994b; Karley et al. 2000a; Leigh and Storey 1993; Leigh and Tomos 1993; Volkov et al. 2004). The best-studied example is the mature grass leaf and the distribution of vacuolar solutes between the two main tissues, epidermis and mesophyll (for information about bundle sheath, see Koroleva et al. 1997). Potassium, nitrate and Na are present at similar concentrations in the epidermis and mesophyll. Depending on plant nutrition, this distribution can change (Fricke et al. 1996). Calcium, at osmotically significant concentrations (> 5–10 mM), is found almost exclusively in the epidermis and is absent from the mesophyll; P distributes the opposite (Fricke et al. 1994a). Chloride concentrations are higher in the epidermis and increase in this tissue in particular in response to salinity (Fricke et al. 1996). Notably, the tissue distribution of vacuolar P and Ca between epidermis and mesophyll is opposite in leguminous species, possibly in dicotyledonous species in general (for a review, see Leigh and Tomos 1993; for the distribution in *Arabidopsis*, see Volkov et al. 2004).

The available data suggest that the epidermal vacuole uses predominantly inorganic ions (K, Na, Cl, Ca, nitrate) for the generation of osmolality, whereas the mesophyll vacuole uses a mixture of both, inorganic and organic (sugars, amino acids) solutes. There are some exceptions concerning absence of organic solutes from epidermal vacuoles. Within the barley leaf epidermis, cells closest to stomatal pores can accumulate large (> 100 mM) concentrations of malic acid. Within the epidermis of *Thellungiella*, a halophytic close relative of *Arabidopsis*, almost 400 mM of S accumulates in epidermal vacuoles (as determined by energy-dispersive X-ray analysis). This represents most likely 200 mM of glucosinolates.

Vacuolar solute concentrations can differ also between cells within one tissue, in a non-random way (Fig. 1). In barley, solutes are distributed unevenly between the adaxial (upper) and abaxial (lower) epidermal layer and between different cell types, particularly within the anatomically more complex adaxial epidermis (Fricke et al. 1994c, 1995).

Fig. 1 **A–C** Compartmentation of solutes between leaf tissues. Differences in solute content or concentration are symbolized by different *shades of grey*. **A** Possible distribution patterns between the two major leaf tissues, epidermis and mesophyll: **a** Solutes are distributed differentially between mesophyll and epidermis. In the epidermis, solutes distribute between cells according to a reduced pattern; in the mesophyll, solutes distribute evenly. **b** Solutes distribute evenly between leaf tissues and there exist no differences in solute concentrations. **c** Solute distribution within the epidermis follows a complex pattern; similarly, within the mesophyll, solute concentrations differ between cells bordering the adaxial or abaxial epidermis or bordering only other mesophyll cells. **d** as in **a**, except that upper (adaxial) and lower (abaxial) epidermis differ in solute concentrations (see also **B**). **e** There are no systematic differences in solute concentrations between tissues and cells but solutes distribute randomly. **B** Nitrate and chloride distribute opposite between the adaxial and abaxial epidermal layer in fully expanded leaves of barley (Fricke et al. 1995). In the adaxial epidermis, nitrate concentrations increase with time (days) and level off at 200–250 mM, while Cl concentrations remain below 100 mM; in the abaxial epidermis, the distribution is opposite. The micrograph shows a cross-section of a barley leaf. **C** Solute concentrations differ in a systematic way between cells of the upper (adaxial) epidermis of barley leaves. The distribution appears to be related to the proximity of cells to stomatal pores or the top of ridges. The micrograph shows a surface view of a double-leaf replica of the upper epidermis of a mature barley leaf ▶

The above data were obtained for mature, transpiring leaf tissue. Growing leaf tissues compartmentalize solutes in a similar way. The main difference is an inability of growing tissue to accumulate large (> 50 mM) concentrations of Ca in epidermal cells (Fricke 2004b). In response to salt, both growing and non-growing tissues of the developing leaf three of barley accumulate Cl and Na, and loose K, but their K : Na ratios are affected differently (Fricke 2004a). Sugars, which are negligible in the epidermis, contribute less than 20% to bulk osmolality in the leaf elongation zone, but may have an important role during osmotic adjustment to salinity and drought in the mesophyll and bundle sheath (Barlow 1986; Delane et al. 1982; Hu and Schmidhalter 1998). When salt is added to growth media and growth stops transiently (for 20–30 minutes), solutes start to accumulate in a strict base-to-tip pattern, aiding first osmotic water uptake and growth in the basal leaf elongation zone before allowing mature tissue to recover turgor and adjust osmotically to the low-water potential environment (Fricke et al. 1994b).

We do not know the molecular mechanisms through which a differential accumulation of solutes between tissues (epidermis versus mesophyll; growing versus mature tissues) is achieved. There exist two principle mechanisms: differential supply of solutes to tissues; or differential transport properties of tissues, at the plasma membrane or tonoplast (for a review, see Karley et al. 2000b; Leigh and Tomos 1993). In the case of Ca, the absence of a transpiration stream that passes radially through the grass leaf elongation zone (which is enclosed in sheaths of older leaves) may explain the abundance of epidermal Ca in emerged compared to growing tissue: Ca is carried with the transpiration stream to the site of evaporation, where it is left behind (see also

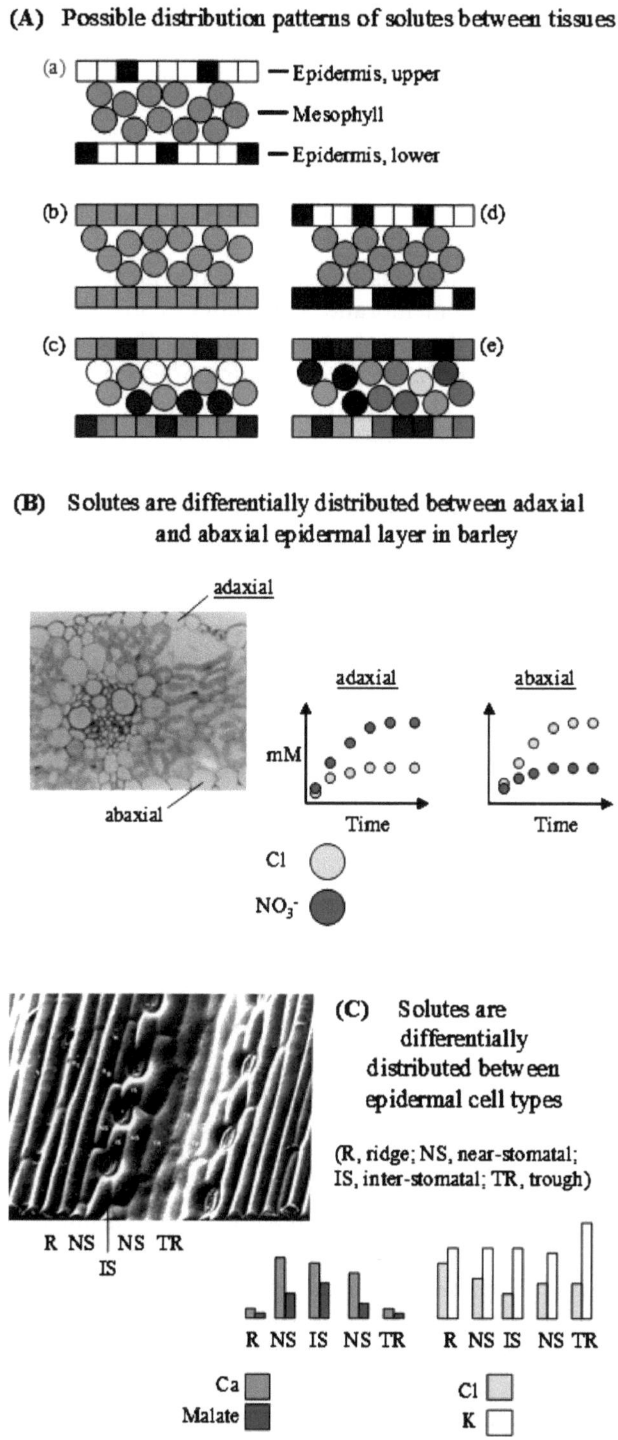
(A) Possible distribution patterns of solutes between tissues
(a)
Epidermis, upper
Mesophyll
Epidermis, lower
(b)
(d)
(c)
(e)
(B) Solutes are differentially distributed between adaxial and abaxial epidermal layer in barley
adaxial
abaxial
adaxial
abaxial
mM
Time
Time
Cl
NO3-
(C) Solutes are differentially distributed between epidermal cell types
(R, ridge; NS, near-stomatal; IS, inter-stomatal; TR, trough)
R NS NS TR
IS
R NS IS NS TR
R NS IS NS TR
Ca
Malate
Cl
K

Storey and Leigh 2004). In contrast, Karley et al. (2000) using a patch-clamp approach, concluded that differences in Na accumulation between barley leaf epidermal and mesophyll protoplasts could be explained by differences in transport properties of Na at the plasma membrane.

2.4 Solute Transport: Plasma Membrane and Tonoplast

What determines concentrations of solutes in the vacuole, processes at the tonoplast or at the plasma membrane? The truth lies probably somewhere between. The tonoplast is the membrane delineating the vacuole and as such would be expected to exert a prime controlling function. The plasma membrane constitutes the boundary between protoplasm and apoplast and as such controls what enters and exits the cell. The driving force for movement of solutes is provided by metabolic energy stored as a pH gradient between two compartments (e.g. sucrose) and by the electrochemical gradient of a particular solute (e.g. K, Ca). Gradients in pH between cell compartments are established through activity of H^+-pumps (ATPases and pyrophosphatases; Hasegawa et al. 2000). The pH gradient along the apoplast-cytosol-vacuole path is mirror-image like. The apoplastic pH is 5 – 6 and 1 – 2 units smaller (H^+ concentration 10 – 100 times higher) than the cytosolic pH (pH 7.0 – 7.4), but similar to the vacuolar pH. Different mechanisms must operate for solutes such as sucrose to move across the plasma membrane into the cell (up-ward gradient in pH) and to move across the tonoplast from cytosol to vacuole (down-ward gradient in pH). For the movement of ionic solutes such as K and Ca, the membrane potential is paramount.

Membrane potential is a driving force for ion movement and regulates channel activity through voltage-gating. Membrane potential between apoplast and cytosol (more negative) is typically in the range – 120 to – 200 mV, but can be as little as – 70 mV in grass leaf and root cells (Carden et al. 2003; Cuin et al. 2003). Therefore, membrane potential is sufficient to accumulate K from an apoplastic few mM to close to 100 mM in the cytosol through facilitated diffusion through K-channels.

There exists a range of channels (Maser et al. 2001; Pilot et al. 2003; Very and Sentenac 2002), which could potentially function in facilitating K-uptake into growing leaf tissues. Probably the best characterized candidate channel involved in growth associated K-accumulation is ZmK1 from maize (*Zea mays*). ZmK1 operates at the plasma membrane, the cellular control point for entry and exit of solutes, and has been proposed to play a key role in growth-associated K uptake in the coleoptile epidermis of maize and in the auxin-mediated growth response to gravity (Bauer et al. 2000; Philippar et al. 1999). The coleoptile is not a true leaf but it is related to a leaf through its ontogeny. Furthermore, ZmK1 is inward-rectifying and allows uptake of K into cells. It has also high homology to AtAKT1, a K channel that has been characterized in the model plant *Arabidopsis thaliana* and that plays a key role in the K nutrition of plants at low external concentrations of K (Hirsch et al. 1998) and accounts for 50% of K uptake currents in leaf mesophyll cells (Dennison et al. 2001). The outward-rectifying channel KCO may function in the retrieval of K from the vacuole of cells (van den Wijngaard et al. 2005). Together with K transporters, KCO would enable the vacuolar compartment, particularly in the epidermis, to buffer demands of the mesophyll and growing tissues.

In elongating cotton fiber cells, elongation is accompanied by an increased expression of sucrose and K transporters (Ruan et al. 2001). In maize leaves, different developmental zones show different surface fluxes of K, Ca and protons (Zivanovic et al. 2005), and light-stimulated growth of poplar depends on ion transport mechanisms, which are possibly specific to growing tissue (Stiles and Van Volkenburgh 2002). Stiles and Van Volkenburgh (Stiles et al. 2003) concluded that light-dependent uptake of K into growing tobacco leaf tissues is not so much required for increase in cell solute load and generation of osmotic force, but for extrusion of protons, which in turn aids cell expansion through modifying wall properties.

Different mechanisms must operate at the tonoplast. The electric potential difference between cytosol and vacuole is close to zero (Carden et al. 2003; Cuin et al. 2003). If K moved only by diffusion (through channels) across the tonoplast, it would not be possible to establish a concentration difference across the tonoplast, yet it has been shown that vacuolar K exceeds cytosolic K by more than 100 mM or is lower by 20–40 mM (Carden et al. 2003; Cuin et al. 2003). Transporters must be responsible for this differential accumulation of K. The most likely candidates are transporters belonging to the KUP/HAK/HKT family of K transporters (Maser et al. 2001; Vallejo et al. 2005; Santa-Maria et al. 1997; Schachtman 2000). Differences in HAK1 expression have been reported for root developmental zones in barley (Vallejo et al. 2005).

Transport of Na into the vacuole can be achieved through Na^+/H^+ antiporters of the NHX family (Xue et al. 2004; Zhu 2003). Loading of the vacuole with Na through NHX is a mechanism to cope with salinity and maintain cytosolic levels of Na low, but it is also potentially a way to provide the osmotic force for water uptake in environments with lower, non-toxic levels of Na.

3 Aquaporins

3.1 Aquaporins and Water Movement Through Cellular Membranes

As detailed above, cell expansion requires continuous uptake of water to maintain turgor pressure. This water movement is driven by a gradient of water potential between cellular membranes established through accumulation of solutes. Water molecules move from cell to cell through the phospholipid bilayer by diffusion. However, the high water permeability found in most biological membranes, including the tonoplast, cannot be explained by purely diffusional processes but by the presence of water channels (aquaporins). Aquaporins represent an important selective pathway for water (and/or small neutral solutes) movement across cellular membranes and a large number of aquaporins has been found in plants (Chaumont et al. 2001; Johanson et al. 2001; Sakurai et al. 2005). This probably reflects the importance of aquaporins in maintaining sufficient water movement through membranes in physiological processes such as long-distance water transport from roots to leaves or cell osmoregulation (reviewed in Chaumont et al. 2005; Hachez et al. 2006a; Luu and Maurel 2005; Maurel et al. 2002; Tyerman et al. 2002).

Aquaporins that are present in the plasma membrane (PIPs) and tonoplast (TIPs) are likely to play essential roles in cell expansion. Vacuole biogenesis and enlargement require the transport of osmotically active substances across the tonoplast, followed by a rapid influx of water. Water transport measurements on tonoplast vesicles isolated from tobacco suspension cultures and wheat root cells (Maurel et al. 1997; Niemietz and Tyerman 1997) and on isolated vacuoles (Morillon and Lassalles 1999) showed generally a very high osmotic water permeability coefficient ($P_f > 200\ \mu m\ s^{-1}$) that was inhibited by mercury, an aquaporin inhibitor. These data support an aquaporin-mediated water movement across the tonoplast that probably plays an important role in cell expansion but also in water homeostasis (see below). In contrast, water permeability of the plasma membrane determined for isolated plasma membrane vesicles from tobacco suspension cultures and wheat root cells (Maurel et al. 1997; Niemietz and Tyerman 1997) or from swelling assay of protoplasts from different plant species (Chaumont et al. 2005) was on average much lower ($P_f < 30\ \mu m\ s^{-1}$) than permeability of the tonoplast, although high P_f values were obtained for some cell types and developmental stages (Chaumont et al. 2005; Maurel et al. 2002). It is possible that the comparatively low permeability of the plasma membrane results from experimental approaches. Plasma membrane permeability determined for intact cells within tissues may differ from that determined for isolated vesicles or protoplasts. For example, cell pressure probe measurements suggest that cell permeability values are higher than values of isolated protoplasts (Zhang and Tyerman 1999; Volkov et al. 2006). Despite this uncertainty, it appears justified to conclude that plasma membrane permeability represents the limiting factor in cell-to-cell water movement.

The difference in permeability between plasma membrane and tonoplast is essential for cell water homeostasis. During cell expansion, continuous uptake of solutes and water could perturb cytosol metabolism. Cytosol volume is small relative to vacuole and total cell volume. The volume and osmotic potential of the cytosol have to be rapidly equilibrated (buffered) in response to changes in external osmotic potential, and this can be achieved best through a much higher water permeability of the tonoplast, making vacuole water readily available as cytosol water is lost or taking up water into the vacuole as cytosol water increases (Maurel et al. 2002; Tyerman et al. 2002).

3.2 Tonoplast Aquaporins and Cell Expansion

The first indication of aquaporin involvement in cell elongation came from the analysis of gene expression patterns in plant organs, tissues and cells (reviewed in Maurel et al. 2002). Using in situ hybridization and transcriptional fusion between the promoter of the *Arabidopsis thaliana TIP1;1* gene, encoding a tonoplast aquaporin, and β-glucuronidase gene, Ludevid et al. (1992) detected a high *AtTIP1;1* expression in root and stem elongating tissues. No expression was detected in the meristems (cell division zones) or older parts of organs. Interestingly, *AtTIP1;1* expression was shown to be up-regulated after application of gibberellic acid, a hormone promoting cell expansion, in *Arabidopsis ga1* dwarf mutant impaired in gibberellin synthesis (Phillips and Huttly 1994). The physiological role of *AtTIP1;1* was further investigated *in planta* by an RNA interference approach (Ma et al. 2004). Plants with down-regulated *AtTIP1;1* displayed pleiotropic phenotypes including a reduced growth of varying severity according to the silencing efficiency.

Maize tonoplast aquaporin ZmTIP1;1, a close homologue of AtTIP1;1, was highly expressed in expanding cells in roots, leaves and reproductive organs. Transcript levels were also abundant in dividing cells (Barrieu et al. 1998; Chaumont et al. 1998). Elevated expression of tonoplast aquaporins has been reported also for elongating tissues in hypocotyls of soybean, castor bean and radish seedlings (Eisenbarth and Weig 2005; Higuchi et al. 1998; Maeshima 1990; Suga et al. 2001), during cold-induced stalk elongation in tulip (Balk and de Boer 1999) and pea fruit growth (Ozga et al. 2002). *HvTIP1;1* transcripts were increased in the *slender* mutant of barley, which is characterized by a faster elongation rate of leaves compared to the wild-type (Schunmann and Ougham 1996). Together these studies demonstrate a positive correlation between tonoplast aquaporin expression and cell elongation and indicate that this process requires a high hydraulic permeability of the tonoplast to support water entry into the vacuole and guarantee cellular water homeostasis.

Aquaporin activity in the tonoplast appears to affect not only cell elongation but also final cell size. Cauliflower tonoplast aquaporin BobTIP1;1 was fused to the green fluorescent protein (GFP) and expressed in tobacco suspension cells (Reisen et al. 2003). BobTIP1;1-GFP fusion protein was still an active water channel and localized in the vacuolar membrane. The fusion protein did not affect the growth rate of cell suspensions but increased the size and surface of cells two-fold, parallel to a swelling of the vacuole. It is possible that over-expression of BobTIP1;1 induced a concomitant solute transport which increased the osmotic gradient and water entry into the vacuole (Reisen et al. 2003).

3.3 Plasma Membrane Aquaporins and Cell Expansion

The involvement of plasma membrane aquaporins in the growth of cells in leaves and roots has been deduced from pressure probe and osmotic swelling experiments. Osmotic water permeability of elongating epidermal and mesophyll cells from barley leaf was 31 to 55% higher than that of non-expanding cells (Volkov et al. 2006). Interestingly, the increased P_f in epidermal elongating cells correlated with the expression of barley *HvPIP1;6* gene encoding an active plasma membrane water channel (Fricke et al. 2006). Treatment of maize roots with mercury chloride, an aquaporin inhibitor, reduced maize root elongation by around 75% as well as the hydraulic conductivity of growing cells in the distal region of the elongation zone (Hukin et al. 2002). Recently an extensive study of plasma membrane PIP gene and protein expression has been performed in maize roots grown aeroponically (Hachez et al. 2006b). Twelve of the 13 maize *PIP* genes identified (Chaumont et al. 2001) were expressed in primary roots. Expression was found to be dependent on the developmental stage of the root with an increase in expression towards either the elongation or mature zone (Hachez et al. 2006b). Aquaporins present in the plasma membrane of expanding root cells might have a dual function; they facilitate water entry into cells and maintain turgor pressure as the mechanical force driving wall expansion, and they participate in radial movement of water from soil to xylem vessels.

Although most PIPs are widely expressed in plant tissues, preferential expression in elongating tissues has been reported (Maurel et al. 2002). *Arabidopsis* PIP1;2 is expressed in expanding and differentiating cells comprising the root elongation zone, vascular bundle sheaths, filaments of stamen and young siliques (Kaldenhoff et al. 1995). Plasma membrane aquaporin expression has also been reported for the elongation zone of tobacco roots (Otto and Kaldenhoff 2000), castor bean and radish hypocotyls (Eisenbarth and Weig 2005; Suga et al. 2002), expanding cells of reproductive tissues (Bots et al. 2005; O'Brien et al. 2002), and for the elongation zone of barley leaves (Hollenbach and Dietz 1995; Fricke et al. 2006; Wei et al., 2006, personal communication). De-regulation of plasma membrane aquaporins by gene silencing, gene knock-out or over-expression leads to several phenotypes related to water relations but no exhaustive study on cell expansion has been conducted yet (reviewed in Hachez et al. 2006a).

There is accumulating evidence that aquaporin activity is regulated through many different post-transcriptional and post-translational mechanisms, and this provides another means through which cell expansion can be modified (Chaumont et al. 2005; Luu and Maurel 2005). Differential elongation of cells at the upper and lower side of gravitropically bending roots in pea was caused by a difference in the water-uptake rate and cell or tissue hydraulic conductivity rather than a difference in the driving force (Miyamoto et al. 2002, 2005). No significant difference in the levels of putative aquaporins between the upper and lower side of roots was observed using general aquaporin antibodies. The increased water conductivity measured in the (faster) elongating part of the root might have resulted from activation of pre-existing aquaporins, for example through aquaporin heteromerization (Fetter et al. 2004), phosphorylation (Johansson et al. 1998; Maurel et al. 1995; Weaver and Roberts 1991), deprotonation (Tournaire-Roux et al. 2003) and/or subcellular trafficking (Vera-Estrella et al. 2004). All these regulatory mechanisms have been extensively described in recent reviews (Chaumont et al. 2005; Luu and Maurel 2005).

4 Conclusions

Solute transport has received the least attention of the three main biophysical variables—walls, water and solutes—potentially limiting growth. This surprises given the importance of solute transport for osmotically driven water uptake by cells. In contrast to water, which accumulates in each cell and tissue in the same chemical form, solute composition and concentration differs between cells and tissues. Study of the role of solute transport in growth must take this heterogeneity into consideration through analyses at cell- and tissue level. The two basic ways through which solutes accumulate differentially between tissues or accumulate preferentially in growing compared to non-growing tissues (Fricke 2004a) are (i) differential supply of solutes to cells or (ii) differential transport properties of cells. In the first instance, we will have to focus on loading and un-loading of xylem and phloem; in the second instance, we need to focus on solute transporters and channels of growing cells.

To the best of our knowledge, there exists not a single study in which solute transport properties specific to growing leaf or root tissue has been studied at the molecular level. The best characterized system is the maize coleoptile, for which the shaker-type, inward-rectifying K-channel, ZmK1 has been shown to be involved in growth-associated uptake of K (Bauer et al. 2000; Philippar et al. 1999). Until we have further evidence for other tissues and organs, we can only speculate about candidate channels and transporters. Mechanisms are expected to differ between tonoplast and plasma membrane. At the plasma membrane, a significant trans-membrane electrical potential difference exists, and K could move into cells through either, active transport by transporters or facilitated diffusion through channels.

Channels can be voltage gated and show characteristic voltage-current relationships. Therefore, it is possible to predict based on electrophysiological analyses (patch-clamping) which channels are involved. Small (20–40 mV) changes in membrane potential can cause several-fold changes in K-uptake, and we need to obtain information on membrane potential in growing tissues and how it differs from that in non-growing tissue. At the tonoplast, where transmembrane potential between cytosolic and vacuolar compartments is close to zero, little driving force exists for (facilitated) diffusion through channels; active transport is required to accumulate vacuolar K above cytosolic K concentration. This transport is aided by the proton gradient across the tonoplast.

Expansion growth of cells must be matched by adequate rates of solute supply or solute uptake (Van Volkenburgh 1999), and either rate can become growth-limiting. This applies in particular to situations where environmental stress imposes extra demand on solute provision. For example, in barley exposed to high external NaCl (Fricke and Peters 2002), high salt overloads the capacity of epidermal cells to maintain osmolality during growth-dilution and to adjust osmotically to the large decrease in external water potential. The reduction in leaf cell expansion may not be so much a detrimental effect of salinity on the plant, but a mechanism through which the plant assures that solutes accumulate sufficiently and guarantee osmotic adjustment in an expanding cell. Similarly, Frensch (1997) concluded that growth in osmotically stressed maize roots is limited by solute supply. It appears that limitation of growth by solute supply occurs particularly in plants exposed to large decreases in external water potential, or in tissues that depend on a high rate of radial transport of phloem-borne solutes (discussed in Cosgrove 1993).

There is accumulating evidence that tonoplast and plasma membrane aquaporins mediate water uptake into growing cells and tissues. Tonoplast osmotic water permeability is by a factor 10 to 100 larger than plasma membrane osmotic water permeability. This assures that vacuole and cytosol are in osmotic equilibrium and suggests that the plasma membrane constitutes the main hydraulic barrier and is therefore the prime target for increasing growth and yield of plants through (genetic) modification of aquaporin activity. Several studies support a role of specific TIP or PIP isoforms in growth, based on expression profiles. "Hard" evidence that water channel activity of these isoforms is actually limiting water uptake into cells is scarce, particularly since the generally short half time of water exchange of plant cells questions the possibility that water transport limits cell and organ growth in the first place! Several technical approaches need to be combined to obtain unequivocal evidence. This will involve analyzing osmotic water permeability of cells and protoplasts/vesicles, monitoring expression levels and tissue localization of PIPs and TIPs, testing their post-translational regulation and, ultimately, testing the significance of altered expression levels of a candidate gene in mutant plants. The high homology of PIP and TIP isoforms within a species makes this task even more difficult, particularly when attempting to monitor protein levels through immunological approaches. Having said this, the above will also provide us with information on the redundancy of particular aquaporin isoforms and on the degree to which the impaired function of one PIP, can be compensated by the 11 to 12 other PIPs present in a species. Aquaporins may also have osmosensing functions and it is these functions which could provide a direct link between growth-associated water uptake and solute uptake. In other words, aquaporins may be involved in regulating the driving force (osmotic gradient) and the resulting flux (water). This would equip plant cells with a means to control cell size and volume.

Technical Box: Single-Cell Sampling

Sampling and analysis of contents of individual cells is essential for the study of water and solute relations of cells (Tomos and Sharrock 2001). Outlaw and colleagues developed techniques for manual micro-dissection, weighing and analyses of individual cells, more than 30 years ago (Outlaw and Zhang 2001). More recently, manual dissection of cells has been replaced by laser micro-dissection of tissues, followed by chemical or molecular (PCR) analyses (Kehr 2001; Tomos and Sharrock 2001). An alternative to dissection of tissue is sampling of single-cell contents, through inserting a silicon-oil filled micro-capillary into cells and extracting 10–20% of cell sap, which shoots into the capillary driven by cell turgor. This technique is particularly suited for analysis of surface tissues such as the leaf epidermis, since these can easily be accessed (for reviews see Tomos et al. 1994; Tomos and Leigh 1999). Deeper-lying tissues such as leaf mesophyll and bundle sheath or root cortex and stele can also be sampled, with some modification of the sampling approach. The extracted cell sap consists to almost 100% of vacuolar sap, in the case of an epidermal cell, but contains significant portions of cytoplasm, in the case of a mesophyll or bundle sheath cell (Fricke et al. 1994a). The volume of extracted sap is in the lower pl (mesophyll) range or reaches 100 pl and more (epidermis). Because of the small volume, further handling of sap is done under (stereo)microscopes, using glass capillaries attached to a micromanipulator; assays have to be carried out in a microenvironment which reduced evaporative loss of water.

Analysis for osmolality is done by picoliter-osmometry. A droplet (5–20 pl) of sap is placed under a drop of liquid paraffin onto a small (1 cm^2) copper stage on which a cover slip with black-and-white background is attached using heat-conducting paste. Standards (NaCl) of known osmolality are placed nearby and the stage is cooled to – 40 °C. Thereafter, the stage is reheated slowly and the melting of ice crystals is observed under the stereomicroscope. The temperature where the last ice crystal in a particular droplet melts is recorded and used, together with values of standards, to calculate the osmolality of the sample.

Solute contents can be analyzed using three types of techniques: EDX, micro-fluorospectrophotometry and capillary-zone electrophoresis. For energy-dispersive X-ray (EDX) analysis, sample droplets (10–20 pl) are pipetted onto a electron-microscope copper folding grid (100 and 200 mesh), which is coated with a film, ideally Pioloform. The same glass constriction pipette is used for samples and an internal standard (typically $RbNO_3$), which is placed in a 1 : 1 ratio together with sample droplets. During pipetting, the grid is submerged in liquid paraffin; this is removed by successive washes in hexane and isopentane, the latter having also a freeze-drying effect on droplets to obtain as amorphous as possible samples. Grids with samples are stored over silica gel, before they are analyzed with a scanning electron microscope (SEM) equipped with an X-ray analyzer. Element specific emission of X-ray is captured and peak integrals are related to the integral of the internal standard, Rb. This approach avoids difficulties from varying X-ray yield due to differences in the topology of droplets. Using standards of known concentrations of, for example, K, Cl and P that have been processed in the same way as the samples, their concentrations can be determined. For an element such as K, it can be assumed that all K measured existed as K^+. For an element such as P, it cannot be said with certainty how much of this P existed as phosphate and how much as part of larger molecules (e.g. nucleotides).

Additional solutes can be analyzed by microspectrofluorophotometry, in particular solutes such as sucrose, glucose, fructose and nitrate. Basically, the same approach is used as for spectrophotometric, enzymatic analyses of solute contents of standard (1-ml) samples, except that samples and cocktail components are pipetted with constriction pipettes of approximate volume (ranging from 10 pl to 0.5 nl) and assays are carried out in an oil-filled well on a glass-microscope slide, under a fluorescence microscope. Any solute for which an assay can be linked to the consumption or production of a fluorescent compound (such as NAD(P)H) can be analyzed, provided the assay does not require other manipulations (e.g. heating or solvent extraction) which might be difficult to carry out at the nano-scale. The same approach can be used to determine enzymatic activities in cell sap samples (Fricke et al. 1994a).

Capillary zone electrophoresis offers the most extensive range of solutes which can be analyzed in extracts of individual cells (Bazzanella et al. 1998). Considering its wide applicability, it surprises that capillary zone electrophoresis has been used by less than a handful laboratories for the analysis of single-cell extracts (Kehr 2001). Possibly, this is due to difficulties and extra facilities required to inject very small (pl) and reproducible volumes into capillaries used for electrophoresis.

Acknowledgements This work was supported by grants from the Biotechnology and Biological Sciences Research Council (BBSRC, UK), the Leverhulme Trust (UK), The Royal Society London and The Royal Society Edinburgh (to W.F.) and by grants from the Belgian National Fund for Scientific Research (FNRS), the Interuniversity Attraction Poles Programme-Belgian Science Policy and the Communauté française de Belgique-Actions de Recherches Concertées (to F.C.). Wieland Fricke would like to thank Vadim Volkov (Paisley and Glasgow University), Alexandre Boscari (Paisley and Nice University), Mathilde Clément (Paisley and Nice University), Anna Amtmann (Glasgow University), Tony Miller (Rothamsted, Harpenden), Tim Flowers (Sussex University), Charles Hachez (Université catholique de Louvain, Belgium) and the entire aquaporin team at Lund University (in particularly Per Kjellbom and Erik Alexandersson) for their help and lots of discussions.

References

Balk PA, de Boer AD (1999) Rapid stalk elongation in tulip (Tulipa gesneriana L. cv. Apeldoorn) and the combined action of cold-induced invertase and the water-channel protein gammaTIP. Planta 209:346–354

Barlow EWR (1986) Water relations in expanding leaves. Aust J Plant Physiol 13:45–58

Barrieu F, Chaumont F, Chrispeels MJ (1998) High expression of the tonoplast aquaporin ZmTIP1 in epidermal and conducting tissues of maize. Plant Physiol 117:1153–1163

Bauer CS, Hoth S, Haga K, Philippar K, Aoki N, Hedrich R (2000) Differential expression and regulation of K^+ channels in the maize coleoptile: molecular and biophysical analysis of cells isolated from cortex and vasculature. Plant J 24:139–145

Bazzanella A, Lochmann H, Tomos AD, Bachmann K (1998) Determination of inorganic cations and anions in single plant cells by capillary zone electrophoresis. J Chrom A 809:231–239

Biela A, Grote K, Otto B, Hoth S, Hedrich R, Kaldenhoff R (1999) The Nicotiana tabacum plasma membrane aquaporin NtAQP1 is mercury-insensitive and permeable for glycerol. Plant J 18:565–570

Bots M, Feron R, Uehlein N, Weterings K, Kaldenhoff R, Mariani T (2005) PIP1 and PIP2 aquaporins are differentially expressed during tobacco anther and stigma development. J Exp Bot 56:113–121

Box S, Schachtman DP (2000) The effect of low concentrations of sodium on potassium uptake and growth of wheat. Aust J Plant Physiol 27:175–182

Boyer J (1985) Water transport. Ann Rev Plant Physiol. 36:473–516

Boyer JS, Silk WK (2004) Hydraulics of plant growth. Func Plant Biol 31:761–773

Carden DE, Walker DJ, Flowers TJ, Miller AJ (2003) Single-cell measurements of the contributions of cytosolic Na^+ and K^+ to salt tolerance. Plant Physiol 131:676–683

Chaumont F, Barrieu F, Herman EM, Chrispeels MJ (1998) Characterization of a maize tonoplast aquaporin expressed in zones of cell division and elongation. Plant Physiol 117:1143–1152

Chaumont F, Barrieu F, Wojcik E, Chrispeels MJ, Jung R (2001) Aquaporins constitute a large and highly divergent protein family in maize. Plant Physiol 125:1206–1215

Chaumont F, Moshelion M, Daniels MJ (2005) Regulation of plant aquaporin activity. Biol Cell 97:749–764

Cherel I (2004) Regulation of K^+ channel activities in plants: from physiological to molecular aspects. J Exp Bot 55:337–351

Cosgrove DJ (1993) How do plant cell walls extend? Plant Physiol 102:1–6
Cosgrove DJ, Li LC, Cho HT, Hoffmann-Benning S, Moore RC, Blecker D (2002) The growing world of expansins. Plant Cell Physiol 43:1436–1444
Cramer GR (1992) Kinetics of Maize Leaf Elongation. 2. Responses of a Na-Excluding Cultivar and a Na-Including Cultivar to Varying Na/Ca Salinities. J Exp Bot 43:857–864
Cuin TA, Miller AJ, Laurie SA, Leigh RA (2003) Potassium activities in cell compartments of salt-grown barley leaves. J Exp Bot 54:657–661
de Souza IRP, MacAdam JW (1998) A transient increase in apoplastic peroxidase activity precedes decrease in elongation rate of B73 maize (Zea mays) leaf blades. Physiologia Plantarum 104:556–562
Delane R, Greenway H, Munns R, Gibbs J (1982) Ion concentration and carbohydrate status of the elongating leaf tissue of Hordeum vulgare growing at high external NaCl. I. Relationship between solute concentration and growth. J Exp Bot 33:557–573
Dennison KL, Robertson WR, Lewis BD, Hirsch RE, Sussman MR, Spalding EP (2001) Functions of AKT1 and AKT2 potassium channels determined by studies of single and double mutants of Arabidopsis. Plant Physiol 127:1012–1019
Dordas C, Chrispeels MJ, Brown PH (2000) Permeability and channel-mediated transport of boric acid across membrane vesicles isolated from squash roots. Plant Physiol 124:1349–1362
Eisenbarth DA, Weig AR (2005) Dynamics of aquaporins and water relations during hypocotyl elongation in Ricinus communis L. seedlings. J Exp Bot 56:1831–1842
Fetter K, Van Wilder V, Moshelion M, Chaumont F (2004) Interactions between Plasma Membrane Aquaporins Modulate Their Water Channel Activity. Plant Cell 16:215–228
Frensch J (1997) Primary responses of root and leaf elongation to water deficits in the atmosphere and soil solution. J Exp Bot 48:985–999
Fricke W (1997) Cell turgor, osmotic pressure and water potential in the upper epidermis of barley leaves in relation to cell location and in response to NaCl and air humidity. J Exp Bot 48:45–58
Fricke W (2002a) Biophysical limitation of cell elongation in cereal leaves. Ann Bot 90:157–167
Fricke W (2002b) Biophysical limitation of leaf cell elongation in source-reduced barley. Planta 215:327–338
Fricke W (2004a) Rapid and tissue-specific accumulation of solutes in the growth zone of barley leaves in response to salinity. Planta 219:515–525
Fricke W (2004b) Solute sorting in grass leaves: the transpiration stream. Planta 219:507–514
Fricke W, Akhiyarova G, Wei W, Alexandersson E, Miller A, Kjellbom PO, Richardson A, Wojciechowski T, Schreiber L, Veselov D, Kudoyarova G, Volkov V (2006) The short-term growth response to salt of the developing barley leaf. J Exp Bot 57:1079–1095
Fricke W, Flowers TJ (1998) Control of leaf cell elongation in barley. Generation rates of osmotic pressure and turgor, and growth-associated water potential gradients. Planta 206:53–65
Fricke W, Hinde PS, Leigh RA, Tomos AD (1995) Vacuolar Solutes in the Upper Epidermis of Barley Leaves - Intercellular Differences Follow Patterns. Planta 196:40–49
Fricke W, Leigh RA, Tomos AD (1994a) Concentrations of Inorganic and Organic Solutes in Extracts from Individual Epidermal, Mesophyll and Bundle-Sheath Cells of Barley Leaves. Planta 192:310–316
Fricke W, Leigh RA, Tomos AD (1994b) Epidermal Solute Concentrations and Osmolality in Barley Leaves Studied at the Single-Cell Level - Changes Along the Leaf Blade, During Leaf Aging and Nacl Stress. Planta 192:317–323

Fricke W, Leigh RA, Tomos AD (1996) The intercellular distribution of vacuolar solutes in the epidermis and mesophyll of barley leaves changes in response to NaCl. J Exp Bot 47:1413–1426
Fricke W, Peters WS (2002) The biophysics of leaf growth in salt-stressed barley. A study at the cell level. Plant Physiol 129:374–388
Fricke W, Pritchard E, Leigh RA, Tomes AD (1994c) Cells of the Upper and Lower Epidermis of Barley (Hordeum-Vulgare L) Leaves Exhibit Distinct Patterns of Vacuolar Solutes. Plant Physiol 104:1201–1208
Fry SC (1998) Oxidative scission of plant cell wall polysaccharides by ascorbate-induced hydroxyl radicals. Biochem J 332(Pt 2):507–515
Gerbeau P, Guclu J, Ripoche P, Maurel C (1999) Aquaporin Nt-TIPa can account for the high permeability of tobacco cell vacuolar membrane to small neutral solutes. Plant J 18:577–587
Hachez C, Zelazny E, Chaumont F (2006a) Modulating the expression of aquaporin genes in planta: A key to understand their physiological functions? Biochim Biophys Acta 1758:1142–1156
Hachez C, Moshelion M, Zelazny E, Cavez D, Chaumont F (2006b) Localization and quantification of plasma membrane aquaporin expression in maize primary root: a clue to understanding their role as cellular plumbers. Plant Mol Biol 62:305–323
Hanba YT, Shibasaka M, Hayashi Y, Hayakawa T, Kasamo K, Terashima I, Katsuhara M (2004) Overexpression of the barley aquaporin HvPIP2;1 increases internal CO_2 conductance and CO_2 assimillation in the leaves of transgenic rice plants. Plant Cell Physiol 45:521–529
Hasegawa PM, Bressan RA, Zhu JK, Bohnert HJ (2000) Plant Cellular and Molecular Responses to High Salinity. Annu Rev Plant Physiol Plant Mol Biol 51:463–499
Henzler T, Steudle E (2000) Transport and metabolic degradation of hydrogen peroxide in Chara corallina: model calculations and measurements with the pressure probe suggest transport of H_2O_2 across water channels. J Exp Bot 51:2053–2066
Higuchi T, Suga S, Tsuchiya T, Hisada H, Morishima S, Okada Y, Maeshima M (1998) Molecular cloning, water channel activity and tissue specific expression of two isoforms of radish vacuolar aquaporin. Plant Cell Physiol 39:905–913
Hill AE, Shachar-Hill B, Shachar-Hill Y (2004) What are aquaporins for? J Memb Biol 197:1–32
Hirsch RE, Lewis BD, Spalding EP, Sussman MR (1998) A role for the AKT1 potassium channel in plant nutrition. Science 280:918–921
Hollenbach B, Dietz KJ (1995) Molecular-Cloning of Emip, a Member of the Major Intrinsic Protein (Mip) Gene Family, Preferentially Expressed in Epidermal-Cells of Barley Leaves. Bot Acta 108:425–431
Hsiao TC, Xu LK (2000) Sensitivity of growth of roots versus leaves to water stress: biophysical analysis and relation to water transport. J Exp Bot 51:1595–1616
Hu Y, Schmidhalter U (1998) Spatial distributions of inorganic ions and sugars contributing to osmotic adjustment in the elongating wheat leaf under saline soil conditions. Aust J Plant Physiol 25:591–597
Huang J, Takano T, Akita S (2000) Expression of alpha-expansin genes in young seedlings of rice (Oryza sativa L.). Planta 211:467–473
Hukin D, Doering-Saad C, Thomas CR, Pritchard J (2002) Sensitivity of cell hydraulic conductivity to mercury is coincident with symplasmic isolation and expression of plasmalemma aquaporin genes in growing maize roots. Planta 215:1047–1056
Johanson U, Karlsson M, Johansson I, Gustavsson S, Sjovall S, Fraysse L, Weig AR, Kjellbom P (2001) The complete set of genes encoding major intrinsic proteins in Ara-

bidopsis provides a framework for a new nomenclature for major intrinsic proteins in plants. Plant Physiol 126:1358–1369
Johansson I, Karlsson M, Johanson U, Larsson C, Kjellbom P (2000) The role of aquaporins in cellular and whole plant water balance. Biochim Biophys Acta 1465:324–342
Johansson I, Karlsson M, Shukla VK, Chrispeels MJ, Larsson C, Kjellbom P (1998) Water transport activity of the plasma membrane aquaporin PM28A is regulated by phosphorylation. Plant Cell 10:451–459
Kaldenhoff R, Kolling A, Meyers J, Karmann U, Ruppel G, Richter G (1995) The blue light-responsive AthH2 gene of Arabidopsis thaliana is primarily expressed in expanding as well as in differentiating cells and encodes a putative channel protein of the plasmalemma. Plant J 7:87–95
Karley AJ, Leigh RA, Sanders D (2000a) Differential ion accumulation and ion fluxes in the mesophyll and epidermis of barley. Plant Physiol 122:835–844
Karley AJ, Leigh RA, Sanders D (2000b) Where do all the ions go? The cellular basis of differential ion accumulation in leaf cells. Trends Plant Sci 5:465–470
Kehr J (2001) High resolution spatial analysis of plant systems. Curr Opin Plant Biol 4:197–201
Koroleva OA, Farrar JF, Tomos AD, Pollock CJ (1997) Patterns of solute in individual mesophyll, bundle sheath and epidermal cells of barley leaves induced to accumulate carbohydrate. New Phytol 136:97–104
Leigh RA, Storey R (1993) Intercellular Compartmentation of Ions in Barley Leaves in Relation to Potassium Nutrition and Salinity. J Exp Bot 44:755–762
Leigh RA, Tomos AD (1993) Ion Distribution in Cereal Leaves - Pathways and Mechanisms. Philos T Roy Soc B 341:75–86
Leigh RA, Wyn Jones RG (1986) Cellular compartmentation in plant nutrition: The selective cytoplasm and the promiscuous vacuole. Adv Plant Nutr 2:249–279
Loque D, Ludewig U, Yuan L, von Wiren N (2005) Tonoplast Intrinsic Proteins AtTIP2;1 and AtTIP2;3 Facilitate NH3 Transport into the Vacuole. Plant Physiol 137:671–680
Ludevid D, Hofte H, Himelblau E, Chrispeels MJ (1992) The Expression Pattern of the Tonoplast Intrinsic Protein Gamma-Tip in Arabidopsis-Thaliana Is Correlated with Cell Enlargement. Plant Physiol 100:1633–1639
Luu DT, Maurel C (2005) Aquaporins in a challenging environment: molecular gears for adjusting plant water status. Plant Cell Env 28:85–96
Ma JF, Tamai K, Yamaji N, Mitani N, Konishi S, Katsuhara M, Ishiguro M, Murata Y, Yano M (2006) A silicon transporter in rice. Nature 440:688–691
Ma S, Quist TM, Ulanov A, Joly R, Bohnert HJ (2004) Loss of TIP1;1 aquaporin in Arabidopsis leads to cell and plant death. Plant J 40:845–859
Maeshima M (1990) Development of Vacuolar Membranes During Elongation of Cells in Mung Bean Hypocotyls. Plant Cell Physiol 31:311–317
Martre P, Bogeat-Triboulot MB, Durand JL (1999) Measurement of a growth-induced water potential gradient in tall fescue leaves. New Phytol 142:435–439
Marty F (1997) The biogenesis of vacuoles: insights from microscopy. In: Leigh RAaS, D. (ed) The Plant Vacuoles. Academic Press, San Diego CA, p 1–42
Maser P, Thomine S, Schroeder JI, Ward JM, Hirschi K, Sze H, Talke IN, Amtmann A, Maathuis FJ, Sanders D, Harper JF, Tchieu J, Gribskov M, Persans MW, Salt DE, Kim SA, Guerinot ML (2001) Phylogenetic relationships within cation transporter families of Arabidopsis. Plant Physiol 126:1646–1667
Maurel C (1997) Aquaporins and Water Permeability of Plant Membranes. Annu Rev Plant Physiol Plant Mol Biol 48:399–429

Maurel C, Javot H, Lauvergeat V, Gerbeau P, Tournaire C, Santoni V, Heyes J (2002) Molecular physiology of aquaporins in plants. Int Rev Cytol 215:105–148
Maurel C, Kado RT, Guern J, Chrispeels MJ (1995) Phosphorylation regulates the water channel activity of the seed-specific aquaporin alpha-TIP. Embo J 14:3028–3035
Maurel C, Tacnet F, Guclu J, Guern J, Ripoche P (1997) Purified vesicles of tobacco cell vacuolar and plasma membranes exhibit dramatically different water permeability and water channel activity. Proc Natl Acad Sci USA 94:7103–7108
Miyamoto N, Katsuhara M, Ookawa T, Kasamo K, Hirasawa T (2005) Hydraulic conductivity and aquaporins of cortical cells in gravitropically bending roots of Pisum sativum L. Plant Prod Sci 8:515–524
Miyamoto N, Ookawa T, Takahashi H, Hirasawa T (2002) Water uptake and hydraulic properties of elongating cells in hydrotropically bending roots of Pisum sativum L. Plant Cell Physiol 43:393–401
Morillon R, Lassalles JP (1999) Osmotic water permeability of isolated vacuoles. Planta 210:80–84
Niemietz CM, Tyerman SD (1997) Characterization of Water Channels in Wheat Root Membrane Vesicles. Plant Physiol 115:561–567
Nonami H, Wu Y, Boyer JS (1997) Decreased Growth-Induced Water Potential (A Primary Cause of Growth Inhibition at Low Water Potentials). Plant Physiol 114:501–509
O'Brien M, Bertrand C, Matton DP (2002) Characterization of a fertilization-induced and developmentally regulated plasma-membrane aquaporin expressed in reproductive tissues, in the wild potato Solanum chacoense Bitt. Planta 215:485–493
Otto B, Kaldenhoff R (2000) Cell-specific expression of the mercury-insensitive plasma-membrane aquaporin NtAQP1 from Nicotiana tabacum. Planta 211:167–172
Outlaw WH, Zhang SQ (2001) Single-cell dissection and microdroplet chemistry. J Exp Bot 52:605–614
Ozga JA, van Huizen R, Reinecke DM (2002) Hormone and seed-specific regulation of pea fruit growth. Plant Physiol 128:1379–1389
Palmer SJ, Davies WJ (1996) An analysis of relative elemental growth rate, epidermal cell size and xyloglucan endotransglycosylase activity through the growing zone of ageing maize leaves. J Exp Bot 47:339–347
Peters WS, Tomos AD (1996) The history of tissue tension. Ann Bot 77:657–665
Philippar K, Fuchs I, Luthen H, Hoth S, Bauer CS, Haga K, Thiel G, Ljung K, Sandberg G, Bottger M, Becker D, Hedrich R (1999) Auxin-induced K^+ channel expression represents an essential step in coleoptile growth and gravitropism. Proc Natl Acad Sci USA 96:12186–12191
Phillips AL, Huttly AK (1994) Cloning of two gibberellin-regulated cDNAs from Arabidopsis thaliana by subtractive hybridization: expression of the tonoplast water channel, gamma-TIP, is increased by GA3. Plant Mol Biol 24:603–615
Pilot G, Gaymard F, Mouline K, Cherel I, Sentenac H (2003) Regulated expression of Arabidopsis shaker K^+ channel genes involved in K^+ uptake and distribution in the plant. Plant Mol Biol 51:773–787
Pritchard J (1994) The Control of Cell Expansion in Roots. New Phytol 127:3–26
Pritchard J (1996) Aphid stylectomy reveals an osmotic step between sieve tube and cortical cells in barley roots. J Exp Bot 47:1519–1524
Reidy B, Nosberger J, Fleming A (2001) Differential expression of XET-related genes in the leaf elongation zone of F. pratensis. J Exp Bot 52:1847–1856
Reisen D, Loborgne-Castel N, Ozalp C, Chaumont F, Marty F (2003) Expression of a cauliflower tonoplast aquaporin tagged with GFP in tobacco suspension cells correlates with an increase in cell size. Plant Mol Biol 52:387–400

Rose JK, Braam J, Fry SC, Nishitani K (2002) The XTH family of enzymes involved in xyloglucan endotransglucosylation and endohydrolysis: current perspectives and a new unifying nomenclature. Plant Cell Physiol 43:1421–1435

Ruan YL, Llewellyn DJ, Furbank RT (2001) The control of single-celled cotton fiber elongation by developmentally reversible gating of plasmodesmata and coordinated expression of sucrose and K^+ transporters and expansin. Plant Cell 13:47–60

Sakurai J, Ishikawa F, Yamaguchi T, Uemura M, Maeshima M (2005) Identification of 33 rice aquaporin genes and analysis of their expression and function. Plant Cell Physiol 46:1568–1577

Santa-Maria GE, Rubio F, Dubcovsky J, Rodriguez-Navarro A (1997) The HAK1 gene of barley is a member of a large gene family and encodes a high-affinity potassium transporter. Plant Cell 9:2281–2289

Schachtman DP (2000) Molecular insights into the structure and function of plant K^+ transport mechanisms. BBA-Biomembranes 1465:127–139

Schaffner AR (1998) Aquaporin function, structure, and expression: are there more surprises to surface in water relations? Planta 204:131–139

Schnyder H, Seo S, Rademacher IF, Kuhbauch W (1990) Spatial-Distribution of Growth-Rates and of Epidermal-Cell Lengths in the Elongation Zone During Leaf Development in Lolium-Perenne L. Planta 181:423–431

Schunmann PHD, Ougham HJ (1996) Identification of three cDNA clones expressed in the leaf extension zone and with altered patterns of expression in the slender mutant of barley: A tonoplast intrinsic protein, a putative structural protein and protochlorophyllide oxidoreductase. Plant Mol Biol 31:529–537

Schunmann PHD, Smith RC, Lang V, Matthews PR, Chandler PM (1997) Expression of XET-related genes and its relation to elongation in leaves of barley (Hordeum vulgare L.). Plant Cell Env 20:1439–1450

Steudle E, Peterson CA (1998) How does water get trough roots? J Exp Bot 49:775–788

Stiles KA, McClintick A, Van Volkenburgh E (2003) A developmental gradient in the mechanism of K^+ uptake during light-stimulated leaf growth in Nicotiana tabacum L. Planta 217:587–596

Stiles KA, Van Volkenburgh E (2002) Light-regulated leaf expansion in two Populus species: dependence on developmentally controlled ion transport 10.1093/jxb/erf015. J Exp Bot 53:1651–1657

Storey R, Leigh RA (2004) Processes modulating calcium distribution in citrus leaves. An investigation using X-ray microanalysis with strontium as a tracer. Plant Physiol 136:3838–3848

Suga S, Imagawa S, Maeshima M (2001) Specificity of the accumulation of mRNAs and proteins of the plasma membrane and tonoplast aquaporins in radish organs. Planta 212:294–304

Suga S, Komatsu S, Maeshima M (2002) Aquaporin isoforms responsive to salt and water stresses and phytohormones in radish seedlings. Plant Cell Physiol 43:1229–1237

Thompson JE, Smith RC, Fry SC (1997) Xyloglucan undergoes interpolymeric transglycosylation during binding to the plant cell wall in vivo: evidence from 13C/3H dual labelling and isopycnic centrifugation in caesium trifluoroacetate. Biochem J 327(Pt 3):699–708

Tomos AD, Hinde P, Richardson P, Pritchard J, Fricke W (1994) Microsampling and measurements of solutes in single cells. In: Harris N, Oparka KJ (eds) Plant Cell Biology—A practical approach. IRL Press, Oxford, p 297–314

Tomos AD, Leigh RA (1999) The pressure probe: A versatile tool in plant cell physiology. Ann Rev Plant Physiol Plant Mol Biol 50:447–472

Tomos AD, Sharrock RA (2001) Cell sampling and analysis (SiCSA): metabolites measured at single cell resolution. J Exp Bot 52:623–630

Tornroth-Horsefield S, Wang Y, Hedfalk K, Johanson U, Karlsson M, Tajkhorshid E, Neutze R, Kjellbom P (2006) Structural mechanism of plant aquaporin gating. Nature 439:688–694

Tournaire-Roux C, Sutka M, Javot H, Gout E, Gerbeau P, Luu DT, Bligny R, Maurel C (2003) Cytosolic pH regulates root water transport during anoxic stress through gating of aquaporins. Nature 425:393–397

Tyerman SD, Niemietz CM, Bramley H (2002) Plant aquaporins: multifunctional water and solute channels with expanding roles. Plant Cell Env 25:173–194

Uehlein N, Lovisolo C, Siefritz F, Kaldenhoff R (2003) The tobacco aquaporin NtAQP1 is a membrane CO_2 pore with physiological functions. Nature 425:734–737

Vallejo AJ, Peralta ML, Santa-Maria GE (2005) Expression of potassium-transporter coding genes, and kinetics of rubidium uptake, along a longitudinal root axis. Plant Cell Env 28:850–862

van den Wijngaard PWJ, Sinnige MP, Roobeek I, Reumer A, Schoonheim PJ, Mol JNM, Wang M, De Boer AH (2005) Abscisic acid and 14-3-3 proteins control K^+ channel activity in barley embryonic root. Plant J 41:43–55

Van Volkenburgh E (1999) Leaf expansion—an integrating plant behaviour. Plant Cell Env 22:1463–1473

Vera-Estrella R, Barkla BJ, Bohnert HJ, Pantoja O (2004) Novel Regulation of Aquaporins during Osmotic Stress. Plant Physiol 135:2318–2329

Very AA, Sentenac H (2002) Cation channels in the Arabidopsis plasma membrane. Trends Plant Sci 7:168–175

Volkov V, Wang B, Dominy PJ, Fricke W, Amtmann A (2004) Thellungiella halophila, a salt-tolerant relative of Arabidopsis thaliana, possesses effective mechanisms to discriminate between potassium and sodium. Plant Cell Env 27:1–14

Volkov V, Hachez C, Moshelion M, Draye X, Chaumont F, Fricke W (2006) Water permeability differs between growing and non-growing barley leaf tissues. J Exp Bot (in press)

Walker DJ, Leigh RA, Miller AJ (1996) Potassium homeostasis in vacuolate plant cells. Proc Natl Acad Sci USA 93:10510–10514

Weaver CD, Roberts DM (1991) Phosphorylation of Nodulin-26 by a Calcium-Dependent Protein-Kinase. Faseb J 5:A426–A426

Wolfe J, Steponkus PL (1981) The stress-strain relation of the plasma membrane of isolated plant protoplasts. Biochim Biophys Acta 643:663–668

Xue ZY, Zhi DY, Xue GP, Zhang H, Zhao YX, Xia GM (2004) Enhanced salt tolerance of transgenic wheat (Tritivum aestivum L.) expressing a vacuolar Na^+/H^+ antiporter gene with improved grain yields in saline soils in the field and a reduced level of leaf Na^+. Plant Science 167:849–859

Yokoyama R, Rose JKC, Nishitani K (2004) A surprising diversity and abundance of xyloglucan endotransglucosylase/hydrolases in rice. Classification and expression analysis. Plant Physiol 134:1088–1099

Zhang WH, Tyerman SD (1999) Inhibition of water channels by HgCl2 in intact wheat root cells. Plant Physiol 120:849–858

Zhu JK (2003) Regulation of ion homeostasis under salt stress. Curr Opin Plant Biol 6:441–445

Zivanovic BD, Pang J, Shabala S (2005) Light-induced transient ion flux responses from maize leaves and their association with leaf growth and photosynthesis. Plant Cell Env 28:340–352

Plant Cell Monogr (5)
J.-P. Verbelen · K. Vissenberg: The Expanding Cell
DOI 10.1007/7089_2006_070/Published online: 27 October 2006
© Springer-Verlag Berlin Heidelberg 2006

Cellulose and Cell Elongation

Kian Hématy · Herman Höfte (✉)

Laboratoire de biologie cellulaire, Institut Jean-Pierre Bourgin,
INRA centre de Versailles, Route de St Cyr (RD 10), 780026 Versailles cedex, France
herman.hofte@versailles.inra.fr

Abstract The cell wall (CW) is a strong but dynamic exoskeleton, which determines the wide variety of cell shapes (for review see Martin et al. 2001) and provides a mechanical barrier against pathogens (Vorwerk et al. 2004). CW structure and composition vary according to cell type and growth stage: the primary CW is a thin and constantly modified structure allowing cell growth driven by turgor pressure (see Verbelen and Vissenberg, in this volume), whereas secondary CWs are thick and rigid structures, which are laid down as soon as the cell has reached its final size.

Cellulose microfibrils (CMFs) constitute the fibre component of the composite material that makes up the plant CW. Cellulose represents 10–14% of the dry weight of primary CWs, 40–60% of secondary CWs and up to 98% in specialized cells, such as cotton fibres. CMFs are highly oriented and in this way influence the mechanical properties and viscoplasticity of the wall (see Burgert and Fratzl, in this volume). Recent data suggest that, besides controlling cell shape, cellulose synthesis also plays a critical role in the transition between cellular growth stages.

In this chapter we focus on the question of how the synthesis and deposition of cell wall material is coordinated with cell expansion in different cell types. Although several actors involved in cellulose synthesis have been identified, the mechanism and regulation of deposition remain largely unknown. Also, how cellulose associates with other CW polymers and how the cells monitor the status of the cell wall is not understood. We will first describe how genetic screens allowed the isolation of key components of the cellulose synthesis machinery in primary and secondary CWs. Next we will discuss recent findings on the role of CW synthesis in the control of cell expansion.

Abbreviations

CESA	Cellulose synthase catalytic subunit
CSC	Cellulose synthase complex
CMFs	Cellulose microfibrils
CW	Cell wall

1 Cellulose Biosynthesis

1.1 Cellulose: A Simple Structure Made by Complex Machinery

The remarkable tensile strength of cellulose micofibrils (CMFs), which is comparable to that of steel, reflects their unique structure. They are composed of multiple β-1,4-linked glucan chains with a parallel orientation, which are linked to each other by hydrogen bonds and van der Waals forces, thus forming a paracrystalline structure referred to as cellulose I. The parallel orientation is thermodynamically metastable: upon denaturation of the microfibrils, the glucan chains reassemble with an antiparallel orientation, yielding cellulose II. This metastable structure can only be understood in the context of the biosynthesis mode: glucans associate upon extrusion from high molecular weight cellulose synthase complexes (CSC) in the plasma membrane (for a review see Doblin et al. 2002). This stands in contrast to chitin, which is a polymer of β-1,4-linked *N*-acetylglucosamine (GlcNAc) that forms, in most organisms, type II crystals with antiparallel chains that assemble after bulk release into the extracellular space (Carlstrom 1957). In higher plants, CSCs consist of six distinguishable globules that assemble in 25 nm hexameric rosettes (Brown and Montezinos 1976). The dimensions of the microfibril in primary cell walls are compatible with the presence of 36 glucan chains, in which case each globule may produce six glucan chains (Doblin et al. 2002). Immunolabelling has confirmed the presence of the catalytic subunit of cellulose synthase in the rosettes (Kimura et al. 1999). Attempts to purify the CSC have remained unsuccessful so far and their exact composition is unknown.

Cellulose synthase catalytic subunits (CESAs) are members of the glycosyltransferase (GT) family 2 based on the CAZy classification (Coutinho et al. 2003). They consist of eight predicted transmembrane domains (TM) and, between TM 2 and 3, a large catalytic domain exposed to the cytosol with the characteristic D,D,D,QXXRW motif of processive glycosyl transferases. The N-terminal cytosolic region contains a ring-finger domain, which is involved in the homo- or heterodimerization of CESAs through disulfide bonds, and may act as a ubiquitin E3 ligase and mediate the targeted degradation of the monomeric form of CESA (Kurek et al. 2002).

The *Arabidopsis* genome contains ten *CESA* genes and over 30 more distantly related cellulose synthase-like (*CSL*) genes (Richmond and Somerville 2000, 2001). Among the ten *CESAs*, two functional groups can be distinguished based on expression patterns and mutant phenotypes. For *CESA1*, 3 and 6, transcripts are present in most tissues, mutants are dwarfed or seedling-lethal and show cellulose defects in primary cell walls. *CESA4*, 7 and 8 are expressed in developing xylem and interfascicular fibre cells. Mutants lack the characteristic secondary thickenings in xylem cells, which causes the collapse of the xylem. Interestingly, orthologues for each of the six isoforms can be found in other species including the monocots rice, barley and maize and the Gymnosperm *Pinus taeda* (Table 1). This indicates that the six isoforms are evolutionary conserved and that their origin preceded the divergence between dicots and monocots (Fig. 1) and, at least for *CESA4*, 7 and 8, the divergence between Angiosperms and Gymnosperms. The requirement for CESA triplets in the same cell types suggests that they are part of the same complex. This has been confirmed in co-immunoprecipitation experi-

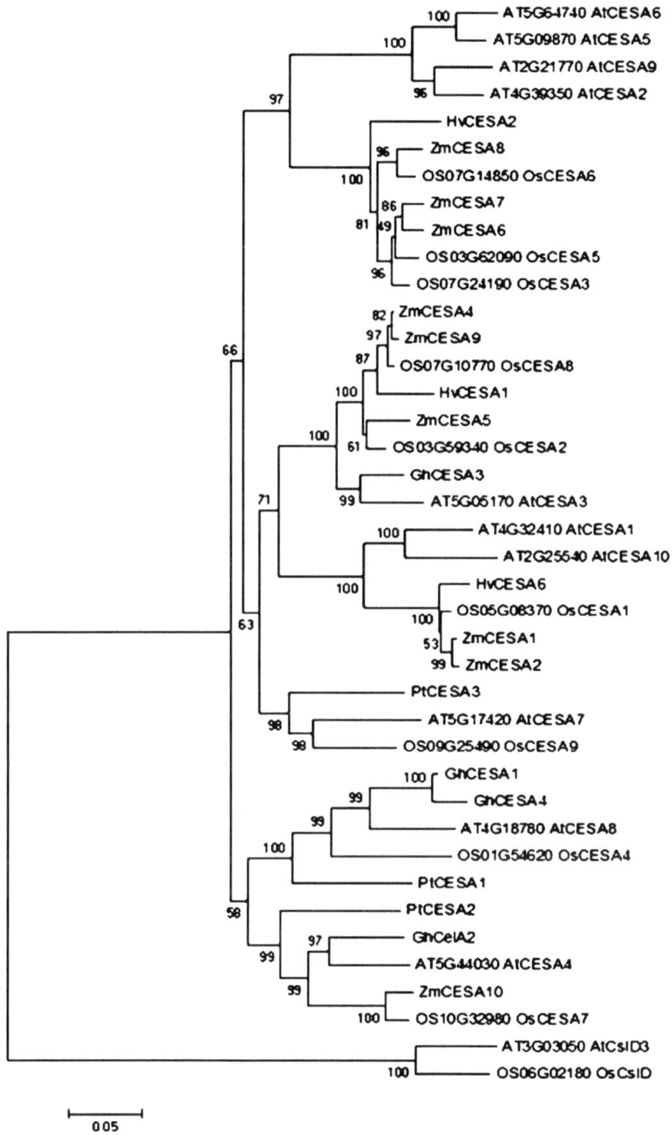

Fig. 1 Neighbor-joining dendrogram of relationship among CESAs of different plant species, rooted with Arabidopsis and Rice CslD/KJK (most closely related CSL family to CESA). The phylogenetic tree is based on ClustalW 1.83 alignment of the full-length protein sequence of *Arabidopsis thaliana* At, *Oryza sativa* Os, *Gossypium hirsutum* Gh, *Hordeum vulgare* Hv, and *Pinus taeda* Pt. The GhCesA2 sequence was assembled from two partial EST covering > 95% of the coding sequence. Sequences have been retrieved from http://cellwall.stanford.edu/, The bootstrapped phylogenetic tree (sampling 1000 times) was generated using MEGA 3.1 (Kumar et al. 2004) based on a ClustalW alignment. The bootstrap value is indicated. Gene numbers for Arabidopsis and Rice are from FLAGdb (http://urgv.evry.inra.fr/projects/FLAGdb++/HTML/index.html/)

Table 1 Orthologous CESA triplets in different species

	Arabidopsis thaliana	*Oryza sativa*	*Hordeum vulgare*	*Zea mays*	*Gossypium hirsutum*	*Pinus taeda*
Primary CW	AtCESA1,10	OsCESA1	HvCESA6	ZmCESA1,2	N.A.	N.A.
	AtCESA3	OsCESA2,8	HvCESA1	ZmCESA4,5,9	GhCESA3	N.A.
	AtCESA6,5,2,9	OsCESA3,5,6	HvCESA2	ZmCESA6,7,8	N.A.	N.A.
Secondary CW	AtCESA4	OsCESA7	N.A.	ZmCESA10	GhCESA2	PtCESA2
	AtCESA7	OsCESA9	N.A.	ZmCESA3?	N.A.	PtCESA3
	AtCESA8	OsCESA4	N.A.	ZmCESA3?	GhCESA1,4	PtCESA1

N.A. full length sequence not available

ments for CESA4, 7 and 8 (Gardiner et al. 2003; Taylor et al. 2003) and *CESA1*, 3 and 6 (Gonneau M and Desprez T, unpublished data). Additional genetic evidence corroborates the direct in vivo interaction, at least between CESA3 and CESA6 subunits. Indeed, mutations in either CESA3 (*ixr1-1* and *2*; Scheible et al. 2001) or CESA6 (*ixr2-1*; Desprez et al. 2002) confer increased resistance to the cellulose inhibitor isoxaben. The simplest explanation for the existence of two non-redundant resistance loci is that isoxaben recognizes an epitope associated with the CESA3- and CESA6-containing complex.

The sequence comparisons suggest that *CESA2*, *5*, *9* and *10* correspond to more recent gene duplications (Fig. 1, Table 1). CESA2, 5 and 9 are most closely related to CESA6, whereas CESA10 is most similar to CESA1. The transcript profiles show that *CESA2*, *5*, *9* and *10* are expressed during embryogenesis. They are partially redundant, as shown by the absence of an observable phenotype for loss of function mutations (Juraniec M and Gonneau M, unpublished results). Interestingly, *cesA6* mutants show a short hypocotyl phenotype only when grown in the dark. Light, through the activation of phytochrome, rescues the cellulose defect and the growth phenotype (Desnos et al. 1996). The expression of *CESA5* is redundant in the light with *CESA6* (Fig. 2, AtGE_7: green part 7 days). CESA5 may therefore replace CESA6 in light-grown tissues, which would explain the absence of a light-grown phenotype in *cesA6* mutants.

1.2 Other Proteins Involved in Cellulose Synthesis

1.2.1 Other Proteins Required for Cellulose Synthesis in Primary Walls

A common phenotype of mutants with cellulose defects in primary cell walls is the reduced cell elongation in dark-grown hypocotyls associated with an

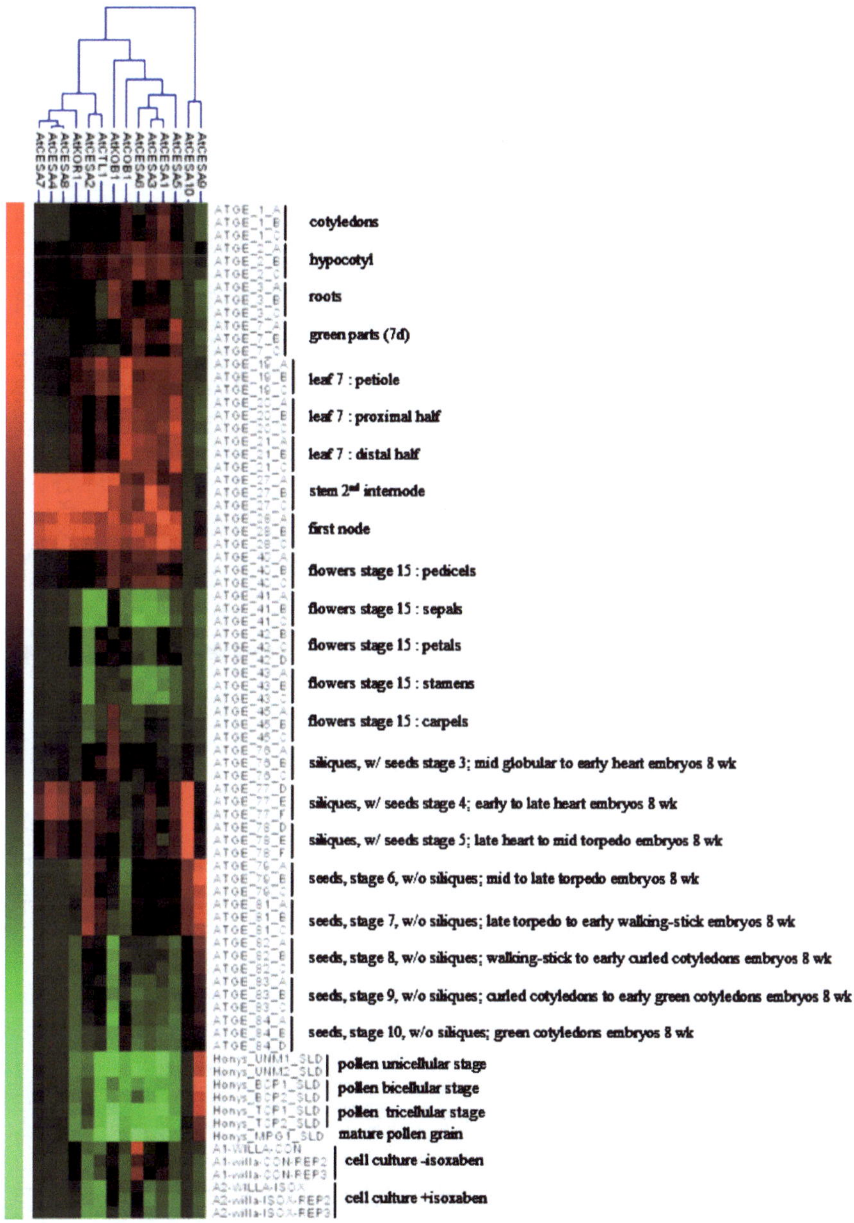

Fig. 2 Hierarchical clustering of several actors in cellulose synthesis (see Table 2 for gene numbers). Genes were clustered based on their expression levels across 79 selected ATH1 microarray data sets from (Honys and Twell 2004; Manfield et al. 2004; Schmid et al. 2005) hosted by NASC http://affymetrix.arabidopsis.info/narrays/experimentbrowse.pl. Clustering was performed using Genesis 1.5.0. software (Sturn et al. 2002). Gene expression levels have been normalized per gene prior to hierarchical clustering. *Green* and *red colour scales* represent low and high expression levels, respectively

Table 2 Arabidopsis cellulose-deficient mutants

AGI number	Function	Name alleles	Refs.
At4g32410	Cellulose synthase catalytic subunit, CESA1	*rsw1*	Arioli et al. 1998
At5g05170	Cellulose synthase catalytic subunit, CESA3	*eli1* *cev1* *ixr1*	Cano-Delgado et al. 2000, 2003 Ellis and Turner 2001; Ellis et al. 2002 Scheible et al. 2001
At5g64740	Cellulose synthase catalytic subunit, CESA6	*prc1* *ixr2* *YFP-CESA6*	Desnos et al. 1996; Fagard et al. 2000 Desprez et al. 2002 Paredez et al. 2006
At5g60920	GPI anchored protein	*cob1*	Schindelman et al. 2001 Roudier et al. 2005
At3g08550	Novel type II intrinsic membrane protein or secreted serine rich protein	*kob1* *eld1* *abi8*	Pagant et al. 2002 Lertpiriyapong and Sung 2003 Brocard-Gifford et al. 2004
At1g05850	Putative secreted basic chitinase-like protein (AtCTL1), GH19	*pom1* *elp1*	Hauser et al. 1995 Zhong et al. 2002b
At5g49720	Membrane-bound endo-β-1,4-glucanase, GH9 activity against CMC and non-crystalline cellulose, not against xyloglucan (Molhoj et al. 2002)	*kor1* *rsw2* *acw1* *tsd1* *irx2* *GFP-KOR1*	Nicol et al. 1998 Lane et al. 2001 Sato et al. 2001 Frank et al. 2002 Szyjanowicz et al. 2004 Robert et al. 2005
At5g44030	Cellulose synthase catalytic subunit, CESA4	*irx5*	Taylor et al. 2003

Table 2 (continued)

AGI number	Function	Name alleles	Refs.
At5g17420	Cellulose synthase catalytic subunit, CESA7	*irx3* *fra5*	Taylor et al. 1999 Zhong et al. 2003
At4g18780	Cellulose synthase catalytic subunit, CESA8	*irx1* *fra6*	Taylor et al. 2000 Zhong et al. 2003
At5g47820	Kinesin	*fra1*	Zhong et al. 2002a
At2g28110	Xylan glucuronyltransferase, GT47	*fra8* *irx7*	Zhong et al. 2005 Brown et al. 2005
At5g15630	COBRA-Like4	*Cobl4/irx6*	Brown et al. 2005
At5g22130	Mannosyltransferase GPI-anchor synthesis	*pnt1*	Gillmor et al. 2005
At2g39770	Mannose-1-phosphate guanylyltransferase	*Cyt1*	Lukowitz et al. 2001
At1g67490	Glucosidase I	*gcs1* *knf*	Boisson et al. 2001 Gillmor et al. 2002
At5g63840	Glucosidase II	*rsw3*	Burn et al. 2002
At5g13710	C-24 methyltransferase	*smt1/cph*	Reviewed in Schrick et al. 2004
At3g52940	C-14 reductase	*fk*	
At1g20050	C-7,8 isomerase	*hyd1*	
At1g12840	Vacuolar ATP-ase subunit	*det3*	Schumacher et al. 1999

Grey and *black bars* represent genes involved in cellulose synthesis in the primary and secondary CW, respectively. *Light grey bar* represents "housekeeping genes" required for normal cellulose synthesis exaggerated radial expansion. Genetic screens for mutations affecting dark-grown hypocotyl elongation yielded numerous cellulose-deficient mutants (Mouille et al. 2003; Robert et al. 2004). Other screens for temperature-sensitive *radial swelling* (*rsw*) mutants (Williamson et al. 2001) or sugar-sensitive *COnditional Root Expansion* (*CORE*) mutants (Hauser et al. 1995), provided other cellulose-deficient mutants and showed the importance of cellulose for cell expansion. Other cellulose–deficient mutants were found in screens for embryo-defective phenotypes (Gillmor et al. 2002, 2005). All these screens identified main actors in cellulose synthesis in the primary CW, like CESAs but also several non-GT proteins (Table 2).

KORRIGAN1 (KOR1). Mutant alleles have been identified with phenotypes ranging from seedling-lethal with cytokinesis defects; uncontrolled cell proliferation in the meristem; temperature-sensitive root-swelling; and dwarf to normal size with collapsed xylem. These phenotypes show that KOR1 is required for cellulose synthesis at all growth stages. KOR1 encodes a membrane-bound cellulase (Nicol et al. 1998). The protein produced in *Pichia pastoris* shows Ca^{2+}-dependent activity on non-crystalline cellulose and not on xyloglucan (Molhoj et al. 2001, Master et al. 2004). KOR1 cycles through the Golgi apparatus, endosomal compartments and the plasma membrane (Robert et al. 2005). All cellulose-producing organisms investigated also express cellulases. In *Agrobacterium* and *Rhizobium*, a cellulase is part of the celllulose synthase operon and mutant analysis showed that the protein has an essential role in cellulose synthesis (Hayashi et al. 2005). The exact role in cellulose synthesis is not known. It has been proposed that KOR1 may play a role in the recycling of the sitosterol-glucoside primer (Peng et al. 2002), remove uncorrectly assembled glucan chains or release the cellulose synthase complex from the growing glucan chain.

COBRA (COB1). Strong mutants are seedling-lethal, a weaker allele shows radial expansion of the root in the presence of high concentrations of sucrose or at high temperatures. COB1 is a GPI-anchored extracellular protein, which in the root is localized above the cortical microtubules in elongating cells. It may play a role in linking CESA complexes to cortical microtubules (Schindelman et al. 2001; Roudier et al. 2005).

POMPOM1(POM1)/ECTOPIC LIGNIN IN PITH (ELP1). Mutants are dwarfed, cellulose-deficient (Mouille et al. 2003) and produce ectopic lignin (Hauser et al. 1995; Zhong et al. 2000). POM1 shows sequence similarity to basic chitinases, but lacks critical catalytic residues, which suggests that the protein does not have chitinase activity (Zhong et al. 2002b). Plants do not contain chitin. N-linked glycans and arabinogalactan proteins, and perhaps NOD-related oligosaccharides, contain instead the chitin building block GlcNAc. It is conceivable that POM1 interacts with one of those and thus regulates cellulose synthesis in an unknown way.

KOBITO1 (KOB1)/ELONGATION DEFECTIVE (ELD1)/ABSCISSIC ACID INSENSITIVE 8 (ABI8). This locus was identified in screens for dwarf mutants (*eld1*), cellulose-deficient dwarf mutants (*kob1*), or mutants that show abscisic acid-insensitive germination (*abi8*). In mutant roots at early growth stages, the cell division zone appears normal and the elongation zone is strongly reduced. Field emission SEM of the innermost cell wall layers in root cells showed the presence of transversely oriented microfibrils in the cell division zone of both wild type and mutant. In the rudimentary elongation zone in *kob1*, only an amorphous matrix was observed. This matrix consisted of pectins since treatment with pectate lyase removed the layer and uncovered an underlying network of randomly oriented microfibrils. The randomization of the remaining microfibrils may be the result of their reorientation during cell expansion or the perturbation of the oriented microfibril deposition. At later growth stages, the mutant phenotype also propagated apically with the disappearance of the root meristem (Brocard-Gifford et al. 2004). Interestingly, the wild type phenotype is rescued in the elongation zone in the presence of 1% glucose.

KOB1 is a type II membrane protein of unknown function. GFP-KOB1 expressed from the 35S promoter was observed in intracellular compartments in the dividing cells and in the plasma membrane in elongating cells. A translational GUS-fusion from its own promoter showed an intracellular punctate staining. An ELD1-GFP fusion instead accumulated in the cell wall. These differences may be related to differences in the fusion proteins: GFP fused to the N-terminus (KOB1) or the C-terminus (ELD1), or GUS fused to the C-terminus (ABI8). Alternatively, they could be due to the promoter used (35S vs endogenous promoter). Resolving this discrepancy will require immunolocalization of the endogenous protein. The relation between the growth defect, the cellulose defect, the ABA-insensitivity and the effect of glucose on the phenotype is not understood, but reflects cross-talk between cellulose metabolism, soluble sugars and hormone signalling.

KOR1, POM1, COB and KOB1 are all members of multigene families with 3, 2, 12 and 3 members, respectively. The function of the other family members remains to be determined.

A number of cellulose-deficient mutants show embryo-defective phenotypes. The corresponding genes encode enzymes involved in *N*-glycan modification or sterol synthesis (Table 2). The cell walls of these mutants show a reduced cellulose content without important changes in other polysaccharides. This suggests that, at least during embryogenesis, cellulose synthesis is particularly sensitive to *N*-glycan-mediated quality control in the ER (Boisson et al. 2001; Lukowitz et al. 2001; Gillmor et al. 2002) and to changes in sterol content, in contrast to enzymes involved in matrix polysaccharide synthesis. The sensitivity of cellulose synthesis to variations in sterol composition may be related to the requirement for sterol-glucoside as a primer for cellulose synthesis, or the requirement of sterol-containing lipid rafts for the correct targeting of cellulose synthase complexes to the plasma membrane (Peng et al. 2002).

1.2.2 Other Proteins Required for Cellulose Synthesis in Secondary Walls

Genes involved in the synthesis of secondary cell walls have been identified in screens for *irregular xylem (irx)*, (Turner and Somerville 1997) and *fragile fibre (fra)*, (Zhong et al. 2001) mutants. These mutants in general do not show growth defects but lack the secondary cell wall thickenings that line the lumen of xylem elements and interfascicular fibres. As for the primary CW, a triplet of CESAs (CESA4, 7 and 8) is essential for cellulose synthesis. Interestingly, KOR1 and POM1 seem to be involved in both primary and secondary CWs whereas a COB1 paralogue (COBL4/IRX6) instead is involved in cellulose synthesis in secondary CWs only (Brown et al. 2005).

1.2.3 The Function of CSL Genes

The CSL family has been classified into eight subfamilies according to sequence similarities. Typically, monocots possess CSLA, C, D, F, H and dicots CSLA, B, C, D and G. The function of most of the CSL proteins is unknown except for the subclass CSLA and CSLF, members of which have been shown to be involved in the synthesis of (gluco)mannans (Dhugga et al. 2004; Liepman et al. 2005) and of mixed-linked β1,3;1,4 glucan (Burton et al. 2006).

At least one member of the CSLD family (CSLD3/KOJAK) is required for cell elongation, specifically in tip growing cells. Mutants in this gene have a normal phenotype except for the root hairs, which are initiated normally but subsequently swell and lyse (Favery et al. 2001). Another family member NaCSLD1 is strongly expressed in pollen tubes of *Nicotiana alata* (Doblin et al. 2001). Both pollen tubes and root hairs are unaffected in *cesA* mutants and express CESA isoforms at basal levels whereas they strongly express CSLD isoforms. Given the high sequence similarity between CSLD and CESA and the presence of the N-terminal ring-finger domain, which is absent in other CSL isoforms, they also might multimerize to form rosettes and synthesize cellulose.

1.3 Transcriptional Co-Regulation Identifies Novel Genes Potentially Involved in Cellulose Synthesis

As expected, the transcripts of the three *CESA* isoforms that are part of the same complex are co-regulated, as shown by the analysis of public transcriptome data sets (Manfield et al. 2004; Persson et al. 2005; Jen et al. 2006, see groups clustered in Fig. 2). Given this co-regulation, it is reasonable to assume that other potential protein partners can be identified by searching for co-regulated genes using the available microarray data for *Arabidopsis*. Figure 2 shows expression profiles for the above-mentioned genes extracted from 79 Affymetrix chip experiments. *CESA1*, 3 and 6 are indeed co-regulated and form a cluster with *CESA5* and *COB1*. *CESA2*, *POM1/CTL1* and *KOB1* form a distinct cluster. The profiles of *CESA4*, 7 and 8 form a cluster with *KOR1*, confirming its involvement also in secondary wall synthesis. Finally *CESA10* is exclusively expressed during embryogenesis and *CESA9* in the developing embryo and in microspores. Using a similar approach, Brown et al. (2005) identified a series of genes with expression patterns that are correlated with *CESA4*, 7 and 8 using *CESA7* as bait. Interestingly, among 19 co-regulated genes tested, ten showed a collapsed xylem phenotype when mutated, confirming their role in secondary cell wall synthesis.

1.4
Regulation of Cellulose Synthesis

The synthesis of cellulose microfibrils involves a large number of steps, which are all potential regulatory targets. As shown above, CESAs are regulated at the transcript level during cellular differentiation by redox-dependent dimerization of CESA monomers and perhaps through glycosylation and quality control in the ER. Other potential levels of regulation are the assembly of dimers into globules and into hexameric complexes, regulated intracellular trafficking from the Golgi apparatus to the cell surface, supply of the substrate UDP-glucose, priming of glycan synthesis, movement of the complexes in the plasma membrane, extrusion and assembly of glucan chains into microfibrils, termination of polymerization, endocytosis and turnover of the complex.

1.4.1
Intracellular Trafficking

Recently, evidence was obtained showing an important role for intracellular trafficking of CESA6 and KOR1 in the regulation of cellulose synthesis. Using functional YFP-CESA6 fusions expressed from the endogenous promoter, Paredez et al. (2006) studied epidermal cells of dark-grown hypocotyls and showed that YFP-CESA6 was present in subcellular compartments, including the Golgi apparatus and the plasma membrane. Interestingly, whereas plasma membranes were strongly labelled in smaller cells at the top of the hypocotyl, larger cells towards the hypocotyl basis showed a much stronger labelling of intracellular compartments. Treatment of seedlings with isoxaben also led to the rapid disappearance of YFP-CESA6 from the cell surface. Together, these observations show that the insertion into and the retrieval from the plasma membrane of the complexes is highly regulated. The cellulase KOR1 also accumulated in intracellular compartments (Robert et al. 2005), including Golgi apparatus and early endosomes and tonoplast, but could not be detected at the cell surface. Several observations suggest, however, that KOR1 cycles between intracellular compartments and the plasma membrane. Interestingly, intracellular KOR1-containing compartments were motile and this motility required intact microtubules, in contrast to the motility of the Golgi apparatus, which depends on actin and not on microtubules. One role for KOR1 could be in the removal of the complexes from the glucan chains prior to their endocytosis.

1.4.2 Surface Movement of CESA6

In YFP-CESA6 transformants, one can observe linear arrays of fluorescent dots in epidermal cells of the hypocotyl. These are presumably the cellulose synthase complexes, which are aligned above the microtubules in the plasma membrane. These objects display linear bidirectional trajectories, which follow microtubules, also when they are reoriented after illumination of the seedling or upon treatment for short periods with the microtubule depolymerizing drug oryzalin. Interestingly, upon complete removal of microtubules by longer treatments with oryzalin, YFP-CESA6 re-adopted transverse trajectories (Paredez et al. 2006). This shows that microtubules are not required for the motility of YFP-CESA6, which must be propelled by the polymerization of glucan chains. The rigidity of the microfibrils most likely ensures the maintenance of linear trajectories. In addition, these results show that, at least for the trajectories of CESA6-containing complexes, two levels of control exist: on one hand a default self-organizing behaviour, which leads to defined trajectories and on the other hand, the coupling to cortical microtubules, which in certain conditions overrides the default organization and imposes a direction, presumably via direct interaction between the cellulose synthase complex and cortical microtubules. Evidence for such a default organization exists for instance during secondary cell wall deposition in *Equisetum* root hairs, in which microfibrils are deposited in a helicoidal pattern, whereas microtubules adopt an axial orientation (Traas et al. 1985). It will be interesting to see whether such uncoupling occurs every time microtubules reorient towards an axial orientation after cessation of elongation of other cell types. A model has been proposed that predicts the orientation of the microfibrils based on the geometry of the cell and the density of the complexes (Emons and Mulder 1998).

1.4.3 Phosphorylation

Recent phosphoproteomics studies on cytoplasmic domains of plasma membrane proteins from *Arabidopsis* cell suspensions identified phosphopeptides that correspond to CESA3, CESA5 and KOR1 (Nuhse et al. 2004). The Ser/Thr phophorylation sites in CESA3 and CESA5 are located in conserved residues in the isoform-specific "hypervariable region" within the N-terminal cytoplasmic domain. Also in KOR1, the three phosphorylation sites in the cytoplasmic tail are highly conserved among different species. It will be interesting to see at what level (activity, assembly, targeting or turnover) phosphorylation plays a regulatory role.

2 Coordination Between Cell Wall Synthesis and Cell Elongation

As explained extensively in the chapter by Fricke and Chaumont in this volume, plants have a hydrostatic skeleton, consisting of cells filled with water and solutes under pressure, which are surrounded by a highly resistant cell wall. Burgert and Fratzl (in this volume) explained that the oriented microfibrils with their high tensile strength play a key role in the resistance of the thin cell wall (0.1–0.5 μm) to the extreme tensile forces imposed by the cell's turgor pressure (0.5–1 MPa). The drawback of this building plan is that the tough cell wall imposes important constraints on all aspects of growth and development.

A long-standing question in plant biology in this respect is how plant cells manage to expand despite the presence of the wall. Early models proposed that cells could grow through intussusception of new wall material within the existing cell wall (Nageli 1858). With such a mode of cell expansion, one expects a perfect correlation between wall incorporation and cell elongation. Subsequent studies on excised oat coleoptile segments (Heyn and Van Overbeek 1931) and on Avena coleoptiles (Bonner 1934) showed that the addition of "growth substance" (later christened auxin) could in certain conditions stimulate cell elongation without a change in the wall synthesis rate. Heyn and Van Overbeek (1931) showed instead that auxin stimulated the plastic extensibility of oat coleoptile walls. These findings were confirmed in intact plants by Roland et al. (1982), who observed a thick helicoïdal wall at early growth stages of epidermal cells of mungbean hypocotyls and a much thinner wall in more elongated cells. Kutschera and Briggs (1987) compared cell wall dynamics in light and dark-grown sunflower hypocotyls. Both light- and dark-grown hypocotyls produced the same amount of wall material despite the dramatic differences in growth rate. They also observed substantial thinning of the wall in more extensively growing cells of dark-grown seedlings. Finally, consistent with previous findings, the extensibility of the wall was greatly reduced in the light compared to the dark. Together, these and other observations led to the conclusion that cell expansion is regulated by the plastic extensibility of the wall rather than by the synthesis rate of wall material. This explains why wall extensibility has been and still is more intensively studied than wall synthesis.

In this context, the acid growth theory plays a central role. This theory states that auxin causes acidification of the apoplast, which in turn stimulates viscoplastic deformation of the wall through the regulation of pH-sensitive wall-modifying agents (for a review see Rayle and Cleland 1992). This theory was corroborated with the discovery of expansins as pH-dependent cell wall-lubricating proteins (see McQueen-Mason et al, in this volume). The acid growth theory requires a mechanism for acidification, via the activation of H^+-ATPases in the plasma membrane, as well as the presence of expansins and perhaps other wall-modifying agents in the cell wall. This theory does not require changes in the composition of the cell wall polymers. Indeed, Kutschera and Briggs (1987) did not observe differences in the relative proportion of cellulose, hemicellulose and pectin between walls of sunflower seedlings grown in the dark or in the light. Brummell and Hall (1985) suggested instead a critical role for the incorporation of matrix polysaccharides in auxin-promoted cell elongation in growing etiolated pea hypocotyl segments. Interestingly, the ionophore monensin, which inhibits secretion of proteins and matrix polysaccharides, inhibited auxin-stimulated elongation, whereas the cellulose synthesis inhibitor DCB did not influence this process.

Along the same lines, Takeda et al. (2002) showed that the incorporation into the cell wall of xyloglucan oligosaccharides by xyloglucan endotransglycosylase/hydrolase (XTH; see Nishitani and Vissenberg, in this volume) promoted cell expansion in pea stem segments, whereas grafting larger XG fragments inhibited cell expansion. The specific action of XTH observed in the root elongation zone of *Arabidopsis* (Vissenberg et al. 2000) is consistent with a role for XG grafting in growth control. Interestingly, expression of a fungal XGase in transgenic poplar caused increased growth and cellulose content, suggesting that the incorporation of XG oligomers in the cell wall may not only control cell elongation but also may feed back to cellulose synthesis (Park et al. 2004). Finally, the pectin composition also appears to change during cell elongation (see Verhertbruggen and Knox, in this volume). The anti-β-1,4-galactan monoclonal antibody LM5 specifically labels elongating cells (McCartney et al. 2003). It is not clear whether this transient galactan accumulation plays a critical role in the control of cell elongation.

2.1
Coupling Cellulose Synthesis and Cell Elongation in the Arabidopsis Hypocotyl

We are using the dark-grown *Arabidopsis* hypocotyl as a model to study the control of cell elongation (Gendreau et al. 1997). Post-embryonic growth occurs in the absence of cell divisions in the epidermis, except for those involved in stomatal guard cell differentiation. The 20 epidermal cells in each cell file of the hypocotyl epidermis measure 10 μm in the embryo and can reach up to 1 mm at the end of elongation. In our standard conditions, seeds germinate 24 h post-imbibition (pi). Between 24 h pi and 48 h pi, all hypocotyl cells grow slowly and synchronously (Refregier et al. 2004), but between 48 h and 52 h pi, cells at the hypocotyl basis show an abrupt fourfold increase in relative elemental growth rate (REGR, a measure for the growth rate per unit length). The REGR diminishes in these cells between 52 and 55 h. To investigate the molecular events in the CW that underly this transient growth acceleration, we used FT-IR microspectroscopy to follow changes in the cell wall architecture before, during and after the growth acceleration (see inset for details; Pelletier S, Renou J-P and Höfte H, unpublished data). This technique provides a molecular fingerprint of the cell walls, reflecting the relative proportions of molecules, but does not provide information on their absolute amounts.

Between 45 h pi and 48 h pi prior to the growth acceleration, we observed major changes in the ratio between cell wall polymers at the hypocotyl basis. As a negative control during the same period we did not observe significant changes in cells at the top of the hypocotyl, in which growth acceleration takes place only at 80 h pi or beyond. Observed changes involved an increase in the relative cellulose content and a reduced protein and pectin content, with no apparent changes in the degree of methylesterification (DM). These changes can be explained by an increased cellulose synthesis without an increase in pectin deposition. During the growth acceleration between 48 h pi and 52 h pi, we observed a further increase in relative cellulose content and a decrease in a specific ester peak (at 1751 cm^{-1}), without an increase in carboxylic acids. This suggests the concomitant de-methyl-esterification and turnover of pectic polysaccharides.

Between 52 h pi and 55 h pi, we did not observe significant changes in relative cell composition, suggesting that the cell walls had reached a new steady state equilibrium. No significant changes were observed at the top of the hypocotyl throughout the 45–55 h pi period. These findings suggest that the ratio between cell wall polymers is stable in cells with a constant REGR (before but also after the growth acceleration) and that the growth acceleration is preceded and accompanied by major changes in the cell wall composition, including an increased relative cellulose content.

Using isoform-specific antisera, we showed that the increase in relative cellulose content coincided with the accumulation of CESA3 and CESA6 cellulose synthase isoforms (Desprez T, Gonneau M and Höfte H, unpublished data). CESA1 has not been analysed yet in this context. An essential role for CESA6 in this process was shown using the loss of function mutant *cesA6*$^{prc1\text{-}1}$. Hypocotyls of this mutant failed to accumulate cellulose and maintained a low relative cellulose/pectin ratio. Interestingly, *cesA6*$^{prc1\text{-}1}$ hypocotyl cells showed a slow elongation phase indistinguishable from that of the wild type, but failed to accelerate their growth and maintained the slow elongation rate instead throughout post-embryonic development (Refregier et al. 2004). An essential role for the increase in relative cellulose content in growth acceleration was further confirmed using isoxaben treatments before and after the growth acceleration. Administration of 4 nM of isoxaben to seedlings at 30 h pi (Refregier et al. 2004) or even at 45 h pi (unpublished data) inhibited subsequent growth acceleration. Interestingly, adding 4 nM isoxaben after 52 h pi did not inhibit further cell elongation. In conclusion, the results show that an increase in the cellulose/pectin ratio occurs prior to and during growth acceleration in cells at the hypocotyl basis. The activity of CESA6 plays a key role in the relative increase in cellulose content and is also required for the growth acceleration to occur, as shown by the *cesA6* mutant phenotype and isoxaben treatments. Once growth has accelerated, cell elongation is not further inhibited by isoxaben. The increase in relative cellulose content may act as a check-point, controlling the transition from cytoplasmic growth to vacuolar cell expansion.

The dwarf phenotype of cellulose-deficient mutants is generally interpreted as being the result of the loss of growth anisotropy as a result of the fragilization of the cellulose-deficient cell wall. Our observations show a more complex picture. The inhibition of the growth acceleration occurs rapidly, within 3 h and before changes in microfibril orientation or radial swelling can be observed. This growth inhibition presumably involves an active process since second-site mutations, corresponding to at least three loci, were found that partially restored the growth acceleration of *cesA6* hypocotyls, without restoring the cellulose defect. One of these genes was cloned and encodes a plasma membrane receptor-like ser/thr kinase (Hématy K and Höfte H, unpublished data). This kinase is a good candidate for a cell wall integrity sensor.

3 Conclusions

Genetic studies and life cell imaging have shed new light on the complexity of the cellulose synthesis machinery. New molecular actors have been identified and an important regulatory role for intracellular trafficking has been demonstrated. Finally, recent evidence shows that cell wall synthesis plays a critical role in the control of early stages of cell elongation.

FT-IR BOX

Chemical Imaging of Plant Cell Walls Using FT-IR Microspectroscopy

Principle

The chemical analysis of cell walls is notoriously difficult due to the complexity and heterogeneity of cell wall polymers and the insolubility in aqueous solutions of many components, such as cellulose or lignin. Spectroscopic techniques combined with a microscope can provide relevant information on the cell wall composition at a cellular level. Fourier transform infra red (FT-IR) spectroscopy is based on the absorption of IR frequencies (10 000 to 100 cm^{-1}) by the vibration of asymmetric chemical bonds, with each

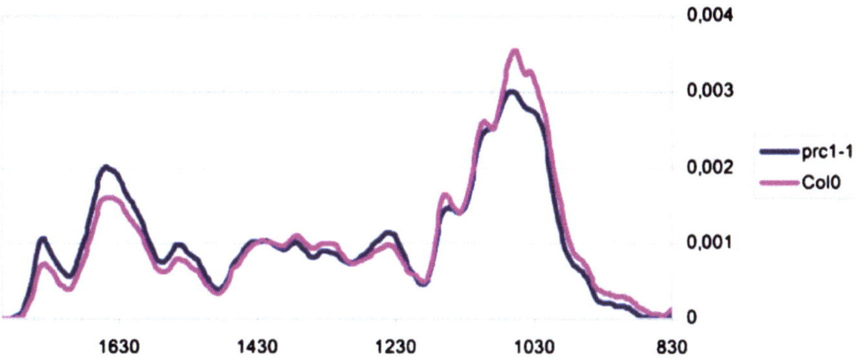

Fig. 3 FT-IR microspectroscopy can be used to distinguish cell wall mutant from wild type. Absorption spectra were obtained in transmission mode from a 50 × 50 μm surface of 4-day-old dark-grown *Arabidopsis* hypocotyls. Comparison between average spectra from cellulose-deficient seedlings *cesa6*$^{prc1\text{-}1}$ and the wild type controls; axis is wavenumber and ordinate is relative intensity (courtesy of G. Mouille)

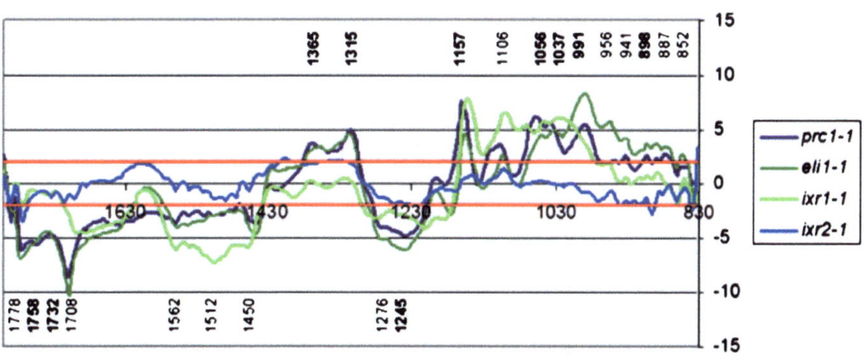

Fig. 4 Student's t test on comparison between average spectra from wild type against *cesa3*$^{eli1\text{-}1}$, and *cesa3*$^{ixr1\text{-}1}$, *cesa6*$^{prc1\text{-}1}$, *cesa6*$^{ixr2\text{-}1}$. Ordinate is t; a value of 2 is significant at $p = 0.01$ (*red line*). Highly significant positive peaks can be assigned to absorption maxima of cellulose (1365, 1315, 1157, 1056, 1037, 991 and 898 cm^{-1} *in bold*). Negative peaks correspond to carboxylic ester (1758, 1732 cm^{-1}) or carboxylic acid (1708 cm^{-1}) bonds and a C – O stretch (1245 cm^{-1}), presumably of pectic polysaccharides. *prc1-1* and *eli1-1* are cellulose-deficient but show increased pectic content. *ixr1-1* and *ixr2-1* are isoxaben-resistant alleles of CESA3 and CESA6, respectively, which do not show a growth phenotype. Note that the cell wall composition of *ixr1-1* is significantly different from the wild type even in the absence of isoxaben, in contrast to the cell walls of *ixr2-1* (courtesy of G. Mouille)

structural feature of a molecule absorbing at characteristic wavenumbers (expressed in cm^{-1}). Digitalization of the interferogram, followed by fast Fourier transformation of the data, quickly provide (within minutes) an absorption spectrum at different wavenumbers (Fig. 3). FT-IR spectroscopy can contribute to the identification of chemical groups, the structure of the molecular backbone and the interaction of the molecule with its environment. Simple unknown compounds can be identified by comparison with a reference database. Complex mixtures like cell walls contain too many chemical bonds with overlapping absorption wavenumbers to allow precise analysis. We use the spectral data between 1830 and 830 cm^{-1} of the mid-IR (4000 to 700 cm^{-1}), which contains information on $C=O$ stretches in carboxylic esters and carboxylic acids, amides and the polysaccharide fingerprint region.

Methodology

Intact organs or homogenized material can be used. Studying intact material preserves the spatial information of the tissue but may generate artefactual information related to differences in cell shapes or in the reflectance of cell surfaces. Analysing homogenized material may provide more easily interpretable data, however, all spatial information is lost. Aiming specific cell types by microspectroscopy allows the separate analysis of primary and secondary cell walls (Sibout et al. 2005). We will focus here on the primary CW analysis developed in our laboratory (Mouille et al. 2003). Four-day-old dark-grown seedlings are squashed, rinsed to eliminate cytoplasmic contents and dried on a BaF_2 slide prior to scanning. Absorption spectra are sampled in transmission mode from a 50×50 μm area on the side of the central cylinder. Given the simple anatomy of this organ, this area corresponds to only two cell types: epidermis and cortex. After base line correction and area normalisation (Fig. 3), significantly different wavenumbers are identified by the Student's t test (Fig. 4), Mutants with related cell wall changes also can be clustered, based on their FT-IR spectra. To this end, discrimant wavenumbers are first selected and used to calculate the Mahalanobis distance between the mutants. These distances are used in a hierarchical clustering procedure. This method is validated by the fact that independently isolated alleles of comparable strength cluster together (Mouille et al. 2003; Robert et al. 2004).

Applications

- High throughput screening for altered CW composition: The method so far has allowed the distinction between mutants with cellulose, pectin, xyloglucan or microtubule defects.
- Chemical screens: The effect of chemicals on the CW can be rapidly assessed.

- In situ analysis with enough precision to allow the visualization of different cell types: In this way we have studied cell walls of the cortex + epidermis, root hairs, mature or germinated pollen grains, parenchyma, xylem and phloem cells on stem sections, envelope and septum in siliques, etc.
- Analysis of CW evolution during growth: We have studied cell wall changes during hypocotyl etiolation in darkness or during cell division, elongation and differentiation in the root.

References

Arioli T, Peng L, Betzner AS, Burn J, Wittke W, Herth W, Camilleri C, Höfte H, Plazinski J, Birch R, Cork A, Glover J, Redmond J, Williamson RE (1998) Molecular analysis of cellulose biosynthesis in Arabidopsis. Science 279:717–720

Boisson M, Gomord V, Audran C, Berger N, Dubreucq B, Granier F, Lerouge P, Faye L, Caboche M, Lepiniec L (2001) Arabidopsis glucosidase I mutants reveal a critical role of N-glycan trimming in seed development. EMBO J 20:1010–1019

Bonner J (1934) Studies on the growth hormone of plants: V. The relation of cell elongation to cell wall formation. Proc Natl Acad Sci USA 20:393–397

Brocard-Gifford I, Lynch TJ, Garcia ME, Malhotra B, Finkelstein RR (2004) The Arabidopsis thaliana ABSCISIC ACID-INSENSITIVE8 encodes a novel protein mediating abscisic acid and sugar responses essential for growth. Plant Cell 16:406–421

Brown DM, Zeef LA, Ellis J, Goodacre R, Turner SR (2005) Identification of novel genes in Arabidopsis involved in secondary cell wall formation using expression profiling and reverse genetics. Plant Cell 17:2281–2295

Brown RM Jr, Montezinos D (1976) Cellulose microfibrils: visualization of biosynthetic and orienting complexes in association with the plasma membrane. Proc Natl Acad Sci USA 73:143–147

Burgert I, Fratzl P (2007) Mechanics of the expanding cell wall. In: Verbelen JP, Vissenberg K (eds) The expanding cell. Plant Cell Monographs, vol 5. Springer, Berlin Heidelberg New York (in press)

Brummell D, Hall J (1985) The role of cell wall synthesis in sustained auxin-induced growth. Physiol Plant 63:406–412

Burn JE, Hurley UA, Birch RJ, Arioli T, Cork A, Williamson RE (2002) The cellulose-deficient Arabidopsis mutant rsw3 is defective in a gene encoding a putative glucosidase II, an enzyme processing N-glycans during ER quality control. Plant J 32:949–960

Burton RA, Wilson SM, Hrmova M, Harvey AJ, Shirley NJ, Medhurst A, Stone BA, Newbigin EJ, Bacic A, Fincher GB (2006) Cellulose synthase-like CslF genes mediate the synthesis of cell wall (1,3;1,4)-beta-D-glucans. Science 311:1940–1942

Cano-Delgado A, Penfield S, Smith C, Catley M, Bevan M (2003) Reduced cellulose synthesis invokes lignification and defense responses in Arabidopsis thaliana. Plant J 34:351–362

Cano-Delgado AI, Metzlaff K, Bevan MW (2000) The eli1 mutation reveals a link between cell expansion and secondary cell wall formation in Arabidopsis thaliana. Development 127:3395–3405

Carlstrom D (1957) The crystal structure of alpha-chitin (poly-*N*-acetyl-D-glucosamine). J Biophys Biochem Cytol 3:669–683

Coutinho PM, Deleury E, Davies GJ, Henrissat B (2003) An evolving hierarchical family classification for glycosyltransferases. J Mol Biol 328:307–317

Desnos T, Orbovic V, Bellini C, Kronenberger J, Caboche M, Traas J, Höfte H (1996) Procustel mutants identify two distinct genetic pathways controlling hypocotyl cell elongation, respectively in dark- and light-grown Arabidopsis seedlings. Development 122:683–693

Desprez T, Vernhettes S, Fagard M, Refregier G, Desnos T, Aletti E, Py N, Pelletier S, Höfte H (2002) Resistance against herbicide isoxaben and cellulose deficiency caused by distinct mutations in same cellulose synthase isoform CESA6. Plant Physiol 128:482–490

Dhugga KS, Barreiro R, Whitten B, Stecca K, Hazebroek J, Randhawa GS, Dolan M, Kinney AJ, Tomes D, Nichols S, Anderson P (2004) Guar seed beta-mannan synthase is a member of the cellulose synthase super gene family. Science 303:363–366

Doblin MS, Kurek I, Jacob-Wilk D, Delmer DP (2002) Cellulose biosynthesis in plants: from genes to rosettes. Plant Cell Physiol 43:1407–1420

Doblin MS, De Melis L, Newbigin E, Bacic A, Read SM (2001) Pollen tubes of Nicotiana alata express two genes from different beta-glucan synthase families. Plant Physiol 125:2040–2052

Ellis C, Turner JG (2001) The Arabidopsis mutant cevl has constitutively active jasmonate and ethylene signal pathways and enhanced resistance to pathogens. Plant Cell 13:1025–1033

Ellis C, Karafyllidis I, Wasternack C, Turner JG (2002) The Arabidopsis mutant cevl links cell wall signaling to jasmonate and ethylene responses. Plant Cell 14:1557–1566

Emons AMC, Mulder BM (1998) The making of the architecture of the plant cell wall: how cells exploit geometry. Proc Natl Acad Sci USA 95:7215–7219

Fagard M, Desnos T, Desprez T, Goubet F, Refregier G, Mouille G, McCann M, Rayon C, Vernhettes S, Höfte H (2000) PROCUSTE1 encodes a cellulose synthase required for normal cell elongation specifically in roots and dark-grown hypocotyls of Arabidopsis. Plant Cell 12:2409–2424

Favery B, Ryan E, Foreman J, Linstead P, Boudonck K, Steer M, Shaw P, Dolan L (2001) KOJAK encodes a cellulose synthase-like protein required for root hair cell morphogenesis in Arabidopsis. Genes Dev 15:79–89

Frank M, Guivarc'h A, Krupkova E, Lorenz-Meyer I, Chriqui D, Schmulling T (2002) Tumorous shoot development (TSD) genes are required for co-ordinated plant shoot development. Plant J 29:73–85

Fricke W, Chaumont F (2007) Solute and water relations of growing plant cells. In: Verbelen JP, Vissenberg K (eds) The expanding cell. Plant Cell Monographs, vol 5. Springer, Berlin Heidelberg New York (in press)

Gardiner JC, Taylor NG, Turner SR (2003) Control of cellulose synthase complex localization in developing xylem. Plant Cell 15:1740–1748

Gendreau E, Traas J, Desnos T, Grandjean O, Caboche M, Höfte H (1997) Cellular basis of hypocotyl growth in Arabidopsis thaliana. Plant Physiol 114:295–305

Gillmor CS, Poindexter P, Lorieau J, Palcic MM, Somerville C (2002) Alpha-glucosidase I is required for cellulose biosynthesis and morphogenesis in Arabidopsis. J Cell Biol 156:1003–1013

Gillmor CS, Lukowitz W, Brininstool G, Sedbrook JC, Hamann T, Poindexter P, Somerville C (2005) Glycosylphosphatidylinositol-anchored proteins are required for cell wall synthesis and morphogenesis in Arabidopsis. Plant Cell 17:1128–1140

Hauser MT, Morikami A, Benfey PN (1995) Conditional root expansion mutants of Arabidopsis. Development 121:1237–1252

Hayashi T, Yoshida K, Park YW, Konishi T, Baba K (2005) Cellulose metabolism in plants. Int Rev Cytol 247:1–34

Heyn ANJ, Van Overbeek J (1931) Weiteres Versuchmaterial zur plastischen and elastischen Dehnbarkeit der Zellmembran. Proc Koninkt Akad v Wetenschap Amsterdam 34:1190

Honys D, Twell D (2004) Transcriptome analysis of haploid male gametophyte development in Arabidopsis. Genome Biol 5:R85

Jen CH, Manfield IW, Michalopoulos I, Pinney JW, Willats WG, Gilmartin PM, Westhead DR (2006) The Arabidopsis co-expression tool (ACT): a WWW-based tool and database for microarray-based gene expression analysis. Plant J 46:336–348

Kimura S, Laosinchai W, Itoh T, Cui X, Linder CR, Brown RM Jr (1999) Immunogold labeling of rosette terminal cellulose-synthesizing complexes in the vascular plant vigna angularis. Plant Cell 11:2075–2086

Kurek I, Kawagoe Y, Jacob-Wilk D, Doblin M, Delmer D (2002) Dimerization of cotton fiber cellulose synthase catalytic subunits occurs via oxidation of the zinc-binding domains. Proc Natl Acad Sci USA 99:11109–11114

Kutschera U, Briggs WR (1987) Rapid auxin-induced stimulation of cell wall synthesis in Pea internodes. Proc Natl Acad Sci USA 84:2747–2751

Lane DR, Wiedemeier A, Peng L, Höfte H, Vernhettes S, Desprez T, Hocart CH, Birch RJ, Baskin TI, Burn JE, Arioli T, Betzner AS, Williamson RE (2001) Temperature-sensitive alleles of RSW2 link the KORRIGAN endo-1,4-beta-glucanase to cellulose synthesis and cytokinesis in Arabidopsis. Plant Physiol 126:278–288

Lertpiriyapong K, Sung ZR (2003) The elongation defective1 mutant of Arabidopsis is impaired in the gene encoding a serine-rich secreted protein. Plant Mol Biol 53:581–595

Liepman AH, Wilkerson CG, Keegstra K (2005) Expression of cellulose synthase-like (Csl) genes in insect cells reveals that CslA family members encode mannan synthases. Proc Natl Acad Sci USA 102:2221–2226

Lukowitz W, Nickle TC, Meinke DW, Last RL, Conklin PL, Somerville CR (2001) Arabidopsis cyt1 mutants are deficient in a mannose-1-phosphate guanylyltransferase and point to a requirement of N-linked glycosylation for cellulose biosynthesis. Proc Natl Acad Sci USA 98:2262–2267

Manfield IW, Orfila C, McCartney L, Harholt J, Bernal AJ, Scheller HV, Gilmartin PM, Mikkelsen JD, Knox JP, Willats WG (2004). Novel cell wall architecture of isoxaben-habituated Arabidopsis suspension-cultured cells: global transcript profiling and cellular analysis. Plant J 40:260–275

Martin C, Bhatt K, Baumann K (2001) Shaping in plant cells. Curr Opin Plant Biol 4:540–549

Master ER, Rudsander UJ, Zhou W, Henriksson H, Divne C, Denman S, Wilson DB, Teeri TT (2004) Recombinant expression and enzymatic characterization of PttCel9A, a KOR homologue from Populus tremula x tremuloides. Biochemistry 43(31):10080–10089

McCartney L, Steele-King CG, Jordan E, Knox JP (2003) Cell wall pectic (1→4)-beta-D-galactan marks the acceleration of cell elongation in the Arabidopsis seedling root meristem. Plant J 33:447–454

McQueen-Mason SJ, Le NT, Brocklehurst D (2007) Expansins. In: Verbelen JP, Vissenberg K (eds) The expanding cell. Plant Cell Monographs, vol 5. Springer, Berlin Heidelberg New York (in press)

Molhoj M, Ulvskov P, Dal Degan F (2001) Characterization of a functional soluble form of a Brassica napus membrane-anchored endo-1,4-beta-glucanase heterologously expressed in Pichia pastoris. Plant Physiol 127:674–684

Mouille G, Robin S, Lecomte M, Pagant S, Höfte H (2003) Classification and identification of Arabidopsis cell wall mutants using Fourier-transform infrared (FT-IR) microspectroscopy. Plant J 35:393–404

Nageli C (1858) Die Starkekbrner: Morphologisch, physiologisch, chemisch-physicalische und systematisch-botanische. Gebrfider Borntraeger, Berlin

Nicol F, His I, Jauneau A, Vernhettes S, Canut H, Höfte H (1998) A plasma membrane-bound putative endo-1,4-beta-D-glucanase is required for normal wall assembly and cell elongation in Arabidopsis. EMBO J 17:5563–5576

Nishitani K, Vissenberg K (2007) Roles of the XTH protein family in the expanding cell. In: Verbelen JP, Vissenberg K (eds) The expanding cell. Plant Cell Monographs, vol 5. Springer, Berlin Heidelberg New York (in press)

Nuhse TS, Stensballe A, Jensen ON, Peck SC (2004) Phosphoproteomics of the Arabidopsis plasma membrane and a new phosphorylation site database. Plant Cell 16:2394–2405

Pagant S, Bichet A, Sugimoto K, Lerouxel O, Desprez T, McCann M, Lerouge P, Vernhettes S, Höfte H (2002) KOBITO1 encodes a novel plasma membrane protein necessary for normal synthesis of cellulose during cell expansion in Arabidopsis. Plant Cell 14:2001–2013

Paredez AR, Somerville CR, Ehrhardt DW (2006) Visualization of cellulose synthase demonstrates functional association with microtubules. Science 312:1491–1495

Park YW, Baba K, Furuta Y, Iida I, Sameshima K, Arai M, Hayashi T (2004) Enhancement of growth and cellulose accumulation by overexpression of xyloglucanase in poplar. FEBS Lett 564:183–187

Peng L, Kawagoe Y, Hogan P, Delmer D (2002) Sitosterol-beta-glucoside as primer for cellulose synthesis in plants. Science 295:147–150

Persson S, Wei H, Milne J, Page GP, Somerville CR (2005) Identification of genes required for cellulose synthesis by regression analysis of public microarray data sets. Proc Natl Acad Sci USA 102:8633–8638

Rayle DL, Cleland RE (1992) The acid growth theory of auxin-induced cell elongation is alive and well. Plant Physiol 99:1271–1274

Refregier G, Pelletier S, Jaillard D, Höfte H (2004) Interaction between wall deposition and cell elongation in dark-grown hypocotyl cells in Arabidopsis. Plant Physiol 135:959–968

Richmond TA, Somerville CR (2000) The cellulose synthase superfamily. Plant Physiol 124:495–498

Richmond TA, Somerville CR (2001) Integrative approaches to determining Csl function. Plant Mol Biol 47:131–143

Robert S, Mouille G, Höfte H (2004) The mechanism and regulation of cellulose synthesis in primary walls: lessons from cellulose-deficient Arabidopsis mutants. Cellulose 11:351–364

Robert S, Bichet A, Grandjean O, Kierzkowski D, Satiat-Jeunemaitre B, Pelletier S, Hauser MT, Höfte H, Vernhettes S (2005) An Arabidopsis endo-1,4-beta-D-glucanase involved in cellulose synthesis undergoes regulated intracellular cycling. Plant Cell 17:3378–3389

Roland J-C, Reis D, Mosiniak M, Vian B (1982) Cell wall texture along the growth gradient of the mung bean hypocotyl: ordered assembly and dissipative processes. J Cell Sci 56:303–318

Roudier F, Fernandez AG, Fujita M, Himmelspach R, Borner GH, Schindelman G, Song S, Baskin TI, Dupree P, Wasteneys GO, Benfey PN (2005) COBRA, an Arabidopsis extracellular glycosyl-phosphatidyl inositol-anchored protein, specifically controls highly anisotropic expansion through its involvement in cellulose microfibril orientation. Plant Cell 17:1749–1763

Sato S, Kato T, Kakegawa K, Ishii T, Liu YG, Awano T, Takabe K, Nishiyama Y, Kuga S, Sato S, Nakamura Y, Tabata S, Shibata D (2001) Role of the putative membrane-bound endo-1,4-beta-glucanase KORRIGAN in cell elongation and cellulose synthesis in Arabidopsis thaliana. Plant Cell Physiol 42:251–263

Scheible WR, Eshed R, Richmond T, Delmer D, Somerville C (2001) Modifications of cellulose synthase confer resistance to isoxaben and thiazolidinone herbicides in Arabidopsis Ixr1 mutants. Proc Natl Acad Sci USA 98:10079–10084

Schindelman G, Morikami A, Jung J, Baskin TI, Carpita NC, Derbyshire P, McCann MC, Benfey PN (2001) COBRA encodes a putative GPI-anchored protein, which is polarly localized and necessary for oriented cell expansion in Arabidopsis. Genes Dev 15:1115–1127

Schmid M, Davison TS, Henz SR, Pape UJ, Demar M, Vingron M, Scholkopf B, Weigel D, Lohmann JU (2005) A gene expression map of Arabidopsis thaliana development. Nat Genet 37:501–506

Schrick K, Fujioka S, Takatsuto S, Stierhof YD, Stransky H, Yoshida S, Jürgens G (2004) A link between sterol biosynthesis, the cell wall, and cellulose in Arabidopsis. Plant J 38:227–243

Schumacher K, Vafeados D, McCarthy M, Sze H, Wilkins T, Chory J (1999) The Arabidopsis det3 mutant reveals a central role for the vacuolar H(+)-ATPase in plant growth and development. Genes Dev 13:3259–3270

Sibout R, Eudes A, Mouille G, Pollet B, Lapierre C, Jouanin L, Seguin A (2005) CINNAMYL ALCOHOL DEHYDROGENASE-C and -D are the primary genes involved in lignin biosynthesis in the floral stem of Arabidopsis. Plant Cell 17:2059–2076

Sturn A, Quackenbush J, Trajanoski Z (2002) Genesis: cluster analysis of microarray data. Bioinformatics 18:207–208

Szyjanowicz PM, McKinnon I, Taylor NG, Gardiner J, Jarvis MC, Turner SR (2004) The irregular xylem 2 mutant is an allele of korrigan that affects the secondary cell wall of Arabidopsis thaliana. Plant J 37:730–740

Takeda T, Furuta Y, Awano T, Mizuno K, Mitsuishi Y, Hayashi T (2002) Suppression and acceleration of cell elongation by integration of xyloglucans in pea stem segments. Proc Natl Acad Sci USA 99:9055–9060

Taylor NG, Laurie S, Turner SR (2000) Multiple cellulose synthase catalytic subunits are required for cellulose synthesis in Arabidopsis. Plant Cell 12:2529–2540

Taylor NG, Scheible WR, Cutler S, Somerville CR, Turner SR (1999) The irregular xylem3 locus of Arabidopsis encodes a cellulose synthase required for secondary cell wall synthesis. Plant Cell 11:769–780

Taylor NG, Howells RM, Huttly AK, Vickers K, Turner SR (2003) Interactions among three distinct CesA proteins essential for cellulose synthesis. Proc Natl Acad Sci USA 100:1450–1455

Traas JA, Braat P, Emons AM, Meekes H, Derksen J (1985) Microtubules in root hairs. J Cell Sci 76:303–320

Turner SR, Somerville CR (1997) Collapsed xylem phenotype of Arabidopsis identifies mutants deficient in cellulose deposition in the secondary cell wall. Plant Cell 9:689–701

Verbelen JP, Vissenberg K (2007) Cell expansion: past, present and perspectives. In: Verbelen JP, Vissenberg K (eds) The expanding cell. Plant Cell Monographs, vol 5. Springer, Berlin Heidelberg New York (in press)

Verhertbruggen Y, Knox JP (2007) Pectic polysaccharides and expanding cell walls. In: Verbelen JP, Vissenberg K (eds) The expanding cell. Plant Cell Monographs, vol 5. Springer, Berlin Heidelberg New York (in press)

Vissenberg K, Martinez-Vilchez IM, Verbelen J-P, Miller JG, Fry SC (2000) In vivo colocalization of xyloglucan endotransglycosylase activity and its donor substrate in the elongation zone of Arabidopsis roots. Plant Cell 12:1229–1237

Vorwerk S, Somerville S, Somerville C (2004) The role of plant cell wall polysaccharide composition in disease resistance. Trends Plant Sci 9:203–209

Williamson RE, Burn JE, Birch R, Baskin TI, Arioli T, Betzner AS, Cork A (2001). Morphogenesis in *rsw1*, a cellulose-deficient mutant of *Arabidopsis thalania*. Protoplasma 215:116–127

Zhong R, Ripperger A, Ye ZH (2000) Ectopic deposition of lignin in the pith of stems of two Arabidopsis mutants. Plant Physiol 123:59–70

Zhong R, Burk DH, Ye ZH (2001) Fibers. A model for studying cell differentiation, cell elongation, and cell wall biosynthesis. Plant Physiol 126:477–479

Zhong R, Burk DH, Morrison WH, Ye ZH (2002a) A kinesin-like protein is essential for oriented deposition of cellulose microfibrils and cell wall strength. Plant Cell 14:3101–3117

Zhong R, Kays SJ, Schroeder BP, Ye ZH (2002b) Mutation of a chitinase-like gene causes ectopic deposition of lignin, aberrant cell shapes, and overproduction of ethylene. Plant Cell 14:165–179

Zhong R, Morrison WH, Freshour GD, Hahn MG, Ye ZH (2003) Expression of a mutant form of cellulose synthase AtCesA7 causes dominant negative effect on cellulose biosynthesis. Plant Physiol 132:786–795

Zhong R, Pena MJ, Zhou GK, Nairn CJ, Wood-Jones A, Richardson EA, Morrison WH, Darvill AG, York WS, Ye ZH (2005) Arabidopsis fragile fiber8, which encodes a putative glucuronyltransferase, is essential for normal secondary wall synthesis. Plant Cell 17:3390–3408

Plant Cell Monogr (5)
J.-P. Verbelen · K. Vissenberg: The Expanding Cell
DOI 10.1007/7089_2006_071/Published online: 14 October 2006
© Springer-Verlag Berlin Heidelberg 2006

Hemicelluloses and Cell Expansion

Nicolai Obel · Lutz Neumetzler · Markus Pauly (✉)

Max-Planck Institute for Molecular Plant Physiology, Am Muehlenberg 1, 14476 Golm, Germany
pauly@mpimp-golm.mpg.de

Abstract Hemicelluloses are defined as wall polysaccharides that are not solubilized from wall materials with buffers, hot water, or chelating agents but only with more or less strong chaotropic agents such as alkali (O'Neill and York 2003). The most abundant hemicelluloses of the wall are xyloglucan, xylan, and mannans in their various forms. All of these polysaccharides have in common that, like cellulose, they consist of a backbone of β-1,4-linked D-pyranosyl residues but, unlike cellulose, they are branched and contain a variety of side chains. It is thought that the backbone structure of the hemicellulose leads to a strong non-covalent H-bond based association with cellulose microfibrils, thus forming a tight cellulose–hemicellulose network. In contrast, the side chains are responsible for a non-crystalline soluble form of parts of the polymer allowing, e.g., access to enzymes and other agents. Incorporation and metabolism of hemicelluloses is thus considered a key factor in the modulation of the load-bearing cellulose–hemicellulose network allowing cell elongation and plant growth to occur.

1 Xyloglucan

1.1 Structure

Xyloglucan (XyG) has been found in the primary cell walls, i.e., the walls of growing cells, of all higher plants (including gymnosperms) that have been investigated to date (Kim et al. 2004; Popper and Fry 2004). However, the percentage of XyG in the cell wall and their fine structure varies among plant species (McNeil et al. 1984). XyG comprises $\sim$ 10–25% of the cell walls of dicots and non-graminaceous monocots and are the most abundant hemicellulose present in these cell walls (McNeil 1984). In contrast, the cell walls of the graminaceous monocots (family Poaceae) contain only 6% XyG (Carpita 1996; Gibeaut et al. 2005).

XyG consists of a 1,4-linked β-DGlc*p* backbone with several α-D-Xyl*p* substituents at O6. These xylosyl residues can be further substituted and the various XyG side chains have been cataloged (Fig. 1). A concise nomenclature was developed to unambiguously describe XyG structures (Fry et al. 1993). According to this nomenclature, a XyG molecule is named after partitioning the backbone into segments of single glucosyl residues and their attached side

Code Letter (Mnemonic)	Represented Structure
G (Glucose)	-β-D-Glcp-
G	Ac(1→6)┐ -β-D-Glcp-
L (gaLactose)	β-D-Galp-(1→2)-α-D-Xylp-(1→6)┐ -β-D-Glcp-
L	Ac(1→6)-β-D-Galp-(1→2)-α-D-Xylp-(1→6)┐ -β-D-Glcp-
F (Fucose)	α-L-Fucp-(1→2)-β-D-Galp-(1→2)-α-D-Xylp-(1→6)┐ -β-D-Glcp-
F	Ac(1→6)┐ α-L-Fucp-(1→2)-β-D-Galp-(1→2)-α-D-Xylp-(1→6)┐ -β-D-Glcp-
J (Jojoba)	α-L-Galp-(1→2)-β-D-Galp-(1→2)-α-D-Xylp-(1→6)┐ -β-D-Glcp-
A (Arabinose)	α-D-Xylp-(1→6)┐ -β-D-Glcp- α-L-Araf-(1→2)┘
B (Beta-Xylose)	α-D-Xylp-(1→6)┐ -β-D-Glcp- β-D-Xylp-(1→2)┘
C	α-D-Xylp-(1→6)┐ -β-D-Glcp- α-L-Araf-(1→3)-β-D-Xylp-(1→2)┘
S (Solanaceae)	α-L-Araf-(1→2)-α-D-Xylp-(1→6)┐ -β-D-Glcp-
S	Ac(1→5)-α-L-Araf-(1→2)-α-D-Xylp-(1→6)┐ -β-D-Glcp-
T (Tomato)	β-L-Araf-(1→3)-α-L-Araf-(1→2)-α-D-Xylp-(1→6)┐ -β-D-Glcp-
U	β-D-Xylp-(1→2)-α-D-Xylp-(1→6)┐ -β-D-Glcp-

Fig. 1 Different known substituted backbone β-D-glucosyl residues in XyGs and their corresponding one-letter codes (Fry et al. 1993)

chain. Depending on the side chain configuration a single-letter code is assigned (Fig. 1, references to this code are printed in **bold letters** throughout the chapter). The structural elucidation of new XyG made it necessary to include additional code letters **T**, **J** (Hantus et al. 1997), and **U** (Ray et al. 2004) to the original list. In addition, XyG can be O-acetylated at various positions (Kiefer et al. 1989). The location of the *O*-acetyl substituent is incorporated into the single-letter code by underlining (Fig. 1, Jia et al. 2005).

Depending on the plant species different side chain configurations occur (example see Fig. 2). The substructure of XyG can be investigated in detail after digestion of the polymer with a β-1,4-endoglucanase (EG), which requires non-substituted Glc*p* units for cleavage (Fig. 2, York et al. 1990). Cleavage sites are marked with an open arrow. Upon digestion XyG oligosaccharides (XyGOs) are released, most of which consist of a cellotetraose backbone. In general, two types of XyG have been classified according to the number of xylosyl substitutions, the "XXXG"-type and the "XXGG"-type, where approximately 75% or 50% of the backbone residues are branched, respectively (Vincken et al. 1997).

The "XXXG"-type XyGs are highly branched polysaccharides that comprise 10–25% of the primary cell walls of the gymnosperms, dicotyledonous plants with the exception of the Laminales (peppermint, basil etc.) and Solanales (potato, tomato, tobacco etc.), and the non-graminaceous monocotyledonous plants (Hoffman et al. 2005 and references therein). The main XyGOs of this type are **XXXG**, **XXFG**, and **XLFG** but their percentage may vary depending on the species. Interestingly, the xylose on the non-reducing end of this type of oligosaccharides has so far never been found to be further substituted. In many of those species the β-Gal*p* residues often contain *O*-acetyl substituents (**L**, **F**) (Kiefer et al. 1989; Lerouxel et al. 2002; Maruyama et al. 1996; York et al. 1988). The *O*-acetyl substituent is mainly found on the O-6 position, but has been shown to migrate in aqueous solutions to the O-3 or O-4 position (Pauly et al. 1999a; Kiefer et al. 1990). The general use of alkali to solubilize XyG from cell walls (Joseleau et al. 1992; O'Neill and Selvendran 1983) hydrolyzes *O*-acetates, obscuring their presence in most species. Despite the preponderance of XyGOs with cellotetraose backbone, minor amounts of XyGOs with a cellotriose backbone have also been observed, e.g., XXG (Pauly et al. 2001a) as well as XyGOs with a cellopentaose backbone, such as in the seeds of the tropical plant *Hymenaea coubaril* L. (**XXXXG**) (Buckeridge et al. 1997) or in morning glory, *Ipomea purpurea* (**XXGGG**) (Hoffman et al. 2005). Such XyG might originate from metabolism in the wall or an altered biosynthetic mechanism (see below) and might have different conformational properties (O'Neill and York 2003).

The "XXGG"-type XyG comprise 10–15% of the walls of the dicots Laminales and Solanales (Hoffman et al. 2005; Vincken et al. 1996, 1997; York et al. 1996). This type of XyG is distinct from the "XXXG"-type also because they can be considered arabino-XyG (**S**, **T**) but not fuco-XyG as no fucosyl

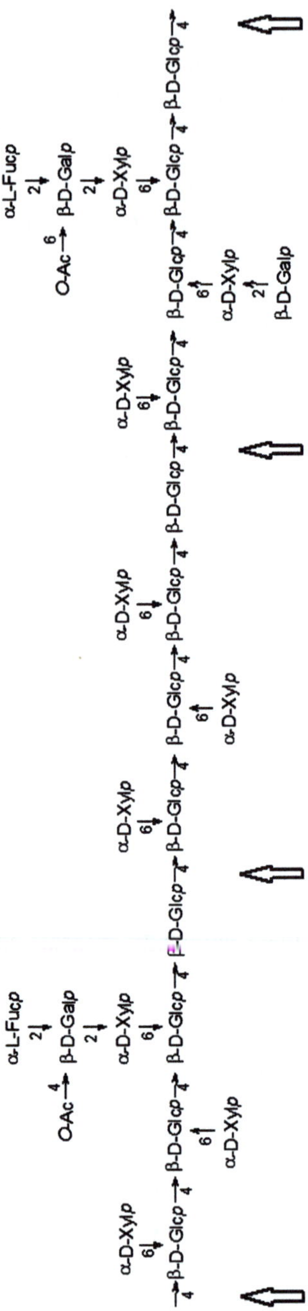

Fig. 2 Example of a general structure of XyG. The one-letter xyloglucan nomenclature for each of the substituted β-D-glucosyl residues is shown below the structure (see also Fig. 1). XyG can be depolymerized with endoglucanases, which hydrolyze the XyG backbone (see *arrows*), yielding XyG oligosaccharides (XyGOs)

residues are present. In addition to the *O*-acetyl substituents on the β-Gal*p*-residue (L), the α-Ara*f*-residue can be O-acetylated at the O-5 position (S) and even some of the backbone β-Glc*p* can be O-acetylated (G) (Jia et al. 2005). The *O*-acetyl group seems to occur most often in the third position from the non-reducing end, as in **XXGG**, suggesting that the *O*-acetyl group might substitute the Xyl residue in the "XXXG"-type XyG. XyGOs such as **XXGGG** have also been found in some of those species (Hoffman. et al. 2005; Jia et al. 2005).

"XXGG"-type XyG are also present in graminaceous monocots such as cereals or grasses (Sims et al. 2000). The XyG are distinct because the only substituents are α-D-Xyl*p* (**X**) or *O*-acetyl substituents (G). Other predominant XyGOs are **XXGGG** and even **XXGGGG** (Jia et al. 2005). Because of their low degree of substitution, these XyGs are insoluble in water and neutral buffers (Akiyama and Kato 1982; Kato et al. 1982a,b). Barley *(Hordeum vulgare L.)* meristematic cells seem to be enriched in XyG before the onset of enhanced β-D-glucan and glucuronoarabinoxylan synthesis during elongation (Carpita 1996; Carpita et al. 2001). Low amounts (10% of branched glucosyl residues) of β-D-Gal*p*-containing side chains (L) have been found in endosperm tissue (Shibuya and Misaki 1978). No α-L-Fuc*p* has been detected in any XyG prepared from the cell walls of whole graminaceous plants.

1.2 Biosynthesis

XyG is synthesized and processed in the Golgi apparatus and then exported via secretory vesicles to the apoplast (exocytosis) (Delmer and Stone 1988; Driouich et al. 1993). XyG has been found intracellularly in the Golgi apparatus using immunogold localization, but not in the endoplasmic reticulum. Partial fractionation of the Golgi membranes of pea stem microsomes by rate-zonal centrifugation has indicated that the glucose–xylose core of the XyG structure is initiated in the *cis* cisternae membranes and completed in the *trans* cisternae membranes, whereas fucosylation occurs in the *trans* cisternae membrane fraction and the Golgi secretory vesicles during transport to the wall (Brummell et al. 1990). Golgi activity increases during primary cell wall deposition, as indicated by hypertrophy of XyG present in the cisternae and an increase in the number of Golgi-derived vesicles during elongation (Lynch and Staehelin 1992). During cell elongation, the number of Golgi-derived vesicles and the amount of XyG present in the vesicles increase, indicating that overall Golgi activity and deposition of XyG in the primary cell wall increase during elongation events.

The very regular structure of the final polymer implies a precise coordination of the enzymes responsible for XyG biosynthesis (Brummell and Maclachlan 1989). Cytosolic nucleotide sugars (NDP-sugars) are transported into the Golgi apparatus, supplying the enzymatic machinery with activated sugars for the biosynthesis of cell wall polysaccharides. Numerous enzymes involved in the conversion of the appropriate nucleotide sugars, as well as Golgi transporters, are necessary (Orellana and Mohnen 1999; Reiter and Vanzin 2001; Seifert 2004). For XyG biosynthesis the presence of both UDP-D-glucose and UDP-D-xylose are required for chain elongation, implying that the addition of side-branch xylosyl residues is essential for the extension of the glucosyl backbone (Hayashi and Matsuda 1981a). However, the precise coordination of the glucosyl- and xylosyltransferases is not known (Gibeaut and Carpita 1994), although the rate of XyG synthesis includes the ratio of UDP-D-glucose to UDP-D-xylose (Hayashi and Matsuda 1981b). One possibility of synthesis regulation is hence the coordinated generation and presence of the appropriate nucleotide sugars. In contrast, the addition of galactosyl and fucosyl residues from UDP-D-galactose and GDP-L-fucose, respectively, onto the XyG backbone appears to be independent of chain elongation (Atalla and Vanderhart 1984; Hayashi 1989) and can also occur during backbone elongation (Atalla and Vanderhart 1984).

Recently, numerous components of the molecular machinery have been identified in *Arabidopsis* with the help of forward genetic approaches such as the *mur*-mutant screen (Reiter et al. 1997) or reverse genetic approaches (for reviews see Keegstra and Raikhel 2001; Scheible and Pauly 2004). The XyG-glucan synthase or synthases have remained elusive so far. However, a polypeptide was identified that acts in vitro as an α-Xyl*p*-transferase (AtXT1) (Faik et al. 2002). The heterologously expressed enzyme was able to transfer UDP-Xyl to an acceptor cello-oligosaccharide molecule with a minimum length of five glucosyl residues in vitro (Faik et al. 2002). Based on the results obtained in the in vitro assay additional xylosyltransferases would be necessary to generate "XXXG"-like XyGOs. Indeed, an additional six putative XT genes have been identified in the *Arabidopsis* genome, but none has been verified biochemically as a xylosyltransferase so far (Faik et al. 2002).

The specific transfer of a β-D-Gal residue to the third xylosyl unit (**XXXG** → **XXLG**) is catalyzed by a galactosyltransferase encoded by the *MUR3* gene in *Arabidopsis* (Madson et al. 2003). In XyG, solubilized by strong alkali from the *mur3* mutant, the Gal content is significantly reduced, causing a reduction of Fuc levels of up to $>90\%$ since the acceptor site for fucosylation (Gal-residues) is missing. However, the loss of **XXLG** is partially compensated by an enhanced biosynthesis of **XLXG** present in the polymer (Madson et al. 2003).

A Gal-transferase that transfers the galactosyl moiety to the middle position (**XXXG**→**XLXG**) has been proposed to be in the same gene family as *MUR3* (Li et al. 2004). An α-Fuc*p*-transferase (*MUR2*, AtFUT1, PsFT1) specifically catalyzes the glycosidic linkage of L-fucose to the third side chain of a building XyGO block (Faik et al. 2000; Madson et al. 2003; Vanzin et al. 2002). Non-fucosylated tamarind XyGOs did not act as fucosyl acceptors, but instead inhibited fucosyl transfer onto tamarind XyG, suggesting that the fucosyltransferase can bind XyGOs, but that only larger XyG fragments (MW > 3000) act as acceptor substrates for fucosylation (Maclachlan et al. 1992). The same size requirement of XyG was also demonstrated for acceptor substrates in the galactosylation reaction catalyzed by galactosyltransferase (Faik et al. 1997). In contrast, overexpression of AtFUT1 (Perrin et al. 2003) did not lead to an increase of fucosylated XyGOs but enhances O-acetylation from 20 to 30%. Thus, there seems to be a dependence of fucosylation and O-acetylation of the Gal residue (Pauly et al. 2001a; Perrin et al. 2003). However, the mechanism of XyG O-acetylation is hitherto unknown. A mutant defective in the synthesis of GDP-Fuc (*mur1*) exhibits a significant decline of L-fucose, which affects not only XyG (Reiter et al. 1993; Zablackis et al. 1996) but also glycoproteins (Rayon et al. 1999) as well as the pectic polysaccharides rhamnogalacturonan I and II (O'Neill et al. 2001; Reuhs et al. 2004). Since the fucosyltransferase is still active in *mur1* the Fuc residue is partially replaced by L-Gal, a homologous sugar, resulting in **XXJG** and **XLJG** oligosaccharides (Pauly et al. 2001a; Zablackis et al. 1996).

Plants carrying strong *MUR1* alleles show a reduction of L-fucose to ~ 40% in roots and to approximately 1% in aerial parts in comparison to wild type levels. These plants display a slightly dwarfed habit and their mechanical strength is halved in elongating inflorescence stems (Reiter et al. 1993). However, these phenotypes were attributed to a structural change in rhamnogalacturonan II rather than XyG (O'Neill et al. 2001). Leaky mutants (*mur1-3* and *mur1-7*) develop normally in regard to strength and habit and all phenotypes can be rescued by supplementing the medium with L-fucose (Reiter et al. 1993). Surprisingly, a more detailed analysis using an antibody (CCRC M1) against fucosylated epitopes present in XyG (Perrin et al. 2003) and rhamnogalacturonan I (Puhlmann et al. 1994) showed these epitopes to be present in the meristematic zones of all primary and lateral roots, as well as in some aboveground tissues of the *mur1* mutant, especially stipules, pollen, and pollen tubes (Freshour et al. 2003). In addition, a polypeptide has been found that can be reversibly glycosylated with UDP-glucose, UDP-xylose and UDP-galactose (Dhugga et al. 1997). The polypeptide is localized in the *trans*-Golgi cisternae and may be involved in XyG biosynthesis (Dhugga et al. 1997). However, recent data indicates that the protein is translocated specifically to the plasmodesmata (Sagi et al. 2005).

The incorporation of XyG into the cell wall is poorly understood. The non-covalent association of secreted XyG and extracellular cellulose (see below) presumably occurs by self-assembly (Hayashi 1989). Pulse-chase experiments using radioactively labeled XyG precursors in suspension-cultured cells indicate that most newly synthesized XyG is tightly bound to the cell wall directly after secretion (Edelmann and Fry 1992). However, in cell suspension cultures it is well known that XyG slough off into the medium (Pauly et al. 2001a). The M_W of alkali-extracted, polymeric XyG chains vary among plant species. A $M_W \leq 240$ kDa was measured for XyG extracted from cultured *Rosa* cells (Thompson and Fry 1997). Similar values could be confirmed for XyGs extracted from pea stem segments (Hayashi and Maclachlan 1984; Talbott and Ray 1992b). In contrast, XyG derived from suspension-cultured maize cells seems to contain XyG with a very high $M_W > 17$ MDa (Kerr and Fry 2003). These findings may represent a fundamental difference in xyloglucan metabolism between dicotyledons and gramineous monocotyledons.

1.3 Macromolecular Organization

Perhaps the most important characteristic feature of XyGs is that they form strong, non-covalent associations with cellulose (Acebes et al. 1993; Hayashi 1989; Hayashi et al. 1994a; Valent and Albersheim 1974). XyG binds to cellulose in vitro in a pH-dependent manner (Hayashi et al. 1987), suggesting that the polymers are associated by hydrogen bonds. In vitro reconstitution studies of XyG and cellulose have shown that XyG binds efficiently to cellulose even in the presence of other β-glucans, arabinogalactans, and pectins (Hayashi et al. 1987). In general, polymers consisting of backbones of β-1,4-linked glucosyl residues like cellulose are water-insoluble unless their self-aggregation is prevented by the presence of side chains such as those in XyG (Akiyama and Kato 1982; Kato et al. 1982a, 1982b; Ring and Selvendran 1981). Several studies, including dynamic conformational simulations, infer that the arrangement and conformation of the side chains of XyG affect their binding to cellulose. It has been proposed that a "twisted" conformation of the XyG backbone is favored in solution, a conformation that is also favored by amorphous unsubstituted glucan chains, as indicated by binding studies with carbohydrate binding proteins (Najmudin et al. 2006). In contrast, a "flat" XyG backbone conformation is adopted when XyG binds to the planar surface of cellulose microfibrils (Levy et al. 1991; Ogawa et al. 1990).

Most XyGOs do not bind to cellulose under physiological conditions (Valent and Albersheim 1974). At least five glucosyl backbone residues are necessary for a XyG that consists only of glucosyl and xylosyl residues (e.g., **GXXXG**) to bind. The extent of the binding increases as the backbone degree of polymerization increases (Hayashi et al. 1994c). When XyG also contains galactosyl residues in the side chains, at least 12 glucosyl backbone residues are necessary for binding (Vincken et al. 1995). Less substituted XyGs give the highest binding yields in in vitro assays, but binding is also dependent on the degree of cellulose crystallinity (Chambat et al. 2005). Tamarind seed XyGs, which possess side chains containing xylosyl and galactosyl residues, bind to cellulose less effectively than cellodextrins (unsubstituted glucans) (Hayashi et al. 1994b,c), suggesting that the side chains of XyG may be important for maintaining XyG solubility. However, enzymatic removal of terminal galactosyl residues results in the formation of a firm gel (Reid et al. 1988), indicating that the presence of monomeric xylosyl side chains is not sufficient to prevent XyG from self-associating.

The availability of *Arabidopsis* mutants that are altered only in XyG side chain composition (*mur2–mur3* see above) offered new tools for studying the effect of specific side-chain residues on the mechanical properties of the wall, as well as their importance in cell elongation and tissue morphology. Absence of the fucosyl residue as in *mur2* does not lead to any significant change in mechanical properties or plant phenotype (Pena et al. 2004; Vanzin et al. 2002). On the contrary, absence of the entire α-L-Fuc*p*-β-Gal*p* side chain, as in *mur3*, leads to significant phenotypic differences such as hypocotyl-base swelling (Pena et al. 2004). In addition, the tensile strength of etiolated hypocotyls is reduced to 50%. Further experiments indicated that both side-chain modifications have little effect on the polymer assembly of the XyG–cellulose network. However, the side-chain modifications present better or worse substrates for XyG-modifying enzymes such as XyG transglycosylases (XTH) (Català et al. 1997; Pena et al. 2004; see also Nishitani and Vissenberg, in this volume).

The addition of fucosyl residues on XyG (as in **XXFG**) might stabilize the "flat" conformation (Levy et al. 1991). This may explain the enhanced cellulose binding and binding rate of fucosylated (pea) XyG to cellulose compared to non-fucosylated (nasturtium) XyG, as shown in vitro with Avicel (microcrystalline cellulose) (Hayashi et al. 1994b; Levy et al. 1997). Antibody studies (Vian et al. 1992) indicate that XyG coat cellulose microfibrils in muro. Electron microscopy of rotary shadowed replicas of rapidly frozen, deep etched cell wall specimens (Itoh and Ogawa 1997; McCann et al. 1990, 1992) and of artificially assembled bacterial cellulose/XyG composites (Whitney et al. 1995) indicates that XyG can cross-link cellulose microfibrils. The cross-links are thought to be XyG because they only vanish after treatment of the walls with strong alkali, which is known to remove XyGs (Itoh and Ogawa 1997; McCann et al. 1990). XyG cross-links were also suggested by antibody studies: labeling cell walls with an XG-specific antibody indicated that the antigen resides mostly between microfibrils (Baba et al. 1994).

The links between cellulose microfibrils are generally shorter (16–40 nm) (McCann et al. 1990; Satiat-Jeunemaitre et al. 1992) than the length of extracted XyG (500 nm for a XyG with a MW of 300 kDa) (McCann et al. 1990), suggesting a lateral interaction between the linking polymers and cellulose in muro. Similar cross-bridge lengths were observed when XyG was mixed with cellulose in vitro (Whitney et al. 1995), showing that the network formation can be an abiotic process. Further evidence for the strong association between cellulose and XyG in the cell wall comes from suspension-cultured tomato cells grown in 2,6-dichlorobenzonitrile (DCB), an herbicide that inhibits cellulose biosynthesis (Shedletzky et al. 1990). The cellulose/XyG complex is virtually absent in the walls of these cells, as nearly all of the XyG produced by these cells is secreted into the medium (Shedletzky et al. 1992), indicating that cellulose is necessary for XyG to be held in the cell wall.

From extraction and solubilization studies it is evident that a XyG molecule can consist of various domains that are distinguished by accessibility to enzymes and solvents, implicating different roles in metabolism (Pauly et al. 1999a). When depectinated cell wall material of pea is digested with a xyloglucan-specific endoglucanase (XEG, Pauly et al. 1999b), only approximately one third of the XyG is released in the form of oligosaccharides. An additional one third is released with strong alkali. The remaining XyG portion can only be solubilized with an endoglucanase due to a concomitant digestion of cellulose microfibrils. Based on this data, it was postulated that XyG occurs in an enzyme-accessible domain (XEG-soluble), a domain that binds to cellulose microfibrils (alkali-soluble) and a domain that is trapped within the microfibril (cellulase-soluble). Molecular weight analysis of alkali-soluble polymeric XyG, with and without prior XEG digestion, clearly indicates that the enzyme-accessible domain and the microfibril-binding domain are connected. Taken together, these results provide evidence that XyGs bind and cross-link cellulose microfibrils in vivo. It has also been suggested that there are covalent links between XyG and pectic polysaccharides (Keegstra et al. 1973) and/or xylan (Coimbra et al. 1995). There is evidence that these complexes occur already in the Golgi apparatus (Cumming et al. 2005), but the structure of such putative covalent links remains to be elucidated.

1.4 Metabolism

XyG is also physiologically active (see below) and appears to have a distinct metabolism (Fry et al. 1993). XyG metabolism is stimulated by low pH (Jacobs and Ray 1975) and IAA (or auxin homologs) (Labavitch and Ray 1974a,b; Masuda 1990), a naturally occurring phytohormone that promotes cell elongation (Rayle and Cleland 1992). IAA-induced growth in pea shoots is accompanied by partial solubilization of XyG (Hoson 1990) and a decrease in average molecular weight (Hayashi and Maclachlan 1984; Labavitch 1981; Labavitch and Ray 1974a; Nishitani and Masuda 1991; Talbott and Ray 1992a). It has been demonstrated that IAA treatment also increases the amount of XyG that can be released from the cell wall by centrifugation (Terry and Bonner 1980; Terry and Jones 1981), including the release of cello-oligosaccharide (Tominaga et al. 1999). These cello-oligosaccharides might be the product of the further degradation of XyGOs or they might originate from the degradation of cellulose. IAA stimulates the metabolic turnover of the insoluble (i.e., cell wall-bound) XyG in this tissue (Nishitani and Masuda 1983; Gilkes and Hall 1977). XyG of elongating pine hypocotyl segments showed a gradual increase in molecular weight that could be suppressed by IAA treatment

(Lorences and Zarra 1987). Depolymerization of cell wall XyG is also accompanied by a reduction in XyG fucosyl content (Hoson et al. 1995). Most of these studies were done on XyG extracted from the whole organ, for example a stem or epicotyl. However, recent results suggest that auxin-induced elongation of stems is controlled by regulating the expansion of epidermal cells, and the accompanying decrease in the molecular weight of XyG is most pronounced in these cells (Hoson et al. 1993; Wakabayashi et al. 1991). Taken together, it is evident that XyG is metabolized during plant cell elongation and thus cell growth.

XyG can be metabolized by several different players, including *endo*-glucanases, glycosidases, expansins, and xyloglucan *endo*transglycosylases/hydrolases.

1.4.1 Endoglucanases

Application of IAA results in an increase in *endo*-β-1,4-glucanase (EG) activity in pea stems (Ferrari and Arnison 1974; Hayashi and Maclachlan 1984). High expression levels for several cloned EGs are also correlated to rapid expansion in plant tissues (Shani et al. 1997) and IAA treatment (Wu et al. 1996). Analysis of an EG purified from pea epicotyls indicated that XyG is its substrate (Hayashi et al. 1984). Some EGs have been shown to be XyG-specific (Tabuchi et al. 1997), and some only cleave specific types of XyG. For example, an EG purified from tobacco callus appears to only hydrolyze the "XXGG"-type XyG, but not seed amyloid XyG ("XXXG"-type) (Truelsen and Wyndaele 1991). Furthermore, an EG activity from tomato has been shown to be increased by XyGOs in vitro (Maclachlan and Brady 1992, 1994). These results led to the hypothesis that plant cell growth is promoted when the cell wall is loosened due to EG-catalyzed cleavage of XyG cross-links (Hensel et al. 1991; Vincken et al. 1995). Recently, a fungal EG has been identified that can induce wall extension (Yuan et al. 2001). Expression of this fungal enzyme in poplar (*Populus alba*) also led to an increased stem growth and increased tissue elasticity (Park et al. 2004). In addition, expression of a plant EG (cellulase from poplar) in *Arabidopsis thaliana* led to an increase in rosette leave size due to an increased cell size (Park et al. 2003). Taken together these results strongly suggest that this class of enzymes indeed plays an important role in cell elongation through modification of the cellulose–XyG network.

Proteins have been identified that inhibit the activity of a fungal XEG (Naqvi et al. 2005; Qin et al. 2003). Sequence homology indicated that such proteins are present in many plant species including four in *Arabidopsis* (Qin et al. 2003). Although a role for these proteins has been shown in the plant defense pathway against fungi (Jones et al. 2006; York et al. 2004), it cannot be excluded at this stage that these proteins might affect the activity of endogenous plant *endo*glucanases, hence modulates the influence of the EGs on plant cell elongation.

1.4.2 Glycosidases

The side chains of XyG or XyGOs can be affected by *exo*glycosidases (Fry 1995). An extract from *Arabidopsis* leaves was able to degrade XyGOs down to **XXG**, demonstrating that α-fucosidase, β-galactosidase, α-xylosidase, and β-glucosidase activities are present in the plant (Iglesias et al. 2006). An α-fucosidase activity could be isolated and cloned from pea epicotyls (Augur et al. 1992) as well as from *Arabidopsis* (AtXFG1) (de la Torre et al. 2002). Both enzymes exhibit fucosidase activity using **XXFG** oligosaccharides as substrate but not from polymeric XyG. Immunocytological studies showed that the pea enzyme is developmentally regulated (Augur et al. 1992). Furthermore, in situ hybridization of α-fucosidase transcripts indicated that it was also expressed in a tissue-specific manner, since the transcripts occurred only in the epidermis of elongating stems and vascular tissue of stem and leaves (Augur et al. 1995). Since the fucosyl residue might be crucial for growth inhibition mediated by XyGOs, the α-fucosidase may play an important role in deactivating these oligosaccharins. A β-galactosidase has been purified from nasturtium seeds that catalyzes the rapid removal of terminal β-D-galactosyl residues from XyGs (Edwards et al. 1986). A XyGO-specific α-xylosidase has been isolated and characterized from germinating nasturtium seeds (Fanutti et al. 1991), pea epicotyls (O'Neill et al. 1989) and *Arabidopsis* (AtXYL1) (Sampedro et al. 2001) that hydrolyzes only a xylosyl residue from the non-reducing end of XyGO. For example, **XXXG** is converted to **GXXG** by the action of this enzyme. **GXXG** can then be further degraded to **XXG** through the action of a β-glucosidase. Such a two-step metabolic stage could occur during maturation of the cell, as upon cessation of cell elongation the relative amount of **XXXG** is reduced whereas the amount of **XXG** is increased (Pauly et al. 2001b).

1.4.3 Expansins

Apoplastic proteins have been identified and cloned that, when added to an abraded, boiled hypocotyl placed under mechanical tension, induce gradual and irreversible elongation, often by more than 40% of its original length. These proteins were termed "expansins" (McQueen-Mason et al. 1992) (see McQueen-Mason et al, in this volume). Experiments carried out with these proteins led to the hypothesis that expansins bind at the interface between cellulose microfibrils and matrix polysaccharides in the walls and induce extension by reversibly disrupting non-covalent bonds within this polymeric network (McQueen-Mason and Cosgrove 1994). However, the precise mechanism by which expansins act on the plant cell wall is not known.

1.4.4
Xyloglucan endotransglycosylase/hydrolases

The hypothesis of wall-loosening through enzyme-mediated XyG metabolism was further substantiated by the discovery of xyloglucan *endo*transglycosylase/hydrolases (XTHs) (see Nishitani and Vissenberg, in this volume). XTHs cleave the β-1,4-linked glucose backbone of a XyG polymer (donor substrate), forming a transient XyG–XTH complex. XTH then transfers its substrate to the non-reducing end of another XyG or XGO (acceptor-substrate) by forming a new β-1,4-glucosyl linkage (Nishitani and Tominaga 1992). Since XyG is thought to cross-link cellulose microfibrils, resisting turgor-driven cell expansion, cleavage of the XyG by XTH and then rejoining to another XyG may cause temporary local loosening of the cell wall, allowing it to expand (Fry 1993; Nishitani 1995). Current models implicate these enzymes as being involved in polymer lengthening by incorporating nascent XyG chains to wall-bound XyG chains.

1.5
Biological Activity of XyGOs

Some of the XyGOs released by EG can be considered oligosaccharins (oligosaccharides with regulatory properties, Albersheim and Darvill 1985; Darvill et al. 1992). It has been shown that **XXFG** at nanomolar concentration inhibits endogenous growth of pea stems (Warneck and Seitz 1993) and its stimulation by 2,4-dichlorophenoxyacetic acid (2,4-D, a synthetic auxin) (Hoson and Masuda 1991; York et al. 1984), protons (Lorences et al. 1990), and gibberellins. Feeding XyG or XyGOs to split pea segments indicated that the polymer addition suppressed cell elongation, whereas the oligosaccharides promoted elongation (Takeda et al. 2002). It was also shown that **XXFG** is released in vivo from polymeric XyG, as it can be found in the medium of suspension-cultured spinach cells (Fry 1986b; McDougall and Fry 1991). Further investigations determined that the α-L-fucosyl residue is necessary for growth inhibition (McDougall and Fry 1989b; Purugganan et al. 1997), but the activity can be abolished by a neighboring galactosyl residue, as in **XLFG** (McDougall and Fry 1989a). **XXFG** in millimolar concentration promotes elongation of pea stems (McDougall and Fry 1990; Takeda et al. 2002) probably by integration into the existing XyG network, most likely through the action of transglycosylase reactions (Takeda et al. 2002) (see above and Nishitani and Vissenberg, in this volume). The α-L-fucosyl residue is not required for this effect as

XLLG and **XXXG** also promote stem elongation when present at millimolar concentration.

The mode of action of these XyGOs is unclear. There is evidence that XyGOs may induce the efflux of protons from the cell into the apoplast (Mutaftschiev et al. 1993) and thus promote the activation of cellulases (Farkaš and Maclachlan 1988) and induce growth of stem segments. Although XyGOs did not show any direct effect on cell wall viscoelasticity, preincubation of the stem segments with XyGOs enhanced their capacity to extend under acidic conditions (Cutillas-Iturralde and Lorences 1997), and decreased the molecular weight of polymeric XyG temporarily loosening the cell wall (Fry et al. 1993). **XXFG** inhibits 2,4-D-induced elongation in a non-competitive manner (Hoson and Masuda 1991). An increase in extractable, cationic cell wall-associated peroxidase isozymes has been correlated with this inhibition, leading to the suggestion that **XXFG** inhibits growth by promoting the peroxidase-catalyzed tightening of the cell wall (Warneck et al. 1996).

The *Arabidopsis* mutants *mur1*, *mur2* and *mur3* are affected in side-chain substitutions but do not exhibit significant growth phenotypes, therefore questioning the XyG-oligosaccharin hypothesis (Reiter 1998), in particular the precise role of the fucosyl residue. First, the aerial parts of *mur1* contain virtually non-fucosylated XyG (Reiter et al. 1997). However, the fucosyl residue is replaced by L-galactose and the L-galactose-containing XyGOs have also been shown to inhibit auxin-induced growth and thus represent a chemical analog (Zablackis et al. 1996). Then, in *mur2* plants no fucosylated XyG has been detected, but trace amounts might still be present (Vanzin et al. 2002) that could be sufficient for the biological activity of the oligosaccharide. However, *mur3* does not contain fucosylated XyG but shows a substantial increase in XLXG (Pena et al. 2004). The plant might compensate for the loss by other means. Nevertheless, it should be pointed out that inhibition of auxin-induced growth has only been demonstrated in pea (*Pisum sativum* L.). Such bioassays have so far never been successfully performed in *Arabidopsis*. There is a possibility that the system "*Arabidopsis*" does not contain XyG-based oligosaccharins, or that the structural requirements for activity on these oligosaccharins is different in this plant species.

XyGOs also affect plant morphogenesis (Pavlova et al. 1992). In the absence of 2,4-D the pentasaccharide **FG** increased the number of adventitious roots in cultured wheat embryos, whereas in the presence of 2,4-D it increased callus formation. XyGOs may also act as elicitors of plant defense responses such as the accumulation of antimicrobial phytoalexins (Pavlova et al. 1996).

In summary, the structure and physical state of XyG in the plant cell wall can vary in response to different physiological circumstances. These variations are a result of biosynthesis and post-biosynthetic processes, such as modification by cell wall proteins, including EGs, XTHs, glycosidases, and expansins.

2 Xylan

2.1 Structure

Xylans, another major group of hemicelluloses, are found in a wide range of plants. The structure, however, varies through different substitutions, which are dependent on the origin of the polymer (Ebringerova and Heinze 2000). In the monocotyledon family Poaceae, glucuronoarabinoxylan (GAX), also named arabinoxylan, is the major hemicellulose. GAX consists of β-(1-4)-linked xylopyranose with α-(1-2) and/or α-(1-3)-linked arabinofuranose (Gruppen et al. 1992). Additionally, the xylan backbone can be O-2 substituted with α-glucuronic acid or 4-*O*-methylglucuronic acid (Gruppen et al. 1992) (see Fig. 3). In grasses, GAX can be O-acetylated at O-2 of the arabinofuranose (Ishii 1991a,b). A unique feature of GAX in the order of Poales is the presence of ferulic acid and *p*-coumaric acid ester linked to the 5′ position of arabinofuranose. These hydroxycinnamates are capable of forming multiple forms of dimers in a similar fashion to monolignols (Grabber et al. 1995). Furthermore, other modifications of the GAX structures exist, such as the presence of xylopyranose β-O-2 linked to ferulated arabinose residues (Wende and Fry 1997) and the presence of shorter arabinan side chains (Schooneveld-Bergmans et al. 1999). In the dicot sycamore, GAX has been detected and the structure is likely to resemble the structure found in grasses (Darvill et al. 1980). However, hydroxy-cinnamic acids have not been reported as being a part of xylan in dicots.

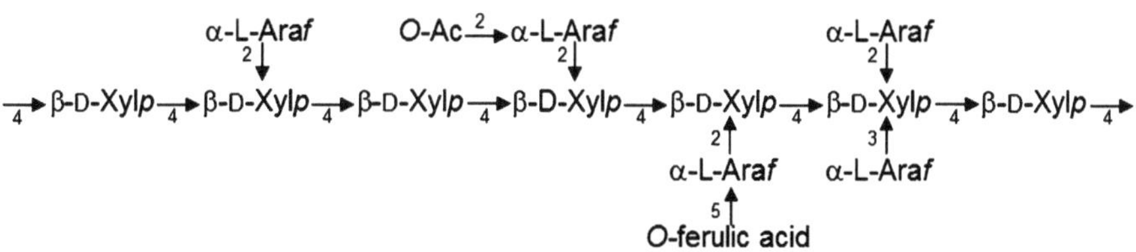

Fig. 3 Example of GAX structure. Xyl*p* refers to xylopyranose, ara*f* is arabinofuranose, while Ac presents an *O*-acetyl substituent

2.2 Biosynthesis and Metabolism

Xylan is synthesized in the Golgi apparatus from UDP-xylose (Feingold et al. 1959). UDP-xylose is generated from UDP-GlcA by a UDP-GlcA decarboxylase (UXS). In barley, higher levels of UXS*i* mRNAs correlate with a higher content of arabinoxylan (Zhang et al. 1996), suggesting that regulation of the synthesis of xylan might lie with the appropriate nucleotide sugar synthesis. Xylan synthase has in vitro a preference for rather long acceptor molecules (Kuroyama and Tsumuraya 2001). The xylan synthase is not yet known, however, it has been speculated that the xylan synthase would be found in the cellulose synthase-like (*Csl*) superfamily since the structure of xylan is similar to the β-glucan chain of cellulose (Richmond and Somerville 2000). Biochemical assays have shown that the xylan arabinosyltransferase uses UDP-arabinofuranose as precursor (Porchia et al. 2002) and the glucuronosyltransferase uses UDP-glucuronic acid (Hobbs et al. 1991; Baydoun and Brett 1997), but the identification of the genes has remained elusive. In the case of arabinoxylan from grasses, feruloylation appears to be localized in the Golgi apparatus, but the exact precursor has not yet been identified. Feruloyl-CoA is a likely candidate (Yoshida-Shimokawa et al. 2001), whereas other precursors such as 1-*O*-feruloyl-glucose (Mock and Strack 1993; Hu et al. 1999; Obel et al. 2003) or feruloyl-4-*O*-glucoside (Lim et al. 2001) cannot be excluded. Presumably the synthesized arabinoxylan is secreted into the apoplast where it becomes incorporated into the existing cell wall matrix. Given the high number of XTHs in rice, where the major hemicellulose is GAX, it can be speculated that some of these proteins might act on GAX as well instead of solely on XyG (Yokoyama et al. 2004).

GAX can be modified through the plant's own hydrolytic enzymes such as xylanases (Benjavongkulchai and Spencer 1986) and arabinofuranosidase (Hrmova et al. 1997). Interestingly, multiple glucuronidases have been isolated from fungi such as *Aspergillus* (Uchida et al. 1992) but no homologs from plants have been described. Xylanase has been located on the pollen coat in maize (Bih et al. 1999) and the activity is tightly controlled through an acidic protease that acts on the deposited "pre-xylanase" to form the active enzyme. This xylanase probably has a function in the hydrolysis of the stigma wall, facilitating pollen entry (Wu et al. 2002). Other xylanases from barley (Banik et al. 1996) and wheat (Cleemput et al. 1997) have been shown to be especially active in the aleurone layer during germination. Transcription of xylanases has also been shown in the stem of barley (Campenhout 2005). The activity of xylanases might be modulated by plant inhibitors such as the *endo*-xylanase inhibitors (Gebruers et al. 2001), which might act on both fungal and endogenous plant xylanases (Elliott et al. 2003). In wheat, rye, barley (Elliott et al. 2003), and rice (Goesaert et al. 2005) xylanase inhibitors against fungal enzymes have been reported and in wheat it has been demonstrated that transcription of some inhibitors can be induced with salicylic acid or methyl jasmonate. Although these inhibitors are mostly found in the seed they might be present in the entire plant (Igawa et al. 2005).

In maize coleoptiles it has been shown that a much higher amount of cellulose is present in the epidermis than in mesophyl cells This is countered by a decrease in GAX. Furthermore it was shown that the ratio of xylan substitution was significantly higher in the epidermis compared to the mesophyl cells (Carpita et al. 2001). Wheat coleoptiles show a gradient in the ratio of arabinose to xylose with the highest value at the base of the coleoptile and the lowest at the top (Obel et al. 2003). During development, a dramatic increase in hydroxycinnamates can be observed, probably reflecting an important role for ferulic acid and *p*-coumaric acid in cell wall stiffening (Obel et al. 2003; Azuma et al. 2005). This idea is supported by growth experiments using hypergravity conditions, where a marked increase in diferulic acids has been detected (Wakabayashi et al. 2005). Taken together, these data indicate a continuous adaptation of arabinoxylan structure to the specific requirements of the cell and it appears likely that as cell elongation ceases the arabinoxylan becomes less substituted with arabinose and has a higher abundance of ferulic acid dimers. This active adaptation of GAX has also been observed in maize where a turnover of GAX in developing maize coleoptiles has been observed (Carpita 1984). A different aspect of GAX in grasses is, furthermore, its potential role in cross-linking of different cell wall polymers after cell elongation has ceased. Since it has been proposed that ferulic acid and *p*-coumaric acid can not only form dimers but can also cross-link to lignin and proteins, they can thus covalently link different independent polymer networks into one molecule (Fry 1986a). The identification of 4-*O*-β- and 8-β-coniferyl alcohol cross-coupled to ferulate strongly supports the idea of a radical cross-coupling of polysaccharides to lignin precursors via ferulic acid (Bunzel et al. 2004).

Xylem from 8 to 15-year old *Alnus incana* Moench. and *Alnus glutinosa* Gaertn was analyzed for its xylan content. While some changes in extractability were observed, no large variation in molecular weight of the xylan was detected (Bikova and Treimanis 2002). In beech (*Fagus crenata*), glucuronoxylan was found to form ester linkages between glucuronic acid and lignin (Imamura et al. 1994). This implies a similar cross-linking of polymers as in grasses. This cross-linking could thus be an important factor for cell wall rigidity. This finding does also lead to the possibility that the methylation of glucuronic acid can be an important regulatory mechanism for cell wall cross-linking and stiffening. Additionally, the O-acetylation level might have a function in controlling how tightly the xylan can bind to cellulose, also providing xylan acetylation with a role in cell wall flexibility (Grondahl et al. 2003).

3 Mannans

3.1 Structure

A wide range of mannans exist in higher plants. Galactomannans and pure mannans have a β-(1-4)-linked mannopyranose backbone and in the case of

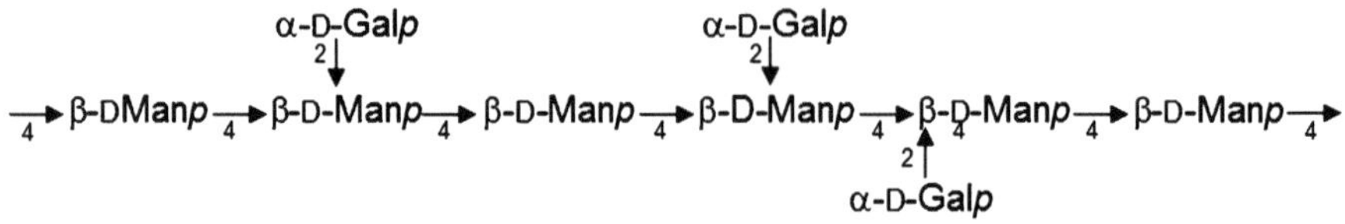

Fig. 4 Galactomannan consists of a mannopyranose, Man*p*, backbone substituted with galactopyranose, Gal*p*

galactomannan α-(1-6)-linked galactopyranose can be attached to the mannosyl residues (see Fig. 4). Galactomannans can often be found in the endosperm of leguminous seeds (Stephen 1983). Glucomannans, on the other hand, have a backbone consisting of β-(1-4)-linked mannopyranose and glucopyranose. When larger amounts of α-(1-6) galactosylation are present this polymer is referred to as galactoglucomannan. In *Arabidopsis* glucomannan has been detected in the thickening xylem fibers using a specific antibody and enzymatic hydrolysis (Handford et al. 2003), but it was not detected in leaves using chemical techniques (Zablackis et al. 1995). However, a mannan synthase activity has been reported for some *CslA* family members from *Arabidopsis* (see below). With the exception of wood, mannans are not very abundant but they are consistently observed in minor proportions in most plants. Species-specific alteration in substitution of the backbone is observed, possibly to adapt the mannan to specific requirements of the given species.

3.2 Metabolism

A mannan synthase from guar has been identified, which belongs to the cellulose synthase-like family, *CslA* (Dhugga et al. 2004). Biochemical evidence shows that some of the homologous *CslA* genes in *Arabidopsis* (*CslA9*, *CslA7*, and *CslA2*) synthesize mannan or glucomannan depending on the activated sugar supplied, implying that the ratio of glucose and mannose in the final glucomannan is dependent on the substrate availability of GDP-mannose and UDP-glucose (Liepman et al. 2005). Plants defective in *AtCslA7* have been shown to be defective in pollen tube growth, and homozygous knock-out plants of this gene were arrested at the globular stage during embryo development (Goubet et al. 2003). *AtCslA9* mutants have been studied and a strong expression of the gene was present in the elongating zone of roots. No major differences in the linkage structure of the non-cellulosic polysaccharides could be traced to the defective *CslA9* gene (Zhu et al. 2003). A galactomannan galactosyl transferase has been identified in fenugreek (*Trigonella foenum-graecum*) (Edwards et al. 1999) and a homolog has been identified in *Lotus japonicus* (Edwards et al. 2004). The fenugreek galactosyltransferase has been shown to recognize six residues of the mannan chain. The enzyme transfers the galactose from UDP-galactose onto the third mannose from the non-reducing end. When expressed in tobacco an elevated level of galactosylation in the endosperm cell wall galactomannan was measured, without any apparent effect on physiology or morphology (Reid et al. 2003).

The galactomannan in fenugreek has a dual role as seed storage polysaccharide and as regulator of the water balance during germination (Buckeridge et al. 2005). Breakdown of the galactomannan is accordingly tightly controlled, with an onset of α-galactosidase (Leubner-Metzger and Meins 2005) and β-*endo*-mannanase activity 24 h after sowing (Dirk et al. 1999; Leubner-Metzger and Meins 2005). A similar function of galactomannan in solanaceous species is possible and this hypothesis is supported by the finding that both *endo*-mannanase and β-mannosidase expression is correlated with seed germination in tomato (Mo and Bewley 2003). An *endo*-mannanase has been located on the surface of pollen, indicating a function for mannan in anther and pollen development and a possible requirement of mannanase activity in the degradation of the female cell walls such as stigma or transmitting tissue in the pistil (Filichkin et al. 2004).

Oligogalactomannans have been demonstrated to inhibit 2,4-dichlorophenoxyacetic acid-stimulated elongation growth of pea (*Pisum sativum* L. cv. Tyrkys) and spruce (*Picea abies* (L.) Karst) stem segments (Auxtova et al. 1995). These effects of oligosaccharides on growth might be understood in the context of the mannan transglycosylase, which is likely to incorporate newly synthesized galactomannan or mannan into the existing polymers in the wall, in a similar way to that observed for XyG (see above and Nishitani and Vissenberg, in this volume). So far, the mannan transglycosylase activity has been described in kiwi and tomato but the gene has yet to be identified (Schroder et al. 2004). It is thus likely that glucomannans are important for cell expansion, based on the finding that the mannan synthase *CslA9* is highly expressed in the root elongation zone. With the discovery of mannan transglycosylase activity, the control might be at multiple levels such as synthesis, targeted hydrolysis, *or* at the level of transglycosylation. Glucomannans are proposed to be tightly associated with cellulose microfibrils and the strength of the association might depend on the level of O-acetylation (Laffend and Swenson 1968). In maize, a much higher content of glucomannan is present in the epidermis of coleoptiles than in the mesophyll cells (Carpita et al. 2001). This could indicate that in grasses glucomannans are important in the load-bearing network of the cell wall since the epidermis is proposed to be the main element in the coleoptile (Kutschera et al. 1987). The finding that the activity of mannan transglycosylase was increasing with the maturation of the tomato fruit, and that all activity was essentially detected in the skin of the fruit, could also point to an important role for mannans in cell wall stiffening (Schroder et al. 2004).

References

Acebes JL, Lorences EP, Revilla G, Zarra I (1993) Pine xyloglucan, occurence, localization and interaction with cellulose. Physiol Plant 89:417–422

Akiyama Y, Kato K (1982) An arabinoxyloglucan from extracellular polysaccharides of suspension-cultured tobacco cells. Phytochem 21:2112–2114

Albersheim P, Darvill AG (1985) Oligosaccharins. Sci Am 253:58–64

Atalla RH, Vanderhart DL (1984) Native cellulose: A composite of two distinct crystalline forms. Science 223:283–285

Augur C, Benhamou N, Darvill A, Albersheim P (1992) Purification, characterization, and cell wall localization of an alpha-fucosidase that inactivates a xyloglucan oligosaccharin. Plant J 3:415–426

Augur C, Stiefel V, Darvill A, Albersheim P, Puigdomenech P (1995) Molecular cloning and pattern of expression of an alpha-L-fucosidase gene from pea seedlings. J Biol Chem 270:24839–24843

Auxtova O, Liskova D, Kakoniova D, Kubackova M, Karacsonyi S, Bilisics L (1995) Effect of galactoglucomannan-derived oligosaccharides on elongation growth of pea and spruce stem segments stimulated by auxin. Planta 196:420–424

Azuma T, Okita N, Nanmori T, Yasuda T (2005) Changes in cell wall-bound phenolic acids in the internodes of submerged floating rice. Plant Prod Sci 8:441–446

Banik M, Garrett TPJ, Fincher GB (1996) Molecular cloning of cDNAs encoding (1→4)-beta-xylan endohydrolases from the aleurone layer of germinated barley (Hordeum vulgare). Plant Mol Biol 31:1163–1172

Baba K, Sone Y, Misaki A, Hayashi T (1994) Localization of xyloglucan in the macromolecular complex composed of xyloglucan and cellulose in Pea stem. Plant Cell Physiol 3:439–444

Baydoun EAH, Brett CT (1997) Distribution of xylosyltransferases and glucuronyltransferase within the Golgi apparatus in etiolated pea (Pisum sativum L) epicotyls. J Exp Bot 48:1209–1214

Benjavongkulchai E, Spencer MS (1986) Purification and characterization of barley-aleurone xylanase. Planta 169:415–419

Bih FY, Wu SSH, Ratnayake C, Walling LL, Nothnagel EA, Huang AHC (1999) The predominant protein on the surface of maize pollen is an endoxylanase synthesized by a tapetum mRNA with a long 5′ leader. J Biol Chem 274:22884–22894

Bikova T, Treimanis A (2002) Solubility and molecular weight of hemicelluloses from Alnus incana and Alnus glutinosa. Effect of tree age. Plant Physiol Biochem 40:347–353

Brummell DA, Camirand A, Maclachlan GA (1990) Differential distribution of xyloglucan glycosyl transferases in pea Golgi dictyosomes and secretory vesicles. J Cell Sci 96:705–710

Brummell DA, Maclachlan GA (1989) Formation and functions of xyloglucan and derivatives. (review) ACS SympSer 399:18–32

Buckeridge MS, Hutcheon IS, Reid JSG (2005) The role of exo-(1→4)-beta-galactanase in the mobilization of polysaccharides from the cotyledon cell walls of Lupinus angustifolius following germination. Ann Bot 96:435–444

Buckeridge MS, Crombie HJ, Mendes CJM, Reid JSG, Gidley MJ, Vieira CCJ (1997) A new family of oligosaccharides from the xyloglucan of Hymenaea coubaril L (Leguminosae) cotyledons. Carbohydr Res 303:233–237

Bunzel M, Ralph J, Lu F, Hatfield RD, Steinhart H (2004) Lignins and ferulate-coniferyl alcohol cross-coupling products in cereal grains. J Agr Food Chem 52:6496–6502

Campenhout SVG (2005) Differential expression of endo-β-1,4-xylanase isoenzymes X-I and X-II at various stages throughout barley development. Plant Sci 169:512–522

Carpita NC (1984) Cell-wall development in maize coleoptiles. Plant Physiol 76:205–212

Carpita NC (1996) Structure and biogenesis of the cell walls of grasses. Annu Rev Plant Physiol Plant Mol Biol 47:445–476

Carpita NC, Defernez M, Findlay K, Wells B, Shoue DA, Catchpole G, Wilson RH, McCann MC (2001) Cell wall architecture of the elongating maize coleoptile. Plant Physiol 127:551–565

Català C, Rose JK, Bennett AB (1997) Auxin regulation and spatial localization of an endo-1,4-β-D-glucanase and a xyloglucan endotransglycosylase in expanding tomato hypocotyls. Plant J 12:417–426

Chambat G, Karmous M, Costes M, Picard M, Joseleau JP (2005) Variation of xyloglucan substitution pattern affects the sorption on celluloses with different degrees of crystallinity. Cellulose 12:117–125

Cleemput G, Van Laere K, Hessing M, Van Leuven F, Torrekens S, Delcour JA (1997) Identification and characterization of a novel arabinoxylanase from wheat flour. Plant Physiol 115:1619–1627

Coimbra MA, Rigby NM, Selvendran RR, Waldron KW (1995) Investigation of the occurence of xylan-xyloglucan complexes in the cell walls of olive pulp (Olea europaea). Carbohyd Polym 27:277–284

Cumming CM, Rizkallah HD, McKendrick KA, bdel-Massih RM, Baydoun EAH, Brett CT (2005) Biosynthesis and cell-wall deposition of a pectin-xyloglucan complex in pea. Planta 222:546–555

Cutillas-Iturralde A, Lorences EP (1997) Effect of xyloglucan oligosaccharides on growth, viscoelastic properties, and long-term extension of pea shoots. Plant Physiol 113:103–109

Darvill A, Augur C, Bergmann C, Carlson RW, Cheong J-J, Eberhard S, Hahn MG, Ló V-M, Marfà V, Meyer B, Mohnen D, O'Neill MA, Spiro MD, van Halbeek H, York WS, Albersheim P (1992) Oligosaccharins-oligosaccharides that regulate growth, development and defence responses in plants. Glycobiol 2:181–198

Darvill JE, McNeil M, Darvill AG, Albersheim P (1980) Structure of plant cell-cell walls. Glucuronoarabinoxylan, a 2nd hemicellulose in the primary cell walls of suspension-cultured sycamore cells. Plant Physiol 66:1135–1139

Dhugga KS, Barreiro R, Whitten B, Stecca K, Hazebroek J, Randhawa GS, Dolan M, Kinney AJ, Tomes D, Nichols S, Anderson P (2004) Guar seed beta-mannan synthase is a member of the cellulose synthase super gene family. Science 303:363–366

de la Torre F, Sampedro J, Zarra I, Revilla G (2002) AtFXG1, an Arabidopsis gene encoding alpha-L-fucosidase active against fucosylated xyloglucan oligosaccharides. Plant Physiol 128:247–255

Delmer DP, Stone BA (1988) Biosynthesis of plant cell walls. In: Preiss J (ed) The biochemistry of plants, a comprehensive treatise. Academic, Orlando, pp 373–420

Dhugga KS, Tiwari SC, Ray PM (1997) A reversibly glycosylated polypeptide (RGP1) possibly involved in plant cell wall synthesis: purification, gene cloning, and trans-Golgi localization. Proc Nat Acad Sci USA 94:7679–7684

Dirk LMA, van der Krol AR, Vreugdenhil D, Hilhorst HWM, Bewley JD (1999) Galactomannan, soluble sugar and starch mobilization following germination of Trigonella foenum-graecum seeds. Plant Physiol Biochem 37:41–50

Driouich A, Faye L, Staehelin LA (1993) The plant Golgi apparatus – a factory for complex polysaccharides and glycoproteins. Trends Biochem Sci 18:210–214

Ebringerova A, Heinze T (2000) Xylan and xylan derivatives – biopolymers with valuable properties, 1 – Naturally occurring xylans structures, procedures and properties. Macromol Rapid Comm 21:542–556

Edelmann HG, Fry SC (1992) Kinetics of integration of xyloglucan into the walls of suspension-cultured Rose cells. J Exp Bot 43:463–470

Edwards M, Bowman YJL, Dea ICM, Reid JSG (1986) A β-D-galactosidase from nasturtium (Tropaeolum majus L). J Biol Chem 263:4333–4337

Edwards ME, Choo TS, Dickson CA, Scott C, Gidley MJ, Reid JSG (2004) The seeds of Lotus japonicus lines transformed with sense, antisense, and sense/antisense galactomannan galactosyltransferase constructs have structurally altered galactomannans in their endosperm cell walls. Plant Physiol 134:1153–1162

Edwards ME, Dickson CA, Chengappa S, Sidebottom C, Gidley MJ, Reid JSG (1999) Molecular characterisation of a membrane-bound galactosyltransferase of plant cell wall matrix polysaccharide biosynthesis. Plant J 19:691–697

Elliott GO, McLauchlan WR, Williamson G, Kroon PA (2003) A wheat xylanase inhibitor protein (XIP-I) accumulates in the grain and has homologues in other cereals. J Cereal Sci 37:187–194

Faik A, Bar-Peled M, DeRocher AE, Zeng WQ, Perrin RM, Wilkerson C, Raikhel NV, Keegstra K (2000) Biochemical characterization and molecular cloning of an alpha-1,2-fucosyltransferase that catalyzes the last step of cell wall xyloglucan biosynthesis in pea. J Biol Chem 275:15082–15089

Faik A, Chileshe C, Sterling J, Maclachlan G (1997) Xyloglucan galactosyl- and fucosyltransferase activities from pea epicotyl microsomes. Plant Physiol 114:245–254

Faik A, Price NJ, Raikhel NV, Keegstra K (2002) An Arabidopsis gene encoding an alpha-xylosyltransferase involved in xyloglucan biosynthesis. Proc Nat Acad Sci USA 99:7797–7802

Fanutti C, Gidley MJ, Reid JSG (1991) A xyloglucan-oligosaccharide-specific α-D-xylosidase from germinated nasturtium (Tropaeolum majus L) seeds: purification, properties and its interaction with a xyloglucan-specific endo-(1→4)-β-D-glucanase and other hydrolases during storage-xyloglucan mobilisation. Planta 184:137–147

Farkaš V, Maclachlan G (1988) Stimulation of pea 1,4-β-glucanase activity by oligosaccharides derived from xyloglucan. Carbohydr Res 184:213–219

Feingold DS, Neufeld EF, Hassid WZ (1959) Xylosyl transfer catalyzed by an asparagus extract. J Biol Chem 234:488–489

Ferrari TE, Arnison PG (1974) Extraction and partial characterization of cellulases from expanding Pea epicotyls. Plant Physiol 54:487–493

Filichkin SA, Leonard JM, Monteros A, Liu PP, Nonogaki H (2004) A novel endo-beta-mannanase associated with anther and gene in tomato LeMAN5 is pollen development. Plant Physiol 134:1080–1087

Freshour G, Bonin CP, Reiter WD, Albersheim P, Darvill AG, Hahn MG (2003) Distribution of fucose-containing xyloglucans in cell walls of the mur1 mutant of Arabidopsis. Plant Physiol 131:1602–1612

Fry SC, York WS, Albersheim P, Darvill A, Hayashi T, Joseleau JP, Kato Y, Lorences EP, Maclachlan GA, McNeil M, Mort AJ, Reid JSG, Seitz HU, Selvendran RR, Voragen AGJ, White AR (1993) An anambiguous nomenclature for xyloglucan-derived oligosaccharides. Phys Plant 89:1–3

Fry SC (1986a) Cross-linking of matrix polymers in the growing cell-walls of Angiosperms. Annu Rev Plant Physiol Plant Mol Biol 37:165–186

Fry SC (1986b) In-vivo formation of xyloglucan nonasaccharide: a possible biologically active cell-wall fragment. Planta 169:443–453

Fry SC (1993) Plant cell expansion: Loosening the ties. Curr Biol 3:355–357
Fry SC (1995) Polysaccharide-modifying enzymes in the plant cell wall. Annu Rev Plant Physiol Plant Mol Biol 46:497–520
Fry SC, Aldington S, Hetherington PR, Aitken J (1993) Oligosaccharides as signals and substrates in the plant cell wall. Plant Physiol 103:1–5
Gebruers K, Debyser W, Goesaert H, Proost P, Van Damme J, Delcour JA (2001) *Triticum aestivum* L. endoxylanase inhibitor (TAXI) consists of two inhibitors, TAXI I and TAXI II, with different specificities. Biochem J 353:239–244
Gibeaut DM, Carpita NC (1994) Biosynthesis of plant-cell wall polysaccharides. Faseb J 8:904–915
Gibeaut DM, Pauly M, Bacic A, Fincher GB (2005) Changes in cell wall polysaccharides in developing barley (Hordeum vulgare) coleoptiles. Planta 221:729–738
Goesaert H, Gebruers K, Courtin CM, Delcour JA (2005) Purification and characterization of a XIP-type endoxylanase inhibitor from Rice (Oryza sativa). J Enzym Inhib Med Ch 20:95–101
Goubet F, Misrahi A, Park SK, Zhang ZN, Twell D, Dupree P (2003) AtCSLA7, a cellulose synthase-like putative glycosyltransferase, is important for pollen tube growth and embryogenesis in Arabidopsis. Plant Physiol 131:547–557
Gilkes NR, Hall MA (1977) The hormonal control of cell wall turnover in Pisum sativum L. New Phytol 78:1–15
Grabber JH, Hatfield RD, Ralph J, Zon J, Amrhein N (1995) Ferulate cross-linking in cell-walls isolated from maize cell-suspensions. Phytochem 40:1077–1082
Grondahl M, Teleman A, Gatenholm P (2003) Effect of acetylation on the material properties of glucuronoxylan from aspen wood. Carbohydr Polym 52:359–366
Gruppen H, Hoffmann RA, Kormelink FJM, Voragen AGJ, Kamerling JP, Vliegenthart JFG (1992) Characterization by H-1-NMR spectroscopy of enzymatically derived oligosaccharides from alkali-extractable wheat flour arabinoxylan. Carbohydr Res 233:45–64
Handford MG, Baldwin TC, Goubet F, Prime TA, Miles J, Yu XL, Dupree P (2003) Localisation and characterisation of cell wall mannan polysaccharides in Arabidopsis thaliana. Planta 218:27–36
Hantus S, Pauly M, Darvill AG, Albersheim P, York WS (1997) Structural characterization of novel L-galactose-containing oligosaccharide subunits of jojoba seed xyloglucans. Carbohydr Res 304:11–20
Hayashi T (1989) Xyloglucans in the primary cell wall. Annu Rev Plant Physiol Plant Mol Biol 40:139–168
Hayashi T, Maclachlan G (1984) Pea xyloglucan and cellulose III: metabolism during lateral expansion of pea epicotyl cells. Plant Physiol 76:739–742
Hayashi T, Matsuda K (1981a) Biosynthesis of xyloglucan in suspension-cultured soybean cells – evidence that the enzyme-system of xyloglucan synthesis does not contain β-1,4-glucan 4-β-D-glucosyltransferase activity. Plant Cell Physiol 22:1571–1584
Hayashi T, Matsuda K (1981b) Sugar nucleotides from suspension-cultured soybean cells. Agricult Biol Chem 45:2907–2908
Hayashi T, Wong YS, Maclachlan G (1984) Pea xyloglucan and cellulose 2 hydrolysis by Pea endo-1,4-beta-glucanases. Plant Physiol 75:605–610
Hayashi T, Marsden MPF, Delmer DP (1987) Pea xyloglucan and cellulose V. Xyloglucan-cellulose interactions in vitro and in vivo. Plant Physiol 83:384–389
Hayashi T, Baba K, Ogawa K (1994a) Macromolecular complexes of xyloglucan and cellulose obtained by annealing. Plant Cell Physiol 35:219–223
Hayashi T, Ogawa K, Mitsuishi Y (1994b) Characterization of the adsorption of xyloglucan to cellulose. Plant Cell Physiol 35:1199–1205

Hayashi T, Takeda T, Ogawa K, Mitsuishi Y (1994c) Effects of the degree of polymerization on the binding of xyloglucans to cellulose. Plant Cell Physiol 35:893–899

Hensel A, Brummell DA, Hanna R, Maclachlan G (1991) Auxin-dependent breakdown of xyloglucan in cotyledons of germinating nasturtium seeds. Planta 183:321–326

Hoffman M, Jia ZH, Pena MJ, Cash M, Harper A, Blackburn AR, Darvill A, York WS (2005) Structural analysis of xyloglucans in the primary cell walls of plants in the subclass Asteridae. Carbohydr Res 340:1826–1840

Hobbs MC, Baydoun EAH, Delarge MHP, Brett CT (1991) Interactions between a xylosyltransferase and a glucuronyltransferase involved in glucuronoxylan synthesis in pea epicotyls. Biochem Soc T 19:S245

Hoson T (1990) Effect of auxin on autolysis of cell walls in Azuki bean epicotyls. Plant Cell Physiol 31:281–287

Hoson T, Masuda Y (1991) Effect of xyloglucan nonasaccharide on cell elongation induced by 2,4-dichlorophenoxyacetic acid and indole-3-acetic acid. Plant Cell Physiol 32(6):777–782

Hoson T, Sone Y, Misaki A, Masuda Y (1993) Role of xyloglucan breakdown in epidermal cell walls for auxin-induced elongation of azuki bean epicotyl segments. Physiol Plant 87:142–147

Hoson T, Tabuchi A, Masuda Y (1995) Mechanism of xyloglucan breakdown in cell walls of Azuki bean epicotyls. J Plant Physiol 147:219–224

Hrmova M, Banik M, Harvey AJ, Garrett TPJ, Varghese JN, Hoj PB, Fincher GB (1997) Polysaccharide hydrolases in germinated barley and their role in the depolymerization of plant and fungal cell walls. Int J Biol Macromol 21:67–72

Hu WJ, Harding SA, Lung J, Popko JL, Ralph J, Stokke DD, Tsai CJ, Chiang VL (1999) Repression of lignin biosynthesis promotes cellulose accumulation and growth in transgenic trees. Nat Biotechnol 17:808–812

Igawa T, Tokai T, Kudo T, Yamaguchi I, Kimura M (2005) A wheat xylanase inhibitor gene, Xip-I, but not Taxi-I, is significantly induced by biotic and abiotic signals that trigger plant defense. Biosci Biotechnol Biochem 69:1058–1063

Iglesias N, Abelenda JA, Rodino M, Sampedro J, Revilla G, Zarra I (2006) Apoplastic glycosidases active against xyloglucan oligosaccharides of Arabidopsis thaliana. Plant Cell Physiol 47:55–63

Imamura T, Watanabe T, Kuwahara M, Koshijima T (1994) Ester linkages between lignin and glucuronic-acid in lignin-carbohydrate complexes from Fagus-crenata. Phytochem 37:1165–1173

Ishii T (1991a) Acetylation at O-2 of arabinofuranose residues in feruloylated arabinoxylan from bamboo shoot cell-walls. Phytochem 30:2317–2320

Ishii T (1991b) Isolation and characterization of a diferuloyl arabinoxylan hexasaccharide from bamboo shoot cell-walls. Carbohydr Res 219:15–22

Itoh T, Ogawa T (1997) Molecular architecture of the cell wall of poplar cells in suspension culture, as revealed by rapid-freezing and deep-etching techniques. Plant Cell Physiol 34:1187–1196

Jacobs M, Ray PM (1975) Promotion of xyloglucan metabolism by acidic pH. Plant Physiol 56:373–376

Jia ZH, Cash M, Darvill AG, York WS (2005) NMR characterization of endogenously O-acetylated oligosaccharides isolated from tomato (Lycopersicon esculentum) xyloglucan. Carbohydr Res 340:1818–1825

Jones RW, Ospina-Giraldo M, Deahl K (2006) Gene silencing indicates a role for potato endoglucanase inhibitor protein in germplasm resistance to late blight. Am J Potato Res 83:41–46

Joseleau J-P, Cartier N, Chambat G, Faik A, Ruel K (1992) Structural features and biological activity of xyloglucans from suspension-cultured plant cells. Biochimie 74:81–88

Kato Y, Ito S, Iki K, Matsuda K (1982a) Xyloglucan and β-glucan in cell walls of rice seedlings. Plant Cell Physiol 23:351–364

Kato K, Shiozawa R, Takeda S, Ito S, Matsuda K (1982b) Structural investigation of a β-D-glucan and a xyloglucan from bamboo-shoot cell-walls. Carbohydr Res 109:233–248

Keegstra K, Raikhel N (2001) Plant glycosyltransferases. Curr Opin Plant Biol 4:219–224

Keegstra K, Talmadge KW, Bauer WD, Albersheim P (1973) The structure of plant cell walls III: a model of the walls of suspension-cultured sycamore cells based on the interconnections of the macromolecular components. Plant Phys 51:188–196

Kerr EM, Fry SC (2003) Pre-formed xyloglucans and xylans increase in molecular weight in three distinct compartments of a maize cell-suspension culture. Planta 217:327–339

Kiefer LL, York WS, Albersheim P, Darvill AG (1990) The structure of plant cell walls XXX: structural characterization of an arabinose-containing heptadecasaccharide enzymatically isolated from sycamore extracellular xyloglucan. Carbohydr Res 197:139–158

Kiefer LL, York WS, Darvill AG, Albersheim P (1989) Xyloglucan isolated from suspension-cultured sycamore cell walls is O-acetylated. Phytochem 28:2105–2107

Kim SW, Ban SH, Chung H, Cho S, Chung HJ, Choi PS, Yoo OJ, Liu JR (2004) Taxonomic discrimination of flowering plants by multivariate analysis of Fourier transform infrared spectroscopy data. Plant Cell Reports 23:246–250

Kuroyama H, Tsumuraya Y (2001) A xylosyltransferase that synthesizes beta-(1→4)-xylans in wheat (*Triticum aestivum* L.) seedlings. Planta 213:231–240

Kutschera U, Bergfeld R, Schopfer P (1987) Cooperation of epidermis and inner tissues in auxin-mediated growth of maize coleoptiles. Planta 170:168–180

Labavitch JM (1981) Cell wall turnover in plant development. Ann Rev Plant Physiol 32:385–406

Labavitch JM, Ray PM (1974a) Relationship between promotion of xyloglucan metabolism and induction of elongation by indolacetic acid. Plant Physiol 54:499–502

Labavitch JM, Ray PM (1974b) Turnover of cell wall polysaccharides in elongating pea stem segments. Plant Physiol 53:669–673

Laffend KB, Swenson HS (1968) Effect of acetyl content of glucomannan on its sorption onto cellulose and on its beater additive properties: 1. Effect on sorption. Tappi 51:118–122

Lerouxel O, Choo TS, Seveno M, Usadel B, Faye L, Lerouge P, Pauly M (2002) Rapid structural phenotyping of plant cell wall mutants by enzymatic oligosaccharide fingerprinting. Plant Physiol 130:1754–1763

Leubner-Metzger G, Meins F (2005) Function and regulation of plant beta-1,3-glucanases(PR2). In: Datta SK, Mathukrishnan S (eds) Pathogenesis-related proteins in plants. CRC, Boca Raton, FL, pp 49–76

Levy S, York WS, Stuike-Prill R, Meyer B, Staehelin LA (1991) Simulations of the static and dynamic molecular conformations of xyloglucan. The role of the fucosylated sidechain in surface-specific sidechain folding. The Plant J 1:195–215

Levy S, Maclachlan G, Staehelin LA (1997) Xyloglucan sidechains modulate binding to cellulose during in vitro binding assays as predicted by conformational dynamics simulations. Plant J 11:373–386

Li XM, Cordero I, Caplan J, Molhoj M, Reiter WD (2004) Molecular analysis of 10 coding regions from arabidopsis that are homologous to the MUR3 xyloglucan galactosyltransferase. Plant Physiol 134:940–950

Liepman AH, Wilkerson CG, Keegstra K (2005) Expression of cellulose synthase-like (Csl) genes in insect cells reveals that CslA family members encode mannan synthases. Proc Natl Acad Sci USA 102:2221–2226

Lim EK, Li Y, Parr A, Jackson R, Ashford DA, Bowles DJ (2001) Identification of glucosyltransferase genes involved in sinapate metabolism and lignin synthesis in Arabidopsis. J Biol Chem 276:4344–4349

Lorences EP, Zarra I (1987) Auxin-induced growth in hypocotyl segments of Pinus pinaster Aiton: changes in molecular weight distribution of hemicellulosic polysaccharides. J Exp Bot 38:960–967

Lorences EP, McDougall GJ, Fry SC (1990) Xyloglucan- and cello-oligosaccharides: antagonists of the growth-promoting effect of H^+. Physiol Plant 80:109–113

Lynch MA, Staehelin LA (1992) Domain-specific and cell type-specific localization of two types of cell wall matrix polysaccharides in the clover root tip. J Cell Biol 118:467–479

Maclachlan G, Brady C (1992) Multiple forms of 1,4-beta-glucanase in ripening tomato fruits include a xyloglucanase activatable by xyloglucan oligosaccharides. Aust J Plant Physiol 19:137–146

Maclachlan G, Brady C (1994) Endo-1,4-beta-glucanase, xyloglucanase, and xyloglucan endotransglycosylase activities versus potential substrates in ripening tomatoes. Plant Physiol 105:965–974

Maclachlan G, Levy B, Farkaš V (1992) Acceptor requirements for GDP-fucose-xyloglucan 1,2-alpha-L-fucosyl-transferase activity solubilized from Pea epicotyl membranes. Arch Biochem Biophys 294:200–205

Madson M, Dunand C, Li X, Verma R, Vanzin GF, Caplan J, Shoue DA, Carpita NC, Reiter WD (2003) The MUR3 gene of Arabidopsis encodes a xyloglucan galactosyltransferase that is evolutionarily related to animal exostosins. Plant Cell 15:1662–1670

Maruyama K, Goto C, Numata M, Suzuki T, Nakagawa Y, Hoshino T, Uchiyama T (1996) O-acetylated xyloglucan in extracellular polysaccharides from cell-suspension cultures of Mentha. Phytochem 41:1309–1314

Masuda Y (1990) Auxin-induced cell elongation and cell wall changes. Bot Mag 103:345–370

McCann M, Wells B, Roberts K (1990) Direct visualization of cross-links in the primary plant cell wall. J Cell Sci 96:323–334

McCann M, Wells B, Roberts K (1992) Complexity in the spatial localization and length distribution of plant cell-wall matrix polysaccharides. J Microsc 166:123–136

McDougall GJ, Fry SC (1989a) Structure–activity relationship for xyloglucan oligosaccharides with antiauxin activity. Plant Physiol 89:883–887

McDougall GJ, Fry SC (1989b) Anti-auxin activity of xyloglucan oligosaccharides: the role of groups other than the terminal alpha-L-fucose residue. J Exp Bot 40:1–11

McDougall GJ, Fry SC (1990) Xyloglucan oligosaccharides promote growth and activate cellulase: evidence for a role of cellulase in cell expansion. Plant Physiol 93:1042–1048

McDougall GJ, Fry SC (1991) Xyloglucan nonasaccharide, a naturally occurring oligosaccharin, arises in vivo by polysaccharide breakdown. J Plant Physiol 137:332–336

McNeil M, Darvill A, Fry SC, Albersheim P (1984) Structure and function of the primary cell walls of plants. Annu Rev Biochem 53:625–663

McQueen-Mason S, Cosgrove DJ (1994) Disruption of hydrogen-bonding between plant-cell wall polymers by proteins that induce wall extension. Proc Natl Acad Sci USA 91:6574–6578

McQueen-Mason S, Durachko DM, Cosgrove DJ (1992) Two endogenous proteins that induce cell-wall extension in plants. Plant Cell 4:1425–1433

McQueen-Mason S, Le NT, Brocklehurst D (2007) Expansins. In: Verbelen JP, Vissenberg K (eds) The expanding cell. Plant Cell Monographs, vol 5. Springer, Berlin Heidelberg New York (in press)

Mo B, Bewley JD (2003) The relationship between beta-mannosidase and endo-beta-mannanase activities in tomato seeds during and following germination: a comparison of seed populations and individual seeds. J Exp Bot 54:2503–2510

Mock HP, Strack D (1993) Energetics of the uridine 5′-diphoglucose-hydroxy-cinnamic acid acyl-glucosyltransferase reaction. Phytochem 32:575–579

Mutaftschiev S, Macaya A, Prat R, Devillers P, Goldberg R (1993) Early effects of plant cell wall fragments on plant cell growth. Plant Physiol Biochem 31:459–467

Najmudin S, Guerreiro CIPD, Carvalho AL, Prates JAM, Correia MAS, Alves VD, Ferreira LMA, Romao MJ, Gilbert HJ, Bolam DN, Fontes CMGA (2006) Xyloglucan is recognized by carbohydrate-binding modules that interact with beta-glucan chains. J Biol Chem 281:8815–8828

Naqvi SMS, Harper A, Carter C, Ren G, Guirgis A, York WS, Thornburg RW (2005) Nectarin IV, a potent endoglucanase inhibitor secreted into the nectar of ornamental tobacco plants: isolation, cloning, and characterization. Plant Physiol 139:1389–1400

Nishitani K (1995) Endo-xyloglucan transferase, a new class of transferase involved in cell wall construction. J Plant Res 108:137–148

Nishitani K, Masuda Y (1983) Auxin-induced changes in the cell wall xyloglucans: effects of auxin on the two different subfractions of xyloglucans in the epicotyl cell wall of Vigna angularis. Plant Cell Physiol 24:345–355

Nishitani K, Masuda Y (1991) Auxin-induced changes in the cell wall structure: changes in the sugar compositions, intrinsic viscosity and molecular weight distributions of matrix polysaccharides of the epicotyl cell wall of Vigna angularis. Physiol Plant 52:482–494

Nishitani K, Tominaga R (1992) Endo-xyloglucan transferase, a novel class of glycosyltransferase that catalyzes transfer of a segment of xyloglucan molecule to another xyloglucan molecule. J Biol Chem 267:21058–21064

Nishitani K, Vissenberg K (2007) Roles of the XTH protein family in the expanding cell. In: Verbelen JP, Vissenberg K (eds) The expanding cell. Plant Cell Monographs, vol 5. Springer, Berlin Heidelberg New York (in press)

Obel N, Porchia AC, Scheller HV (2003) Intracellular feruloylation of arabinoxylan in wheat: evidence for feruloyl-glucose as precursor. Planta 216:620–629

O'Neill MA, Selvendran RR (1983) Isolation and partial characterization of a xyloglucan from the cell walls of Phaseolus coccineus. Carbohydr Res 111:239–255

O'Neill MA, York WS (2003) In: Rose JK (ed) The composition and structure of plant primary cell walls. Blackwell/CRC, Boca Raton, FL, pp 11–31

O'Neill RA, Albersheim P, Darvill AG (1989) Purification and characterization of a xyloglucan oligosaccharidespecific xylosidase from pea seedlings. J Biol Chem 264:20430–20437

O'Neill MA, Eberhard S, Albersheim P, Darvill AG (2001) Requirement of borate cross-linking of cell wall rhamnogalacturonan II for Arabidopsis growth. Science 294:846–849

Ogawa K, Hayashi T, Okamura K (1990) Conformational analysis of xyloglucans. Int J Biol Macromol 12:218–223

Orellana A, Mohnen D (1999) Enzymatic synthesis and purification of [H-3]uridine diphosphate galacturonic acid for use in studying Golgi-localized transporters. Anal Biochem 272:224–231

Park YW, Baba K, Furuta Y, Iida I, Sameshima K, Arai M, Hayashi T (2004) Enhancement of growth and cellulose accumulation by overexpression of xyloglucanase in poplar. FEBS Lett 564:183–187

Park YW, Tominaga R, Sugiyama J, Furuta Y, Tanimoto E, Samejima M, Sakai F, Hayashi T (2003) Enhancement of growth by expression of poplar cellulase in Arabidopsis thaliana. Plant J 33:1099–1106

Pauly M, Albersheim P, Darvill A, York WS (1999a) Molecular domains of the cellulose/xyloglucan network in the cell walls of higher plants. Plant J 20:629–639

Pauly M, Andersen LN, Kauppinen S, Kofod LV, York WS, Albersheim P, Darvill A (1999b) A xyloglucan-specific endo-beta-1,4-glucanase from Aspergillus aculeatus: expression cloning in yeast, purification and characterization of the recombinant enzyme. Glycobiol 9:93–100

Pauly M, Eberhard S, Albersheim P, Darvill A, York WS (2001a) Effects of the mur1 mutation on xyloglucans produced by suspension-cultured Arabidopsis thaliana cells. Planta 214:67–74

Pauly M, Qin Q, Greene H, Albersheim P, Darvill A, York WS (2001b) Changes in the structure of xyloglucan during cell elongation. Planta 212:842–850

Pavlova ZN, Ash OA, Vnuchkova VA, Babakov AV, Torgov VI, Nechaev OA, Usov AI, Shibaev VN (1992) Biological activity of a synthetic pentasaccharide fragment of xyloglucan. Plant Sci 85:131–134

Pavlova ZN, Loskutova NA, Vnuchkova VA, Muromtsev GS, Usov AI, Shibaev VN (1996) Xyloglucan oligosaccharins as elicitors of plant defense responses. Russ J Plant Physiol 43:242–246

Pena MJ, Ryden P, Madson M, Smith AC, Carpita NC (2004) The galactose residues of xyloglucan are essential to maintain mechanical strength of the primary cell walls in Arabidopsis during growth. Plant Physiol 134:443–451

Perrin RM, Jia Z, Wagner TA, O'Neill MA, Sarria R, York WS, Raikhel NV, Keegstra K (2003) Analysis of xyloglucan fucosylation in Arabidopsis. Plant Physiol 132:768–778

Popper ZA, Fry SC (2004) Primary cell wall composition of pteridophytes and spermatophytes. New Phytol 164:165–174

Porchia AC, Sorensen SO, Scheller HV (2002) Arabinoxylan biosynthesis in wheat: characterization of arabinosyltransferase activity in Golgi membranes. Plant Physiol 130:432–441

Puhlmann J, Bucheli E, Swain MJ, Dunning N, Albersheim P, Darvill AG, Hahn MG (1994) Generation of monoclonal antibodies against plant cell-wall polysaccharides. Plant Physiol 104:710

Purugganan MM, Braam J, Fry SC (1997) The arabidopsis TCH4 xyloglucan endotransglycosylase. Plant Physiol 115:181–190

Qin Q, Bergmann CW, Rose JKC, Saladie M, Kolli VSK, Albersheim P, Darvill AG, York WS (2003) Characterization of a tomato protein that inhibits a xyloglucan-specific endoglucanase. Plant J 34:327–338

Ray B, Loutelier-Bourhis C, Lange C, Condamine E, Driouich A, Lerouge P (2004) Structural investigation of hemicellulosic polysaccharides from Argania spinosa: characterisation of a novel xyloglucan motif. Carbohydr Res 339:201–208

Rayle DL, Cleland RE (1992) The acid growth theory of auxin-induced cell elongation is alive and well. Plant Physiol 99:1271–1274

Rayon C, Cabanes-Macheteau M, Loutelier-Bourhis C, Salliot-Maire I, Lemoine J, Reiter WD, Lerouge P, Faye L (1999) Characterization of N-glycans from Arabidopsis, application to a fucose-deficient mutant. Plant Physiol 119:725–733

Reid JSG, Edwards M, Dea ICM (1988) Enzymatic modification of natural seed gums. In: Phillips GO, Williams PA, Wedlock DJ (eds) Gums and stabilizers for the food industry. IRL, Oxford, pp 391–398

Reid JSG, Edwards ME, Dickson CA, Scott C, Gidley MJ (2003) Tobacco transgenic lines that express fenugreek galactomannan galactosyltransferase constitutively have structurally altered galactomannans in their seed endosperm cell walls. Plant Physiol 131:1487–1495

Reiter WD (1998) The molecular analysis of cell wall components. Trends Plant Sci 3:27–32

Reiter WD, Vanzin GF (2001) Molecular genetics of nucleotide sugar interconversion pathways in plants. Plant Mol Biol 47:95–113

Reiter WD, Chapple CCS, Somerville C (1993) Altered growth and cell walls in a fucose-deficient mutant of Arabidopsis. Science 261:1032–1035

Reiter WD, Chapple C, Somerville CR (1997) Mutants of Arabidopsis thaliana with altered cell wall polysaccharide composition. Plant J 12:335–345

Reuhs BL, Glenn J, Stephens SB, Kim JS, Christie DB, Glushka JG, Zablackis E, Albersheim P, Darvill AG, O'Neill MA (2004) L-Galactose replaces L-fucose in the pectic polysaccharide rhamnogalacturonan II synthesized by the L-fucose-deficient mur1 Arabidopsis mutant. Planta 219:147–157

Richmond TA, Somerville CR (2000) The cellulose synthase superfamily. Plant Physiol 124:495–498

Ring SG, Selvendran RR (1981) An arabinogalactoxyloglucan from the cell wall of Solanum tuberosum. Phytochem 20:2511–2519

Sagi G, Katz A, Guenoune-Gelbart D, Epel BL (2005) Class 1 reversibly glycosylated polypeptides are plasmodesmal-associated proteins delivered to plasmodesmata via the Golgi apparatus. Plant Cell 17:1788–1800

Sampedro J, Sieiro C, Revilla G, Gonzalez-Villa T, Zarra I (2001) Cloning and expression pattern of a gene encoding an alpha-xylosidase active against xyloglucan oligosaccharides from Arabidopsis. Plant Physiol 126:910–920

Satiat-Jeunemaitre B, Martin B, Hawes C (1992) Plant cell wall architecture is revealed by rapid-freezing and deep-etching. Protoplasma 167:33–42

Scheible WR, Pauly M (2004) Glycosyltransferases and cell wall biosynthesis: novel players and insights. Curr Opin Plant Biol 7:285–295

Schooneveld-Bergmans MEF, Beldman G, Voragen AGJ (1999) Structural features of (glucurono)arabinoxylans extracted from wheat bran by barium hydroxide. J Cereal Sci 29:63–75

Schroder R, Wegrzyn TF, Bolitho KM, Redgwell RJ (2004) Mannan transglycosylase: a novel enzyme activity in cell walls of higher plants. Planta 219:590–600

Seifert GJ (2004) Nucleotide sugar interconversions and cell wall biosynthesis: how to bring the inside to the outside. Curr Opin Plant Biol 7:277–284

Shani Z, Dekel M, Tsabary G, Shoseyov O (1997) Cloning and characterization of elongation specific endo-1,4-beta-glucanase (cel1) from Arabidopsis thaliana. Plant Mol Biol 34:837–842

Shedletzky E, Shmuel M, Delmer DP, Lamport DTA (1990) Adaptation and growth of tomato cells on the herbicide 2,6-dichlorobenzonitrile leads to production of unique cell walls virtually lacking a cellulose-xyloglucan network. Plant Physiol 94:980–987

Shedletzky E, Shmuel M, Trainin T, Salman S, Delmer DP (1992) Cell wall structure in cells adapted to growth on the cellulose-synthesis inhibitor 2,6-dichlorobenzonitrile. Plant Physiol 100:120–130

Shibuya N, Misaki A (1978) Structure of hemicellulose isolated from rice endosperm cell wall: Mode of linkages and sequences in xyloglucan, β-glucan and arabinoxylan. Agricult Biol Chem 42:2267–2274

Stephen AM (1983) Other plant polysaccharides. In: Aspinall GO (ed) The polysaccharides. Academic, Orlando, pp 98–193

Sims IM, Middleton K, Lane AG, Cairns AJ, Bacic A (2000) Characterisation of extracellular polysaccharides from suspension cultures of members of the Poaceae. Planta 210:261–268

Tabuchi A, Kamisaka S, Hoson T (1997) Purification of xyloglucan hydrolase/endotransferase from cell walls of azuki bean epicotyls. Plant Cell Physiol 38:653–658

Takeda T, Furuta Y, Awano T, Mizuno K, Mitsuishi Y, Hayashi T (2002) Suppression and acceleration of cell elongation by integration of xyloglucans in pea stem segments. Proc Natl Acad Sci USA 99:9055–9060

Talbott LD, Ray PM (1992a) Changes in molecular size of previously deposited and newly synthesized pea cell wall matrix polysaccharides Effects of auxin and turgor. Plant Physiol 98:369–379

Talbott LD, Ray PM (1992b) Molecular size and separability features of pea cell wall polysaccharides Implications for models of primary wall structure. Plant Physiol 98:357–368

Terry ME, Bonner BA (1980) An examination of centrifugation as a method of extracting an extracellular solution from peas, and its use for the study of indoleacetic acid-induced growth. Plant Physiol 66:321–325

Terry ME, Jones RL (1981) Soluble cell wall polysaccharides released from pea stems by centrifugation I Effect of auxin. Plant Physiol 68:531–537

Thompson JE, Fry SC (1997) Trimming and solubilization of xyloglucan after deposition in the walls of cultured rose cells. J Exp Bot 48:297–305

Tominaga R, Samejima M, Sakai F, Hayashi T (1999) Occurence of cello-oligosaccharides in the apoplast of auxin-treated pea stems. Plant Physiol 119:249–254

Truelsen TA, Wyndaele R (1991) Cellulase in tobacco callus – regulation and purification. J Plant Physiol 139:129–134

Uchida H, Nanri T, Kawabata Y, Kusakabe I, Murakami K (1992) Purification and characterization of intracellular alpha-glucuronidase from *Aspergillus niger*. Biosci Biotechnol Biochem 56:1608–1615

Valent BS, Albersheim P (1974) The structure of plant cell walls V: oN the binding of xyloglucan to cellulose fibers. Plant Physiol 54:105–108

Vanzin GF, Madson M, Carpita NC, Raikhel NV, Keegstra K, Reiter WD (2002) The mur2 mutant of Arabidopsis thaliana lacks fucosylated xyloglucan because of a lesion in fucosyltransferase AtFUT1. Proc Natl Acad Sci USA 99:3340–3345

Vian B, Temsah M, Reis D, Roland JC (1992) Co-localization of the cellulose framework and cell-wall matrix in helicoidal constuctions. J Microsc 166:111–122

Vincken JP, York WS, Beldman G, Voragen AGJ (1997) Two general branching patterns of xyloglucan, XXXG and XXGG. Plant Physiol 114:9–13

Vincken J-P, De Keizer A, Beldman G, Voragen AG (1995) Fractionation of xyloglucan fragments and their interaction with cellulose. Plant Physiol 108:1579–1585

Vincken J-P, Wijsman AJM, Beldman G, Niessen WMA, Voragen AGJ (1996) Potato xyloglucan is built from XXGG-type subunits. Carbohydr Res 288:219–232

Wakabayashi K, Sakurai N, Kuraishi S (1991) Differential effect of auxin on molecular weight distributions of xyloglucans in cell walls of outer and inner tissues from segments of dark grown squash (Cucurbita maxima Duch) hypocotyls. Plant Physiol 95:1070–1076

Wakabayashi K, Soga K, Kamisaka S, Hoson T (2005) Increase in the level of arabinoxylan-hydroxycinnamate network in cell walls of wheat coleoptiles grown under continuous hypergravity conditions. Physiol Plant 125:127–134

Warneck H, Seitz HU (1993) Inhibition of gibberellic acid-induced elongation-growth of pea epicotyls by xyloglucan oligosaccharides. J Exp Bot 44:1105–1109

Warneck HM, Haug T, Seitz HU (1996) Activation of cell wall-associated peroxidase isoenzymes in pea epicotyls by a xyloglucan-derived nonasaccharide. J Exp Bot 47:1897–1904

Wende G, Fry SC (1997) 2-*O*-*β*-D-xylopyranosyl-(5-*O*-feruloyl)-L-arabinose, a widespread component of grass cell walls. Phytochem 44:1019–1030

Whitney SEC, Brigham JE, Darke AH, Reid JSG, Gidley MJ (1995) In vitro assembly of cellulose/xyloglucan networks: ultrastructural and molecular aspects. Plant J 8:491–504

Wu SC, Blumer JM, Darvill AG, Albersheim P (1996) Characterization of an Endo-beta-1,4-glucanase gene induced by auxin in elongating pea epicotyls. Plant Physiol 110:163–170

Wu SSH, Suen DF, Chang HC, Huang AHC (2002) Maize tapetum xylanase is synthesized as a precursor, processed and activated by a serine protease, and deposited on the pollen. J Biol Chem 277:49055–49064

Yokoyama R, Rose JKC, Nishitani K (2004) A surprising diversity and abundance of xyloglucan endotransglucosylase/hydrolases in rice. Classification and expression analysis. Plant Physiol 134:1088–1099

York WS, Darvill AG, Albersheim P (1984) Inhibition of 2,4-dichlorophenoxyacetic acid-stimulated elongation of pea stem segments by a xyloglucan oligosaccharide. Plant Physiol 75:295–297

York WS, Oates JE, van Halbeek H, Darvill AG, Albersheim P (1988) Location of the O-acetyl substituents on a nonasaccharide repeating unit of sycamore extracellular xyloglucan. Carbohydr Res 173:113–132

York WS, van Halbeek H, Darvill AG, Albersheim P (1990) Structural analysis of xyloglucan oligosaccharides by H1 NMR-spectroscopy and fast-atom bombardment mass-spectrometry. Carbohydr Res 200:9–31

York WS, Kolli K, Orlando R, Albersheim P, Darvill AG (1996) The structures of arabinoxyloglucans produced by solanaceous plants. Carbohydr Res 285:99–128

York WS, Qin QA, Rose JKC (2004) Proteinaceous inhibitors of endo-beta-glucanases. Biochim Biophys Acta, Proteins Proteomics 1696:223–233

Yoshida-Shimokawa T, Yoshida S, Kakegawa K, Ishii T (2001) Enzymic feruloylation of arabinoxylan-trisaccharide by feruloyl-CoA: arabinoxylan-trisaccharide *O*-hydroxycinnamoyl transferase from Oryza sativa. Planta 212:470–474

Yuan S, Wu YJ, Cosgrove DJ (2001) A fungal endoglucanase with plant cell wall extension activity. Plant Physiol 127:324–333

Zablackis E, Huang J, Müller B, Darvill AG, Albersheim P (1995) Characterization of the cell-wall polysaccharides of *Arabidopsis thaliana* leaves. Plant Physiol 107:1129–1138

Zablackis E, York WS, Pauly M, Hantus S, Reiter WD, Chapple CCS, Albersheim P, Darvill A (1996) Substitution of L-fucose by L-galactose in cell walls of Arabidopsis mur1. Science 272:1808–1810

Zabotina OA, Gurjanov OP, Ayupova DA, Beldman G, Voragen AGJ, Lozovaya VV (1996) Biologically-active soluble oligosaccharides from pea stem tissues. Plant Cell Reports 15:954–957

Zhang ZQ, Pierce ML, Mort AJ (1996) Detection and differentiation of pectic enzyme activity in vitro and in vivo by capillary electrophoresis of products from fluorescent-labeled substrate. Electrophoresis 17:372–378

Zhu YM, Nam J, Carpita NC, Matthysse AG, Gelvin SB (2003) Agrobacterium-mediated root transformation is inhibited by mutation of an Arabidopsis cellulose synthase-like gene. Plant Physiol 133:1000–1010

Plant Cell Monogr (5)
J.-P. Verbelen · K. Vissenberg: The Expanding Cell
DOI 10.1007/7089_2006_072/Published online: 16 September 2006
© Springer-Verlag Berlin Heidelberg 2006

Roles of the XTH Protein Family in the Expanding Cell

Kazuhiko Nishitani[1] (✉) · Kris Vissenberg[2]

[1]Graduate School of Life Sciences, Tohoku University, 980-8578 Sendai, Japan
nishitan@mail.tains.tohoku.ac.jp

[2]Biology Dept., Plant Physiology and Morphology, University of Antwerp, Groenenborgerlaan 171, 2020 Antwerp, Belgium

Abstract Since xyloglucan is believed to be an important structural polysaccharide in the cell wall, possibly interconnecting adjacent cellulose microfibrils, enzymes that modify xyloglucan during the cell expansion process receive much attention. One of the enzymes is xyloglucan endotransglucosylase/hydrolase (XTH), a subgroup of glycoside hydrolase family 16. XTH proteins examined so far display endotransglycosylase (XET), hydrolase (XEH), or both activities towards xyloglucans. Genome sequencing of several model plant species has revealed that XTH proteins are encoded by large multigene families. Comprehensive analyses of XTH gene expression, together with functional analyses based on loss-of-function mutants, have provided evidence in support of the hypothesis that each member of these multigene families has its own role. This is reflected by different substrate specificities and pH dependencies of several individual members. Expression of each of these genes is precisely regulated by various hormones as well as by environmental signals. Some members appear to be critical in promoting cell wall expansion and are therefore essential for cell expansion, whereas others are required for construction of cell walls in cells that have completed the expansion process.

1 Introduction

1.1 Early Work that Lead to the Discovery of the XTH Protein Family and Development of a Unified Nomenclature

Two lines of evidence first suggested that xyloglucan plays a role in "the expanding cell", namely, structural analyses of the cell wall in suspension-cultured cells and metabolic analyses of cell wall polysaccharides during auxin-induced cell expansion.

In 1973, Albersheim and colleagues described a structural model for the plant cell wall in which xyloglucan molecules (see Obel et al., in this volume) were strongly attached to, and occasionally cross-linked or indirectly tethered to, cellulose microfibrils via hydrogen bonds (Keegstra et al. 1973). This first molecular model of the plant cell wall assumed that these polysaccharide cross-links between cellulose microfibrils (see Hématy and Höfte, in this volume) serve as load-bearing linkages, so that modification of the linkages would be necessary for cell wall extension, and, hence, cell expansion and morphogenesis (Valent and Albersheim 1974). Meanwhile, Ray and colleagues, who had been studying the dynamic aspects of plant cell walls, showed that the metabolic turnover of xyloglucan, as detected by release of a ^{14}C-labeled xyloglucan fragment from the cell wall, is enhanced during auxin- and acid-induced cell expansion in pea epicotyls (Labavitch and Ray 1974; Jacobs and Ray 1975). Following these pulse-chase experiments by Ray's group, Nishitani and Masuda (1981, 1982, 1983) have found direct evidence that the molecular size of xyloglucan decreases during auxin- and acid-induced cell extension growth in epicotyl sections of azuki bean (*Vigna angularis*). Similar changes in the molecular weight of xyloglucans during cell expansion processes have been observed in various plant species, including monocots (Inouhe et al. 1984) and gymnosperms (Lorences et al. 1990).

These observations lead to the hypothesis that splitting of the load-bearing xyloglucans connecting cellulose microfibrils is the rate-limiting step for changes in the mechanical properties of the cell wall and that it thereby regulates stress-relaxation processes of the cell wall. However, simple cleavage of linkages between cellulose microfibrils without integration of newly deposited microfibrils would merely result in thinning of the cell wall, eventually causing a loss of its mechanical strength and its ability to resist mechanical tension. In fact, the cell wall does not become thinner during typical cell expansion. Thus, a simple xyloglucan cleavage model cannot account for ordinary cell wall expansion in which newly synthesized cellulose microfibrils are integrated into a preexisting cellulose/xyloglucan framework (reviewed by Nishitani 1997).

To explain this paradox, Albersheim (1976) proposed the involvement of "*an endotransglycosylase that would transfer a portion of a polysaccharide to itself*" in the remodeling process of expanding cells, a hypothetical process that, at the time, lacked experimental or even circumstantial evidence. It was not until the 1990s, when the existence of this hypothetical process was suggested by three independent lines of evidence. Finally, this hypothesis was proven in 1992 when an enzyme capable of catalyzing molecular grafting between xyloglucans was purified from *Vigna angularis* and its mode of action was characterized (Nishitani and Tominaga 1992; see Nishitani 1997 for a review of the discovery of the XTH protein family).

Since then, the number of proteins and genes in this family has grown rapidly, resulting in overlapping and contradictory nomenclature and classification. Accordingly, discussion among researchers through the Cell Wall Newsgroups and a discussion forum at the 9th International Cell Wall Meeting held in Toulouse (France) in 2001 led to adoption of the name xyloglucan endotransglucosylase/hydrolase (XTH) for this family of proteins and genes (Rose et al. 2002). In addition, it was decided that the XTH nomenclature should be based on sequence homology. As a result, the genes and proteins in several model plants including *A. thaliana* (Yokoyama and Nishitani 2001b), rice (Yokoyama and Nishitani 2004) and poplar (Geisler-Lee et al. 2006) as well as in other plants were (re)named according to this nomenclature.

In this chapter we focus on the roles of individual members of the XTH family of proteins in cell expansion.

2
XTH Proteins

2.1
XTH as a Subgroup of Glycoside Hydrolase Family 16

According to the carbohydrate active enzyme database (CAZyme; http://www.cazy.org/CAZY/index.html), in which carbohydrate-related enzymes are classified according to structural features of catalytic and carbohydrate-binding modules, XTH belongs to a subgroup of glycoside hydrolase family 16 (GH16), a family of enzymes characterized by Pfam ID: Glyco_hydro_16 (accession number PF00722) (Henrissat 1991; Henrissat and Bairoch 1993, 1996). XTHs exhibit one or both of two enzyme activities: xyloglucan : xyloglucosyl transferase (EC 2.4.1.207) and xyloglucan-specific endo-β-1,4-glucanase (EC 3.2.1.151) activity, which are currently referred to as xyloglucan endotransglucosylase (XET) and xyloglucan endohydrolase (XEH) activity, respectively (Rose et al. 2002). Thus, the XTH family is defined based on the structural features of the proteins but not on their enzymatic activities. Based on the structural features, the XTH family of proteins and, hence, the genes encoding them, are grouped into three classes (I, II, and III). Members of the class I/II subfamily are thought to mediate transferase-directed catalytic (XET) activity, whereas members of the class III subfamily mediate hydrolysis-directed (XEH) activity.

The XTH family of proteins contains the signature motif DEIDFEFLG, which includes the amino acids that mediate catalysis (Nishitani 1995, 1997). A site-specific mutagenesis study of AtXTH22 demonstrated that the first glutamic acid residue in the DEIDFEFLG motif is essential for catalytic activity (Campbell and Braam 1998). XTH proteins are typically N-glycosylated on a threonine or serine residue in the vicinity of the catalytic site. Removal of this *N*-glycosyl group in recombinant AtXTH22 protein reduces the enzyme activity by 98%. A similar effect is found using a recombinant XTH protein from the very primitive plant *Selaginella* (Lycopodiophyte; Van Sandt et al. 2006). This indicates that the *N*-glycosyl group has a critical role in maintaining enzyme activity. Furthermore, the XTHs possess signal peptides targeting them for secretion into the apoplast (Yokoyama and Nishitani 2001a).

In addition to XTH, the family GH16 includes keratan-sulfate endo-1,4-β-galactosidase (EC 3.2.1.103), glucan endo-1,3-β-D-glucosidase (EC 3.2.1.39), endo-1,3(4)-β-glucanase (EC 3.2.1.6), licheninase (EC 3.2.1.73), agarase (EC 3.2.1.81), and κ-carrageenase (EC 3.2.1.83), all of which are bacterial or fungal proteins and are not found in higher plants (http://www.cazy.org/CAZY/index.html). Thus, evolution of the XTH family protein is intrinsic to plants, implying that it has a unique function in plant cell walls and, therefore, plant growth and development.

2.2 XTH Families of Genes in Plants

Genome sequencing of several model plants has revealed that higher plants typically contain dozens of XTH proteins. For example, in the genome of *Arabidopsis thaliana* (L.) Heynh. ecotype Columbia, 33 open reading frames encoding XTH proteins have been detected (Yokoyama and Nishitani 2001b). Transcripts derived from individual XTH genes were identified by RT-PCR using gene-specific primer sets (Yokoyama and Nishitani 2001b), indicating that all 33 *A. thaliana* XTH open reading frames are functional genes. In addition, the completed genomic database for rice (*Oryza sativa* L. Cultivar Nipponbare) contains 29 members of the XTH gene family. Expression of transcripts of most of these rice XTH genes was confirmed by gene-specific microarray analysis and real-time RT-PCR analysis (Yokoyama et al. 2004).

Figure 1 shows a phylogenetic tree of XTH families for the two model plants. Given that xyloglucan is considerably less abundant in cell walls of monocotyledons than dicotyledons and is not typically ascribed an important load-bearing role in the cell wall, it is noteworthy that the number of XTH family members in rice and *A. thaliana* (Yokoyama et al. 2004) is similar. In this respect it can also be noted that using a fluorescent assay to detect XET activity Vissenberg et al. (2003) found very high XET activity in the elongation zone of several grass species. One possible explanation for this apparent paradox is that the xyloglucan in rice walls has a more important structural function than is generally thought. An alternative explanation is that XTHs are involved in the modification of hemicelluloses other than xyloglucan. At present, there is no reason to exclude the possibility that XTHs from rice and other monocotyledons act on other polymers such as (1–3), (1→4)-β-D-glucans or glucuronoarabinoxylans, which are abundant in monocot cell walls (Yokoyama et al. 2004). More recently, sequencing of the genome of *Populus trichocarpa* Torr. and Gray (http://genome.jgi-psf.org/Poptr1/Poptr1.home.html) has revealed that it contains 41 XTH genes (Geisler-Lee et al. 2006). It has been suggested that the higher number of XTH family proteins in woody plants than in *A. thaliana* and rice indicates that XTH plays more important roles in secondary wall construction processes, which are more sophisticated in woody plants.

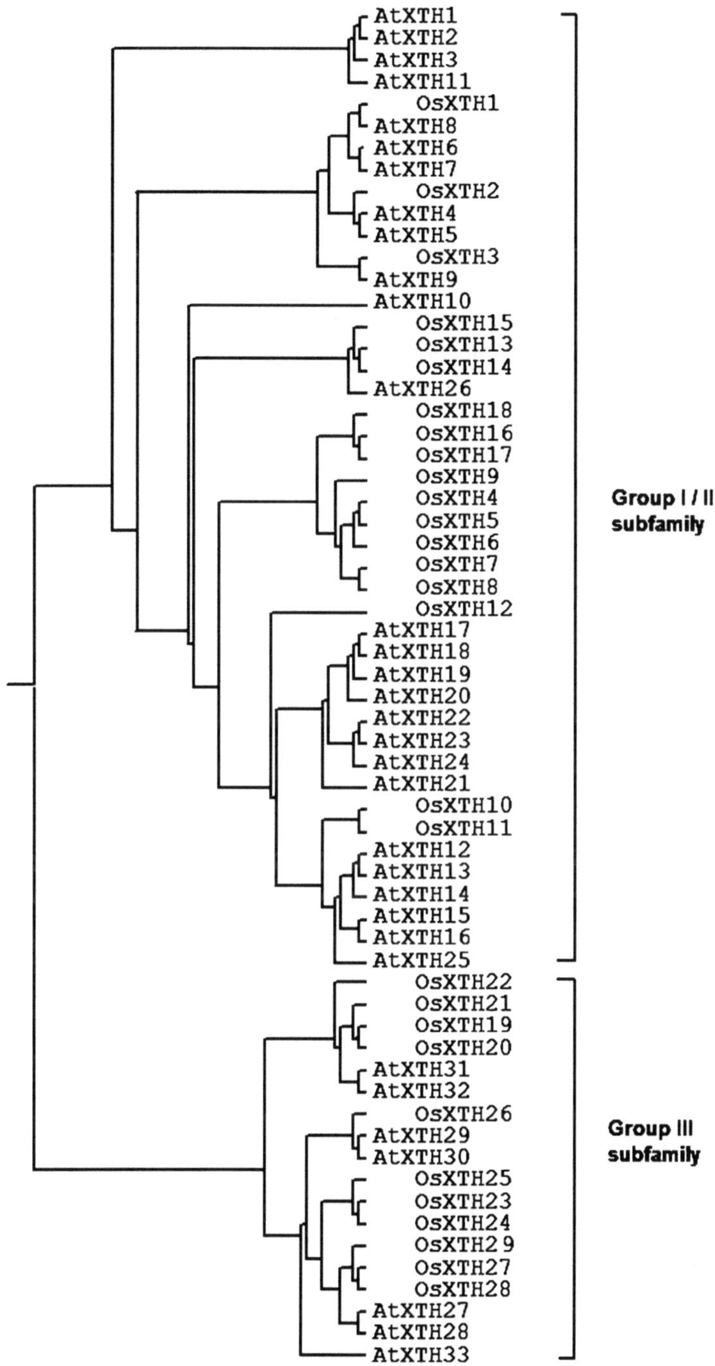

Fig. 1 Phylogenetic tree of XTHs from *A. thaliana* and *O. sativa*. Modified from Yokoyama et al. (2004)

2.3 Crystallography

A recombinant XTH encoded by *Populus tremula* x *tremuloides* XET16A (Ptt-XET16A recently renamed to PttXTH34, Geisler-Lee et al. 2006), which is structurally closest to *A. thaliana* AtXTH5, was heterogeneously expressed in *Pichia pastoris*, and its crystal structure was determined at 1.8-Å resolution (Johansson et al. 2003, 2004). This crystallographic study showed that the overall structure of XTH is a curved β-sandwich, a structural feature also found in other GH16 family proteins; however, the substrate binding cleft has a structure characteristic of the GH7 family proteins (Johannsson et al. 2003, 2004). In addition, crystallographic analysis suggests that glycoside hydrolases in the GH11 family (1,4-β-D-xylan endohydrolases) have similar β-sandwich structures (Strohmeier et al. 2004). In the light of both the crystallographic data for the poplar XTHs and the genomic data for the rice XTH family members, it is likely that XTHs from other plants can act on polysaccharides other than xyloglucans, especially xylans and glucans.

3 Biochemical Characterization of XTH

3.1 Measurement of Enzyme Activity

As mentioned above, all XTH family proteins examined so far have XET, XEH, or both activities towards xyloglucans. Thus, two different assays are required for biochemical characterization of XTH proteins.

3.1.1 XEH Activity (Hydrolase Activity)

Endohydrolytic activity toward xyloglucans can be assayed by measuring changes in specific viscosity and/or reducing power of a xyloglucan solution over time. Based on this assay, the first XTH capable of specifically hydrolyzing xyloglucan was purified from nasturtium (*Tropaeolum majus* L.) (Edwards et al. 1986). At that time, this enzyme was implicated in the degradation of storage xyloglucan in germinating seeds of the plants and was not considered essential in the regulation of primary cell wall architecture (Edwards et al. 1985). So far, all the XTH proteins with hydrolytic activity are identified as members of class III (Fanutti et al. 1993; Farkas et al. 1992; Kaku et al. 2002).

3.1.2 XET Activity (Transferase Activity)

Segment transfer between xyloglucan molecules, also known as the molecular grafting reaction, can be assayed by two different procedures: (i) measurement of changes in the molecular weight distribution profile of xyloglucan molecules, and (ii) measurement of the rate of transfer of a portion of a donor substrate to a radio-actively or fluorescently labeled xyloglucan acceptor (Nishitani 1997).

The first evidence for the existence of enzymes capable of transferring a segment of xyloglucan to another xyloglucan polymer was obtained when a xyloglucan with a defined molecular weight distribution was incubated with the apoplastic enzymes prepared from *V. angularis* (azuki bean) (Nishitani and Tominaga 1991). Specifically, it was shown that an initially uniform population of xyloglucans with a mean molecular weight of 420 kDa diverged to give two populations with mean molecular weights of approximately 149 and 820 kDa, the latter being insoluble. In addition, a purified azuki bean XTH catalyzed the transfer of an approximately 130 kDa xyloglucan from a 230 kDa xyloglucan donor substrate to a fluorescently labeled 15 kDa xyloglucan (Nishitani and Tominaga 1992).

It is not possible, however, to demonstrate molecular transfer between xyloglucans by analyzing this disproportionation reaction. Given that xyloglucans serve as both the donor and acceptor molecules, the substrates should be distinguished either by chemical or isotopic labeling. For example, a 2-pyridylamino xyloglucan oligosaccharide (2PA-XXXG) was used to demonstrate the XTH-mediated molecular transfer reaction between xyloglucans. Also, changes in the molecular weight distribution pattern of fluorescently labeled xyloglucans can be monitored by high-performance gel permeation chromatography coupled with fluorescence detection. This approach made it possible to demonstrate and characterize the XET activity of VaXTH1 purified from an apoplastic solution of azuki bean epicotyl (Nishitani 1992, Nishitani and Tominaga 1992). By using this stable fluorescent probe, the molecular size of xyloglucan segments transferred from the donor to the acceptor molecule was measured. The data indicated that the donor substrate molecules were cleaved randomly in an endo-type fashion to generate a split segment with an average molecular weight half of that of the donor substrate (Nishitani and Tominaga 1992).

XET activity can also be measured quantitatively and with high sensitivity using [*reducing-terminal-Glc-1-*^{3}H]XXFG (Fry et al. 1992). The reaction typically includes a mixture of high molecular weight xyloglucan and the [^{3}H]XXFG. The enzymatic reaction splits the high molecular weight xyloglucan and ligates a portion of it to the non-reducing terminus of [^{3}H]XXFG, thereby producing a high molecular weight [^{3}H]-labeled xyloglucan, which is then adsorbed to filter paper that is measured for adsorbed radioactivity.

Sulová et al. (1995) developed a simple and rapid colorimetric procedure for assaying XET activity by using the blue–green coloration of polymeric xyloglucans. This method was based on the disappearance of the coloration upon depolymerization of xyloglucan as a result of segment transfer to an oligomeric acceptor xyloglucan. This assay, however, cannot distinguish XET and XEH activities.

To visualize XET activity *in muro*, Ito et al. (1999) prepared a fluoresceinyl xyloglucan oligomer by coupling 5-[(2-aminoethyl)thioureidyl]fluorescein to a xyloglucan oligomer. Using this brighter fluorescent probe, they demonstrated that the incorporation of the labeled acceptor into tobacco suspension cultured cells decreased when XTH gene expression is reduced by overexpression of the antisense strand of EXGT-N1 mRNA (Ito et al. 1999). A mixture of sulforhodamine-conjugated xyloglucan oligosaccharides is another bright fluorescent probe that can be used for visualization of XET activity (Fry 1997). This method proved to be effective in "dot-blots", tissue prints (Fry 1997), and zymograms (Iannetta and Fry 1999). Using this fluorescent probe, Vissenberg et al. (2000) developed an *in muro* method for the localization of XET activity in living cells. They showed that the fluorescence was most prominent in the elongation zone of the root in several plant species, from the most primitive up to the most evolved land plants, providing evidence that XET activity correlates with cell expansion in specific cells of the root (Vissenberg et al. 2000, 2001, 2003).

3.2 Mode of Enzyme Action

3.2.1 Substrate Specificity

Azuki bean VaXTH1 is a strictly xyloglucan-specific enzyme that acts efficiently on xyloglucans regardless of whether they have galactosyl or fucosyl side chains (Nishitani and Tominaga 1992), but it does not act on other polysaccharides such as carboxymethylcellulose or β-(1,3);(1,4)-mixed glucan. Furthermore, VaXTH1 has higher reaction rates when xyloglucans with higher molecular sizes are used as the donor substrate, but it has little or no donor substrate activity on xyloglucans smaller than 10 kDa. A kiwifruit XTH (AdXET6) (Schröder et al. 1998) and a nasturtium XTH (NmXET) (Fanutti et al. 1996), in contrast, act efficiently on low molecular weight donor substrates.

The acceptor substrate activity of VaXTH1 does not depend on the molecular size (Nishitani and Tominaga 1992), whereas recombinant *A. thaliana* AtXTH22 protein has a much higher affinity for high molecular weight xyloglucans ($K_m = 0.3\ \mu M$) than for an oligomer, XLLGol ($K_m = 73\ \mu M$) (Purugganan et al. 1997). Thus, among the XTH proteins, there is divergence in the specificity for the molecular size of both the donor and acceptor xyloglucans.

3.2.2
pH Dependence

Azuki bean VaXTH1 exhibits a sharp pH dependence with maximum transferase activity at pH 5.8 (Nishitani and Tominaga, 1992). A similar steep pH dependence curve has been reported for XTHs purified from kiwifruit (Schröder et al. 1998), cauliflower (Hendriksson et al. 2003), and epicotyl of nasturtium (Sulová et al. 2003) as well as some recombinant *A. thaliana* XTHs produced in insect cells (Campbell and Braam 1999), and a recombinant poplar XTH produced in *Pichia* yeast (Saura-Valls et al. 2006). These enzyme activities decline steeply as the pH value increases above 6. The pH of apoplastic solutions derived from epicotyls of azuki bean range from 6.2 to 6.6 (Nishitani and Tominaga 1991), and auxin decreases the pH value by about one unit in several plant tissues (Jacobs and Ray 1976). Given that the transferase activity is optimal at pH 5.7, auxin can up-regulate the XTH activity in the apoplastic space via acidification. These findings imply that the enzyme activities of XTHs in growing tissue are sensitive to and may be controlled by the apoplastic pH. An XTH isozyme from nasturtium seeds, however, shows a different pH dependence to the XTH from the epicotyl; specifically, the seed XTH has maximum activity across a broader pH range (between pH 5.5 and 8) (Sulová et al. 2003). An XTH from the very primitive plant *Selaginella*, Sk-XTH1, displays XET activity over a broad pH range (4.5–7.5). This could reflect that "ancient" XTHs are able to be active over broader pH ranges than "more recent" ones that probably adopted specialized functions requiring a steeper pH dependence (Van Sandt et al. 2006)

3.2.3
Mechanism of Enzyme Reaction

Sulová et al. (1998, 1999) isolated an XTH–xyloglucan complex that was fairly stable under various conditions but decomposed quickly upon addition of an acceptor xyloglucan. This strongly supported the idea that XTH forms a relatively stable glycosyl–enzyme intermediate. Based on this and other data, it appears that XTH acts by first recognizing an unsubstituted β-1-4-glucosyl residue next to the substituted residue of a donor xyloglucan molecule at a random distance from the terminus (Nishitani and Tominaga 1992; Steele et al. 2001). In a second step, the splitting of the glucoside linkage is coupled to the formation of a stable glucosyl intermediate, in which the first glutamic acid residue in the catalytic cleft of XTH is covalently linked to the split end of the donor xyloglucan (Johansson et al. 2004). This xyloglucosyl moiety in the intermediate is transferred to the 4-hydroxyl group of the non-reducing terminus of the acceptor xyloglucan to complete the transfer reaction (XET activity). Alternatively, the xyloglucosyl moiety may be transferred to a water molecule to perform hydrolysis (XEH activity).

Very recently Saura-Valls et al. (2006) measured the kinetic parameters of the recombinant PttXTH34 (PttXET16A) from *Populus tremula* x *tremuloides* (hybrid aspen) using XXXGXXXG as the glycosyl donor and a heptasaccharide derivatized with 8-aminonaphthalene-1,3,6-trisulphonic acid (XXXG-ANTS) as the acceptor substrate. The low molecular weight donor underwent a single transfer reaction to the acceptor substrate under initial rate conditions. Based on these kinetic data, they proposed a ping-pong bi-bi mechanism with substrate inhibition by both donor and acceptor. This model is consistent with that proposed by Baran et al. (2000) for nasturtium seed XTH.

4
Biological Functions of XTHs

4.1
Role of Xyloglucan in Cell Wall Expansion

As described at the beginning of this chapter, cell expansion is typically accompanied by structural changes in cell wall xyloglucans (Nishitani and Masuda 1981). Electron microscopical observation of xyloglucans using a negative staining technique coupled with immunogold labeling shows that most of the xyloglucan molecules are localized between and on cellulose microfibrils (Baba et al. 1994). Dissolution in 24% KOH, which disrupts the crystalline structure of cellulose microfibrils, shows that there are at least two distinct forms of xyloglucan (Nishitani and Masuda 1983). Furthermore, the results indicate that there are two types of non-covalent interactions between xyloglucans and cellulose microfibrils: hydrogen bonding on the surface of cellulose microfibrils, and a stronger interaction by which xyloglucans are entrapped within the paracrystalline core of the cellulose (Hayashi 1989; Edelmann and Fry 1992). In addition to the "adhered" domain of the xyloglucans, there is an "unattached" domain that may function as bridge between cellulose microfibrils. These interactions mediate the interactions in the cellulose/xyloglucan complex that provide a load-bearing structure to the cell wall and constrain the cell shape.

Whitney et al. (1999) showed that a cellulose/xyloglucan composite is less stiff and more elastic than cellulose alone, despite being highly cross-linked. They proposed that domains of cross-linked cellulose behave as mechanical elements and that cellulose/xyloglucan networks provide the balance of elasticity and strength required by primary cell walls. Again, these facts are consistent with the view that molecular modification of xyloglucan molecules is critical for controlling the mechanical properties of the cell wall and, therefore, cell wall elasticity and rigidity.

Using sulforhodamine-labeled xyloglucan oligosaccharides as probes for subcellular localization of XET activity, Vissenberg et al. (2005b) revealed that there are two distinct patterns of XET localization, fibrillar and uniform, which depend on the properties of the cell wall. The fibrillar pattern of XET activity appeared to correlate with the diffuse growth of the cell wall, in which the XTHs act on the xyloglucan domain adhering to the cellulose microfibrils.

4.2 Versatile Functions of XTH in Processing of the Cellulose/Xyloglucan Framework in the Cell Wall

While cellulose microfibrils are produced on the plasma membrane (Doblin 2002), xyloglucans are polymerized within the Golgi apparatus and secreted into the apoplast by exocytosis (Carpita and McCann 2000; Cosgrove 2005). Thus, the cellulose/xyloglucan complex must be assembled outside the plasma membrane. The mechanism by which newly secreted xyloglucans and the nascent cellulose microfibrils are assembled into a cellulose/xyloglucan complex in the cell wall space remains elusive. Once the cellulose/xyloglucan complex is produced outside of the plasma membrane, it is ready to be integrated into the preexisting cell wall framework by the actions of XTH. Figure 2 describes the possible molecular steps by which the cellulose/xyloglucan framework is processed by XTH.

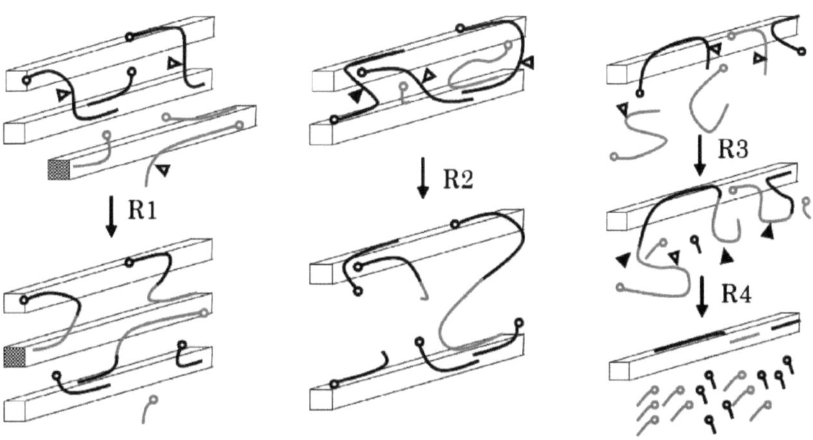

Fig. 2 Versatile actions of XTH in the processing of cellulose/xyloglucan complexes. *Square rods* cellulose microfibrils; *black* and/or *gray strings* xyloglucans; *open circles* nonreducing termini of xyloglucan; *open triangles* XTH with XET activity; *solid triangle* XTH with XEH activity. For R1, R2, R3, and R4, see text. Modified from Nishitani (1998)

Xyloglucan anchored to a preexisting framework may be split and reconnected by XET activity to the non-reducing terminus of a xyloglucan attached to a nascent cellulose microfibril. Conversely, xyloglucan anchored to a nascent cellulose microfibril may be split and reconnected to the non-reducing terminus of a xyloglucan anchored to a preexisting framework (R1 in Fig. 2). These reactions would lead to integration of new cellulose/xyloglucan complexes into the cell wall framework.

When the split end of a xyloglucan cross-link is connected to the free end of an anchored xyloglucan, the cross-link is interchanged. Cleavage of xyloglucan cross-links by hydrolysis would increase mobility of the framework. The cleavage of the cross-links can also be achieved by XET activity if the split end of the xyloglucan cross-link is transferred to a free xyloglucan oligomer (Farkas and MacLachlan 1988; Farkas et al. 1992; R2 in Fig. 2). Repetition of these reactions will render the cellulose/xyloglucan framework "plastic", so that the cell wall can extend upon application of turgor pressure (Nishitani 1997). This process may constitute the molecular basis of chemical creep of the cell wall, although XTH has not been shown to mediate the creeping action *in vitro* (McQueen-Mason et al. 1993; Cosgrove 2005). It should be noted, however, that a negative result in an assay optimized for expansin action does not disprove the ability of XTHs to promote wall loosening. In this respect it needs to be proven whether exogenously applied XTHs effectively permeate the wall matrix.

Cellulose-anchored xyloglucan chains may be extended by the XET activity of XTHs. If a free xyloglucan is split and reconnected to an anchored xyloglucan, then the latter will be elongated (Nishitani and Tominaga 1991, 1992; Thompson et al. 1997). Conversely, an anchored xyloglucan may be split and become linked to the free xyloglucan to extend the reducing terminal domain of the anchored xyloglucan. Thus, cellulose-anchored xyloglucan chains can be extended freely in both directions by the molecular grafting reaction mediated by XTH (Nishitani 1998). Molecular grafting reactions between xyloglucan chains anchored in the same microfibrils will generate a closed loop, which might function as a hook to interact with other polymers. The reverse reaction of the loop formation results in immobilization of the soluble xyloglucan oligomer or polymer by the molecular grafting reaction (R3 in Fig. 2).

Finally, the unattached xyloglucan domain in the cellulose/xyloglucan complex will be degraded by both the XET and XEH activities of XTHs (R4 in Fig. 2). Notably, the XEH and XET activities of XTHs toward xyloglucans can interact synergistically to accelerate cleavage of the cellulose/xyloglucan framework. This might occur when the cell wall is disassembled during rapid cell differentiation processes such as fruit ripening and abscission as well as rapid cell elongation without cell wall synthesis.

Taken together, most of the reactions required for the assembly, rearrangement, and disassembly of the cellulose/xyloglucan framework can be achieved by the single or combined actions of XTHs (Nishitani 1997, 1998). Because of their versatile functions, the XTHs are thought to play a key role in the regulation of the cellulose/xyloglucan framework during various changes associated with cell differentiation, including cell expansion (Nishitani 2002).

4.3 Specific Functions of XTH in Cell Expansion

To gain insight into the *in vivo* action of XTH in cell expansion, the relationship between enzyme activity and cell elongation has been extensively investigated. In some studies, XET activity was shown to correlate with the growth rate (Prichard et al. 1993), but this was not found in other studies, and an inverse correlation has even been reported (Wu et al. 1994). One major complication of this approach is that the XTH family of proteins is encoded by dozens of genes. Given the divergent catalytic actions of XTHs, it is not surprising that a single assay system has not been able to prove a connection between XTH enzyme activity and cell extension growth (Palmer and Davies 1996). Therefore, it is not easy to determine the role of individual XTH proteins in xyloglucan dynamics, which is involved not only in loosening but construction, stiffening, and disassembly of the cell wall. To elucidate the roles of individual XTHs, it is necessary to focus on the function of their proteins and genes in specific cell types or tissues.

In roots of *A. thaliana* and tobacco, elongating cells have higher XET activity than non-elongating cells. Additionally, trichoblasts display high XET activity at the site of future root hair emergence, as revealed by incorporation of sulforhodamine-labeled xyloglucan oligosaccharides (Vissenberg et al. 2000, 2001). Similar patterns of high XET activity in the root elongation zone and specific sites in trichoblasts are observed in various vascular plants, indicating a universal role of XTH in root cell elongation and local loosening of "mature" cells to allow root hair emergence (Vissenberg et al. 2003).

Comprehensive expression analysis in *A. thaliana* showed that the 33 XTH genes exhibit organ- and tissue-specific expression profiles and that they respond differently to various sets of environmental and hormonal signals (Xu et al. 1996; Yokoyama and Nishitani 2000, 2001b; Vissenberg et al. 2005a; Osato et al. 2006). Whereas some genes such as *AtXTH2*, *AtXTH4*, *AtXTH5*, *AtXTH6*, and *AtXTH7* are expressed in the various organs and do not exhibit clear organ-specific expression profiles, expression of most of the genes are organ- or tissue-specific. For example, *AtXTH1* is expressed in the silique, *AtXTH9* in the flower (Hyodo et al. 2003), and *AtXTH17*, *AtXTH18*, *AtXTH19*, and *AtXTH20* in the root (Yokoyama and Nishitani 2001b). Although the four root-specific genes are phylogenetically closely related, they exhibit different tissue specificity within the root: *AtXTH18* is expressed in all cell types in the elongating and differentiating region of the root; *AtXTH19* is expressed in the apical dividing and elongating regions as well as in the differentiation zone and is up-regulated by auxin. In contrast, *AtXTH20* is specifically expressed in vascular tissues in the basal mature region of the root (Vissenberg et al. 2005a). Of the four root-specific XTH genes, *AtXTH18* exhibits the highest level of mRNA expression. Functional analyses of these genes using T-DNA insertion lines and RNA interference technology have revealed that *AtXTH18* plays a primary role in elongation of the primary root (Osato et al. 2006).

To identify the AtXTH genes that are differentially expressed during growth of the inflorescence stem in *A. thaliana*, Imoto et al. (2005) characterized the expression profiles of individual XTH genes along the stem. This analysis showed that *AtXTH9* and *AtXTH16* are preferentially expressed in the apical part of the stem and that *AtXTH24* (*Meri5*) and *AtXTH17* are specific to the lower non-growing part of the stem (Imoto et al. 2005). Because these genes encode proteins that are classified as class II and are assumed to exhibit XET activity, it is likely that XTHs with similar enzyme activities play distinct roles in apparently opposing aspects of cell wall metabolism, namely, cell wall expansion and stiffening.

In juvenile rosette leaves of *A. thaliana*, *AtXTH27* is expressed in elongating immature protoxylem. Its loss-of-function mutants exhibit short-shaped tracheary elements in the tertiary veins and have a reduced number of tertiary veins in the first leaf, indicating an essential role for this gene in the elongation of tracheary elements during vascular development (Matsui et al. 2005). During the expansion process of the immature tracheary elements, xyloglucans in the primary cell wall are thought to be degraded. Because *AtXTH27* encodes a class III XTH, and therefore is assumed to exhibit XEH activity, it is likely that the ATXTH27 protein mediates the degradation of xyloglucans in the primary cell wall of the tracheary elements. Alternatively, it is possible that the AtXTH27 protein mediates the fragmentation of xyloglucans by transferring the large split fragment of the xyloglucan donor molecular to small xyloglucan fragments.

The rice genome possesses 29 XTH genes, most of which show organ- and growth stage-specific expression (Yokoyama et al. 2004). Detailed expression analysis for individual rice XTH genes has revealed that individual OsXTH genes have temporally and spatially controlled expression profiles at particular sites in the rice plant. For example, *OsXTH19*, which encodes a class III XTH, is almost exclusively expressed in the basal 10-mm region and shows the greatest correlation with the rate of blade elongation, implying a role for *OsXTH19* in leaf cell expansion. In the internodes, *OsXTH19* is specifically expressed in the dividing/elongating zone, suggesting that it functions in cell expansion in both leaves and internodes of rice plants (Yokoyama et al. 2004). This is surprising because xyloglucan in rice cell walls is less abundant than in dicotyledonous plants and is not thought to participate in the generation of load-bearing cross-links in rice (Yokoyama et al. 2004; Yokoyama and Nishitani 2004). This would normally suggest a critical role of xyloglucan modification by XTHs in rice cell expansion, but we cannot exclude the possibility that polysaccharides other than xyloglucans are the substrates of XTHs in rice and that their modification plays the key role in cell expansion in grasses.

5
Control of the Expression and Action of XTH

Extensive studies of the expression of XTH genes have shown that individual members of this gene family have different responses to signals. This implies that the actions of XTH genes are differentially regulated by environmental and developmental cues, which are typically mediated by hormonal signals.

5.1
Brassinosteroids

BRU1 was isolated as a transcript whose abundance in soybean (*Glycine max*) epicotyls was increased by treatment with brassinosteroids (Zurek et al. 1994). Sequence analysis of *BRU1* revealed that it encodes a member of the XTH family (Zurek and Clouse 1994). Recombinant BRU1 protein possesses XET activity, and the *BRU1* transcript accumulates in inner epicotyl tissues, particularly in the phloem and paratracheary parenchyma cells (Oh et al. 1998). This was the first evidence that XTH is involved in brassinosteroid-regulated elongation both in vascular and epidermal cells of *G. max*.

In *A. thaliana* seedlings, treatment with brassinosteroids increases expression of *AtXTH22* (*TCH4*), with mRNA accumulation peaking after 2 h (Xu et al. 1995, 1996). *AtXTH22* was isolated as a gene rapidly up-regulated in response to touch, a mechanical stress (Braam 1992). Comprehensive expression analysis of *A. thaliana* XTH genes has shown that 1 μM brassinolide greatly increases the mRNA levels of *AtXTH3*, *AtXTH4*, *AtXTH5*, *AtXTH17*, *AtXTH22*, and *AtXTH23* in 11-day-old seedlings (Yokoyama and Nishitani 2001b). Similar effects of brassinosteroid on XTH genes were observed in 7-day-old seedlings of *A. thaliana*: *AtXTH22* (*TCH4*), *AtXTH23* (*XTR6*), and *AtXTH17* (*BRU8/T8F5.9*) are up-regulated in wild type plants by treatment with 10 nM brassinolide, whereas *AtXTH15* (*XTR7*) is down-regulated in a *det2-1* mutant, which is defective for brassinolide biosynthesis (Goda et al. 2002). Given the essential and versatile roles of brassinosteroids in the growth and development of plants, it is likely that individual brassinosteroid-regulated XTH genes play distinct roles in various aspects of plant development. Furthermore, the expression of *AtXTH22* promoter::GUS gene is observed even in the *det2-1* mutant and the *bri1-2* mutant, which is defective in brassinosteroid perception, indicating that the transcriptional regulation of *AtXTH22* is probably not mediated directly by brassinosteroid signaling (Iliev et al. 2002). Thus, the signaling pathways that link the brassinosteroid receptor to each of the XTH genes are not clear, despite focused studies and the extensive expression data available in public databases.

5.2
Auxin and Hydrogen Ion

Early studies on xyloglucan metabolism showed that molecular weight changes in xyloglucans are closely related to auxin-induced cell expansion (Nishitani and Masuda 1981). Although endo-β-1,4-glucanase has long been considered a promising candidate for mediating the auxin-regulated modification of xyloglucan (Verma et al. 1975), it does not exhibit sufficient activity toward xyloglucans (Ohmiya et al. 1995; Hayashi and Ohsumi 1994). Furthermore, it is difficult to explain how construction and reorganization of the cellulose/xyloglucan framework in the complicated lamellate structure of the cell wall can be mediated by simple hydrolysis of xyloglucan cross-links. This paradox was resolved by the discovery of the XTH family of proteins. Furthermore, this discovery has opened another means of exploring cellulose/xyloglucan reorganization in auxin-regulated cell expansion, and studies on these proteins have furnished new clues for resolving the molecular mechanisms of plant growth (Nishitani 1995).

The tomato XTH gene *SlXTH1* (*LeEXT*) (Okazawa et al. 1993), which was isolated from tomato hypocotyls, is expressed primarily in epidermal and outer cortical cells in elongating regions of the etiolated hypocotyls (Catalá et al. 1997) and in expanding fruit (Catalá et al. 2000). The expression of this gene in the epicotyl section is up-regulated by exogenous application of 5 μM 2,4-dichlorophenoxy acetic acid. Another tomato XTH gene, *SlXTH2* (*LeXET2*), is preferentially expressed in mature non-elongating regions of the hypocotyls. Interestingly, the level of *SlXTH2* mRNA is decreased by auxin (Catalá et al. 2001). Based on both the phylogenetic relationship and the enzyme activity of the recombinant protein, it appears that *SlXTH2* encodes a protein whose enzyme action is similar to that of the *SlXTH1* gene product. Thus, the different expression profiles and responses to auxin of the two tomato XTHs supports the idea that each member of the XTH family of proteins plays a distinct role in cell wall construction, despite similar or identical enzymatic activities. It is likely that some XTH members are specifically involved in auxin-regulated molecular grafting of xyloglucan in expanding cell walls and that other XTH members mediate the grafting of cell walls that are no longer expanding (Nishitani 1997).

Two XTH genes from *V. angularis* (azuki bean), *VaXTH1* and *VaXTH2*, are phylogenetically closely related. Based on their structural similarity as well as studies of VaXTH1 activity, *VaXTH2* is predicted to encode a protein that has XET activity (Nishitani and Tominaga 1992; Nakamura et al. 2003). Although the two genes are predominantly expressed in phloem fibers of growing internodes and are up-regulated by indole-3-acetic acid (IAA), they exhibit temporally distinct expression profiles along the internode and different responses to auxin: *VaXTH1* is expressed nearer to the top of the internode than *VaXTH2*. The promotive effect of IAA on *VaXTH1* expression is blocked by inhibition of cell expansion with 0.25 M mannitol, whereas that of IAA on the *VaXTH2* gene is not affected by mannitol treatment. More interestingly, fusicoccin, a potent stimulator of the plasma membrane ATPase, increases the level of the transcript for *VaXTH1* but not *VaXTH2*. Thus, the two azuki bean XTH genes are involved in distinct aspects of cell wall dynamics in the same cell type during different stages of development (Nakamura et al. 2003). This serves as another example of the division of roles of XTH proteins in auxin-induced cell expansion.

The *A. thaliana* XTH gene *AtXTH19* is expressed in the apical dividing and elongating regions as well as in the differentiation zone of the root in seedlings, and its mRNA expression in intact seedlings is up-regulated by application of 0.1 nM IAA (Vissenberg et al. 2005a). To gain insight into how auxin regulates this XTH gene, expression of *AtXTH19* promoter::GUS fusion genes was examined in both wild-type plants and *axr2-1* mutants in which the domain II of the AXR2/IAA7 protein is not functional (Nagpal et al. 2000). A full-length *AtXTH19* promoter::*GUS* fusion gene (*pAtXTH19::GUS*) was expressed throughout the root. An auxin-responsive element, TGTCTC (Ulmasov et al. 1997), was found between nucleotides – 956 and – 951 upstream of *AtXTH19*. Deletion of the upstream region to nucleotide – 330 of the promoter GUS construct (– *330pAtXTH19::GUS*) eliminated GUS activity in the apical dividing region of the root of the wild-type plant. In addition, in the *axr2-1* mutant plant background, expression of the *pAtXTH19::GUS* gene was drastically reduced in the differentiating regions and was only observed in the apical region of the root. Interestingly, expression of the – *330pAtXTH19::GUS* fusion gene, which lacks the TGTCTC sequence, was also suppressed in the elongation and differentiation zones of the mutant root, implying that transcriptional activity of the – *330pAtXTH19::GUS* fusion gene in the wild-type is still under regulation by auxin signaling via AXR2/IAA7 in the elongating and maturating regions of the root in the wild-type plant (Osato et al. 2006). In this context, it is noteworthy that another potential auxin-responsive sequence, TGTCAC (Okushima et al. 2005), is found between nucleotides – 82 and – 67 upstream of the *AtXTH19* coding region. Although specific auxin-responsive factor proteins inactivated by interaction with the AXR2/IAA protein have not been identified, the auxin signal perceived by the TIR1 protein (Dharmasiri et al. 2005; Kepinski and Leyser 2005) is considered to affect transcriptional activity via the auxin-responsive element (AuxRE).

Several other auxin-responsive *XTH* genes have been identified in *A. thaliana*. External application of 1 μM IAA increases the level of transcripts for *AtXTH4*, *AtXTH22*, and *AtXTH25* and decreases that of *AtXTH30* in 12-day-old seedlings that were grown in liquid culture on a rotary shaker under continuous light (Xu et al. 1996). Comprehensive expression analysis of the 33

AtXTH genes by real-time RT-PCR (Yokoyama and Nishitani 2001b) reveals that application of 1 μM IAA to intact light-grown 11-day-old seedling elevates the expression of *AtXTH3*, *AtXTH17*, *AtXTH19*, *AtXTH22*, and *AtXTH23* and reduces the expression of *AtXTH15*, *AtXTH21*, and *AtXTH27* (Yokoyama and Nishitani 2001b). In addition, microarray analysis has shown that application of 1 μM IAA to in 7-day-old seedlings up-regulates the transcription of *AtXTH22* and *AtXTH23* and down-regulates that of *AtXTH15*, *AtXTH6*, and *AtXTH14* (Goda et al. 2004). The finding of different sets of auxin-responsive genes in these experiments may be partly due to the different growth stages and culture conditions as well as the specific methods used to detect the transcripts. Nevertheless, it is interesting that *AtXTH22* (*TCH4*), which is up-regulated by mechanical stimulus and application of brassinolide, is also up-regulated by auxin in the three independent experimental conditions. Also, an AuxRE element is found in the 5′-upstream region of *AtXTH22* (*TCH4*) (Iliv et al. 2002). Because repetitive mechanical stimulation leads to a delay in shoot growth and differentiation, AtXTH22 may play a role in cell wall strengthening or stiffening as opposed to cell expansion (Braam 2005).

According to the acid growth theory, auxin causes acidification of the cell walls by stimulating hydrogen ion secretion into the cell wall interior, thereby inducing modification of the cell wall, leading to cell expansion (Cleland 1971). It is postulated that auxin and acidification act, at least in part, by separate mechanisms and that their actions are additive (Kutschera 1994). The pH of apoplastic solutions derived from epicotyls of azuki bean range from 6.2 to 6.6 (Nishitani and Tominaga 1991). Auxin decreases the pH value by about one unit in several plant tissues (Jacobs and Ray 1976). XET activities in crude apoplastic solutions derived from azuki bean epicotyls (Nishitani and Tominaga 1991) and pea epicotyls (Fry et al. 1992) have pH optima of 5.4 and 5.5, respectively. Purified azuki bean VaXTH1 exhibits maximal XET activity at pH 5.8 (Nishitani and Tominaga 1992). These XET activities decline steeply as the pH is increased. For example, the crude enzyme preparation from pea was less than half as active at pH 7.0. Taken together, these results show that auxin can enhance the XET activity in the apoplastic space via acidification. This view is consistent with the effect of auxin and acidic pH on cell expansion growth.

5.3 Gibberellin

Gibberellic acids are another class of plant hormones that play a principal role in the regulation of cell expansion during various aspects of plant growth and development. An early report of a correlation between GA_3-induced internode elongation growth and XET activity within internodal tissues (Potter and Fry 1993) suggested that gibberellin regulates the expression of XTH genes. In azuki bean epicotyls, the levels of *VaXTH1* and *VaXTH2* mRNAs are increased by application of GA_3 as well as auxin (Nakamura et al. 2003). In *A. thaliana*, the level of *AtXTH23* transcript in 12-day-old seedlings is increased by treatment with 1 μM gibberellic acid (Yokoyama and Nishitani 2001b). It is not clear whether these gene products are directly involved in cell expansion.

In rice, *OsXTH2* (*OsXTR1*) and *OsXTH23* (*OsXTR3*) are preferentially expressed in the elongation zone of internodes, and their mRNA expression is enhanced by gibberellic acid and brassinosteroid. Furthermore, the levels of mRNAs encoding OsXTH2 and OsXTH23 are lower in Akibare and Waito C dwarf mutants than in wild-type plants (Uozu et al. 2000). Conversely, in a gibberellic acid-insensitive overgrowth mutant, *awaodori*, which grows to two to three times the length of the wild type, the expression levels of four rice XTH genes, including *OsXTH2* and *OsXTH23*, are higher than in wild-type plants (Uozu et al. 2000).

Another rice XTH gene, *OsXTH8*, is preferentially expressed in vascular bundles of leaf sheath and young nodal roots, which is where cell expansion occurs (Jan et al. 2004; this gene was defined as *OsXTH17* in Yokoyama et al. 2004). Its mRNA expression in two-week old seedling is increased by a 24-h treatment with 5 μm GA_3, but it is not affected by other plant hormones. In two mutant rice plants with abnormal heights, there is a positive correlation between the level of the *OsXTH17* mRNA and plant height. Reduction of the level of *OsXTH17* mRNA by RNA interference causes growth repression in rice. These results indicate that *OsXTH17* together with *OsXTH2* and *OsXTH23*, which exhibit temporally distinct expression profiles, may participate in gibberellic acid-stimulated cell elongation in rice.

5.4 Mechanical Stimuli

The expression of *XTH* genes is regulated by various environmental signals, including hypoxia, anoxia, and mechanical stimuli. Application of a simple gentle mechanical stimulus (rubbing plants stems for about 10 s) causes retarded elongation, a response called thigmomorphogenesis (Jaffe 1973; Braam 2005). The *A. thaliana* gene *AtXTH22* (*TCH4*) was originally identified as a touch-induced gene (Braam 1992); however, the signaling pathway by which mechanical stimuli leads to rapid activation of *AtXTH22* expression is not yet fully understood, and the *cis*-acting element responsible for responding to the touch-derived signal has not been identified (Iliev et al. 2002). Regardless, the *AtXTH22* gene product is likely involved in stiffening or reinforcement of the cell wall, and perception of the mechanical signal is expected to lead to strengthening of the shoot.

Plant body weight itself may act as a mechanical signal. Ko et al. (2004) showed that application of 2.5 g of aluminum foil to the top of a 5-cm *A. thaliana* seedling causes a 34.6-fold enhancement of *AtXTH22* expression compared to the unloaded control. Relief of this weight by placing the *A. thaliana* inflorescent stem horizontally reduces the expression levels of most of the cell wall-related genes that were preferentially expressed in the non-elongating basal part of the stem (Imoto et al. 2005, Yokoyama and Nishitani 2006).

Many of the genes that are up-regulated by mechanical stimuli encode calcium-binding proteins and cell wall-related proteins (Braam 2005). The signaling pathways connecting mechanical sensing and transcriptional regulation of cell wall-related genes are not yet known, and molecular dissection of the pathways leading to *AtXTH22* expression should help elucidate the mechanisms by which mechanical stimuli regulate changes in the cell wall architecture and, hence, plant development (Braam 2005).

5.5 Hypoxia and Anoxia

Flooding affects cell elongation, particularly in ground tissues. The *Zea mays* (maize) gene *wusl1005* was isolated as a flooding-induced gene (Peschke and Sachs 1994) and was predicted to encode an XTH protein (Saab and Sachs 1995). The amount of the *XTH* transcript is increased in shoots of maize seedlings subjected to hypoxic stress, and it accumulates in regions of the root and mesocotyl where aerenchyma develops under flooding conditions. Furthermore the mRNA levels are increased by hypoxia but not by other environmental stresses. This induction and aerenchyma formation is blocked by (aminooxy)acetic acid, a potent inhibitor of ethylene synthesis. Furthermore, treatment with ethylene under aerobic conditions causes aerenchyma formation and increases expression of the maize XTH gene. This indicates that induction of *wusl1005* under hypoxic conditions is mediated by ethylene. Interestingly, *wusl1005* expression was also induced under anoxic conditions, which inhibit ethylene production and aerenchyma formation (Saab and Sachs 1996).

Anoxia also up-regulates XTHs and induces shoot growth in arrowhead tubers (*Sagittaria pygmaea* Mig.). Specifically, the levels of *SpXTH1* and *SpXTH4* transcripts are greatly increased by anoxia but not by ethylene or CO_2 (Oogawara et al. 2005). These findings suggest that members of the XTH family participate in cell expansion under anoxic conditions.

6 Concluding Remarks

The fact that XTH proteins are encoded by large multigene families in flowering plants raises the question of whether each of the genes and proteins have specific roles or are functionally redundant. Recent comprehensive analyses of XTH gene expression, particularly in *A. thaliana* and rice, together with functional analyses based on loss-of-function mutants, have provided evidence in support of the hypothesis that each member of the XTH gene family has its own specific role. Some members appear to be critical in promoting cell wall expansion and are therefore essential for cell expansion, whereas others are required for construction of cell walls in cells that have completed the expansion process. Expression of each of these genes, especially at both the transcriptional and post-transcriptional levels, is precisely regulated by various plant hormones, including brassinolide, auxin, and gibberellin, as well as by environmental signals. Dissection of the processes regulating the transcription of individual XTH genes involved in cell expansion should clarify the molecular mechanism by which hormonal and environmental signals regulate cell expansion.

Another important conclusion from the comprehensive studies on the XTH family of proteins is that cell wall expansion is not solely regulated by XTHs, although they carry out an indispensable and rate-limiting step of cell expansion. To address this point, a "cell wall type-specific gene hypothesis" was proposed (Nishitani 2002, 2005), wherein a plant body is composed of dozens of cell types with specific cell wall types. The cell wall dynamics in a given type of cell wall involves different types of enzymes encoded by dozens of gene families, each consisting of several members, referred to as "cell-wall-related gene families". Given the distinct cell-type-specific expression profile for each member of a cell wall-related gene family, specific members of the family are responsible for a given type of cell wall. So far, the results obtained for several cell wall-related gene families are consistent with this hypothesis (e.g., Yokoyama and Nishitani 2001b). The essence of this hypothesis is that there are transcriptional factors that define individual cell wall types and thereby direct individual proteins within the enzyme set to work on specific types of cell wall. Transcriptional factors intrinsic to expanding cells must be regulated, in turn, by a higher class of master genes that would govern morphogenesis in general.

Recently, several candidates for the hypothetical master genes for specific cell wall types have been identified. These include several transcriptional factors that govern cell type-specific regulation of cell wall genes in the root elongation region (Birnbaum et al. 2003) and an NAC transcription factor, which controls secondary wall construction processes (Mituda et al. 2005). Although further studies are needed to clarify the molecular processes by which these transcription factors coordinate the actions of individual cell wall-related genes, these results provide convincing evidence in support of the cell wall type-specific gene hypothesis. Investigation of the signaling networks that link these transcriptional factors and genes involved in cell expansion would provide important clues for understanding the mechanism of cell expansion, particularly that mediated by cell wall loosening and reconstruction.

Acknowledgements This work was supported by a Grant-in-Aid for Scientific Research on Priority Areas (15031202), for Scientific Research (B) (17370012) from the Ministry of Education, Culture, Sports, Science and Technology of Japan, and by the Program for Development of Fundamental Technologies for Controlling the Material Production Process of Plants from New Energy and Industrial Technology Development Organization, Japan. Authors acknowledge the financial support of the Fund for Scientific Research Flanders (FWO – Belgium) and the Institute for Science and Technology (IWT). K.V. is a postdoctoral fellow of the Fund for Scientific Research Flanders (FWO – Belgium).

References

Albersheim P (1976) The primary cell wall. In: Bonner J, Varner JE (ed) Plant biochemistry. Academic, New York, pp 225–274

Baba K, Sone Y, Misaki A, Hayashi T (1994) Localization of xyloglucan in the macromolecular complex composed of xyloglucan and cellulose in pea stems. Plant Cell Physiol 35:439–444

Baran R, Sulová Z, Stratilova E, Farkaš V (2000) Ping-pong character of nasturtium-seed xyloglucan endotransglycosylase (XET) reaction. Gen Physiol Biophys 19:427–440

Birnbaum K, Shasha DE, Wang JY, Jung JW, Lambert GM, Galbraith DW, Benfey PN(2003) A gene expression map of the Arabidopsis root. Science 302:1956–1960

Braam J (1992) Regulated expression of the calmodulin-related TCH genes in cultured Arabidopsis cells: induction by calcium and heat shock. Proc Natl Acad Sci USA 89:3213–3216

Braam J (2005) In touch: plant responses to mechanical stimuli. New Phytol 165:373–389

Campbell P, Braam J (1998) Co- and/or post-translational modifications are critical for TCH4 XET activity. Plant J 15:553–561

Campbell P, Braam J (1999) In vitro activities of four xyloglucan endotransglycosylases from Arabidopsis. Plant J 18:371–382

Carpita N, McCann M (2000) The cell wall. In: Buchanan BB, Gruissem W, Jones RL (ed) Biochemistry and molecular biology of plants. American Society Plant Physiologists, Rockville, Maryland, pp 52–108

Catalá C, Rose JKC, Bennett AB (1997) Auxin regulation and spatial localization of an endo-1,4-β-D-glucanase and a xyloglucan endotransglycosylase in expanding tomato hypocotyls. Plant J 12:417–426

Catalá C, Rose JKC, Bennett AB (2000) Auxin-regulated genes encoding cell wall-modifying proteins are expressed during early tomato fruit growth. Plant Physiol 122:527–534

Catalá C, Rose JKC, York WS, Albersheim P, Darvill AG, Bennett AB (2001) Characterization of a tomato xyloglucan endotransglucosylase gene that is down-regulated by auxin in etiolated hypocotyls. Plant Physiol 127:1180–1192

Cleland R (1971) Cell wall extension. Ann Rev Plant Physiol 22:197–222

Cosgrove DJ (2000) Expansive growth of plant cell walls. Plant Physiol Biochem 38:109–124

Cosgrove DJ (2005) Growth of the plant cell wall. Nature Rev Mol Cell Biol 6:850–861

Davies GJ, Gloster TM, Henrissat B (2005) Recent structural insights into the expanding world of carbohydrate-active enzymes. Curr Opin Struct Biol 15:637–645

Dharmasiri N, Dharmasiri S, Estelle M (2005) The F-box protein TIR1 is an auxin receptor. Nature 435:441–445

Doblin MS, Kurek I, Jacob-Wilk D, Delmer DP (2002) Cellulose biosynthesis in plants: from genes to rosettes. Plant Cell Physiol 43:1407–1420

Edelmann HG, Fry SC (1992) Effect of cellulose synthesis inhibition on growth and the integration of xyloglucan into pea internode cell walls. Plant Physiol 100:993–997

Edwards M, Dea ICM, Bullpin PV, Reid JSG (1985) Xyloglucan (amyloid) mobilization in the cotyledons of *Tropaeolum majus* L. seeds following germination. Planta 163:133–140

Edwards M, Dea ICM, Bullpin PV, Reid JSG (1986) Purification and properties of a novel xyloglucan-specific endo-(1→4)-β-D-glucanase form germinated nasturtium seeds (*Tropaeolum majus* L.). J Biol Chem 261:9489–9494

Fanutti C, Gidley MJ, Reid JSG (1993) Action of a pure xyloglucan endo-transglycosylase (formerly called xyloglucan-specific endo-(1→4)-β-D-glucanase) from the cotyledons of germinated nasturtium seeds. Plant J 3:691–700

Fanutti C, Gidley MJ, Reid JSG(1996) Substrate subsite recognition of the xyloglucan endo-transglycosylase or xyloglucan-specific endo-(1→4)-β-D-glucanase from the cotyledons of germinated nasturtium (*Tropaeolum majus* L.) seeds. Planta 200:221–228

Farkaš V, Maclachlan G (1988) Stimulation of pea 1,4-β-glucanase activity by oligosaccharides derived from xyloglucan. Carbohydr Res 184:213–220

Farkaš V, Sulová Z, Stratilova E, Hanna R, Maclachlan G (1992) Cleavage of xyloglucan by nasturtium seed xyloglucanase and transglycosylation to xyloglucan subunit oligosaccharides. Arch Biochem Biophys 298:365–370

Fry SC (1997) Novel dot-blot assays for glycosyltransferases and glycosylhydrolases: optimization for xyloglucan endotransglycosylase (XET) activity. Plant J 11:1141–1150

Fry SC, Smith RC, Renwick KF, Martin DJ, Hodge S, Matthews KJ (1992) Xyloglucan endotransglycosylase, a new wall-loosening enzyme activity from plants. Biochem J 282:821–828

Geisler-Lee J, Geisler M, Coutinho PM, Segerman B, Nishikubo N, Takahashi J, Aspeborg H, Djerbi S, Master E, Andersson-Gunnerås S, Sundberg B, Karpinski S, Teeri TT, Kleczkowski LA, Henrissat B, Mellerowicz EJ (2006) Poplar carbohydrate-active enzymes: gene identification and expression analyses. Plant Physiol 140:946–962

Goda H, Shimada Y, Asami T, Fujioka S, Yoshida S (2002) Microarray analysis of brassinosteroid-regulated genes in Arabidopsis. Plant Physiol 130:1319–1334

Hayashi T (1989) Xyloglucans in the primary cell wall. Ann Rev Plant Physiol Mol Biol 40:139–168

Hayashi T, Ohsumi C (1994) Endo-1,4-β-glucanase in the cell wall of stems of auxin-treated pea seedlings. Plant Cell Physiol 35:419–424

Hématy K, Höfte H (2007) Cellulose and cell elongation. In: Verbelen JP, Vissenberg K (eds) The expanding cell. Plant Cell Monographs, vol 5. Springer, Berlin Heidelberg New York (in press)

Hendriksson H, Denman SE, Campuzano IDG, Ademark P, Master ER, Teeri TT, Brumer H III (2003) N-linked glycosylation of native and recombinant cauliflower xygloglucan endotransglucosylase 16A. Biochem J 375:61–73

Henrissat B (1991) A classification of glycosyl hydrolases based on amino-acid sequence similarities. Biochem J 280:309–316

Henrissat B, Bairoch A (1993) New families in the classification of glycosyl hydrolases based on amino-acid sequence similarities. Biochem J 293:781–788

Henrissat B, Bairoch A (1996) Updating the sequence-based classification of glycosyl hydrolases. Biochem J 316:695–696

Hyodo H, Yamakawa S, Takeda Y, Tsuduki M, Yokota A, Nishitani K, Kohchi T(2003) Active gene expression of a xyloglucan endotransglucosylase/hydrolase gene, XTH9, in inflorescence apices is related to cell elongation in *Arabidopsis thaliana*. Plant Mol Biol 54:473–482

Iliev EA, Xu W, Polisensky DH, Oh MH, Torisky RS, Clouse SD, Braam J (2002) Transcriptional and posttranscriptional regulation of Arabidopsis TCH4 expression by diverse stimuli. Roles of *cis* regions and brassinosteroids. Plant Physiol 130:770–783

Imoto K, Yokoyama R, Nishitani K (2005) Comprehensive approach to genes involved in cell wall modifications in *Arabidopsis thaliana*. Plant Mol Biol 58:177–192

Inouhe M, Yamamoto R, Masuda Y (1984) Auxin-induced changes in the molecular weight distribution of cell wall xyloglucans in Avena coleoptiles. Plant Cell Physiol 25:1341–1351

Ito H, Nishitani K (1999) Visualization of EXGT-mediated molecular grafting activity by means of a fluorescent-labeled xyloglucan oligomer. Plant Cell Physiol 40:1172–1176

Jacobs M, Ray PM (1975) Promotion of xyloglucan metabolism by acid pH. Plant Physiol 56:373–376

Jacobs M, Ray PM (1976) Rapid auxin-induced decrease in free space pH and its relationship to auxin-induced growth in maize and pea. Plant Physiol 58:203–209

Jaffe MJ (1973) Thigmomorphogenesis: the response of plant growth and development to mechanical stimulation with special reference to *Bryonia dionica*. Planta 114:143–157

Jan A, Yang G, Nakamura H, Ichikawa H, Kitano H, Matsuoka M, Matsumoto H, Komatsu S (2004) Characterization of a xyloglucan endotransglucosylase gene that is up-regulated by gibberellin in rice. Plant Physiol 136:3670–3681

Johansson P, Denman SE, Brumer H, Kallas ÅM, Henriksson H, Bergfors T, Teeri TT, Jones TA (2003) Crystallization and preliminary X-ray analysis of a xyloglucan endotransglycosylase from *Populus tremula* x *tremuloides*. Acta Crystallogr D Biol Crystallogr 59:535–537

Johansson P, Brumer H, Baumann MJ, Kallas ÅM, Henriksson H, Denman SE, Teeri TT, Jones TA (2004) Crystal structures of a poplar xyloglucan endotransglycosylation acceptor binding. Plant Cell 16:874–886

Kaku T, Tabuchi A, Wakabayashi K, Kamisaka S, Hoson T (2002) Action of xyloglucan hydrolase within the native cell wall architecture and its effect on cell wall extensibility in azuki bean epicotyls. Plant Cell Physiol 43:21–26

Keegstra K, Talmadge KT, Bauer WD, Albersheim P (1973) The structure of plant cell walls III: a model of the walls of suspension-cultured sycamore cells based on the interconnections of the macromolecular components. Plant Physiol 51:188–196

Kepinski S, Leyser O (2005) The Arabidopsis F-box protein TIR1 is an auxin receptor. Nature 435:446–451

Ko JH, Ham KH, Park S, Yang J (2004) Plant body weight-induced secondary growth in Arabidopsis and its transcription phenotype revealed by whole-transcriptome profiling. Plant Physiol 135:1069–1083

Kutschera U (1994) The current status of the acid-growth hypothesis. New Phytol 126:549–569

Labavitch JM, Ray PM (1974) Turnover of cell wall polysaccharides in elongating pea stem segments. Plant Physiol 53:669–673

Lorences EP, Revilla G, Acebes JL, Zarra I (1990) Growth capacity and molecular mass distribution of cell wall polysaccharides along the hypocotyl of *Pinus pinaster*. Physiol Plant 79:563–569

Matsui A, Yokoyama R, Ito T, Seki M, Shinozaki K, Takahashi T, Komeda Y, Nishitani K (2005) AtXTH27 plays an essential role in cell wall modification during the development of tracheary elements. Plant J 42:525–534

McQueen-Mason SJ, Fry SC, Durachko DM, Cosgrove DJ (1993) The relationship between xyloglucan endotransglycosylases and in-vitro cell wall extension in cucumber hypocotyls. Planta 190:327–331

Mitsuda N, Seki M, Shinozaki K, Ohme-Takagi M(2005) The NAC transcription factors NST1 and NST2 of Arabidopsis regulate secondary wall thickenings and are required for anther dehiscence. Plant Cell 17:2993–3006

Nagpal P, Walker LM, Young JC, Sonawala A, Timpte C, Estelle M, Reed JW (2000) *AXR2* encode a member of the Aux/IAA protein family. Plant Physiol 123:563–573

Nakamura T, Yokoyama R, Tomita E, Nishitani K (2003) Two azuki bean XTH genes, *VaXTH1* and *VaXTH2*, with similar tissue-specific expression profiles, are differently regulated by auxin. Plant Cell Physiol 44:16–24

Nishitani K (1992) A novel method for detection of endo-xyloglucan transferase. Plant Cell Physiol 33:1159–1164

Nishitani K (1995) Endo-xyloglucan transferase, a new class of transferase involved in cell wall construction. J Plant Res 108:137–148

Nishitani K (1997) The role of endoxyloglucan transferase in the organization of plant cell walls. Int Rev Cytol 173:157–206

Nishitani K (1998) Construction and restructuring of the cellulose–xyloglucan framework in the apoplast as mediated by the xyloglucan-related protein family - a hypothetical scheme. J Plant Res 111:159–166

Nishitani K (2005) Division of roles among members of the XTH gene family in plants. Plant Biosystems 139:98–101

Nishitani K, Masuda Y (1981) Auxin-induced changes in the cell wall structure: changes in the sugar compositions, intrinsic viscosity and molecular weight distributions of matrix polysaccharides of the epicotyl cell wall of *Vigna angularis*. Physiol Plant 52:482–494

Nishitani K, Masuda Y (1982) Acid pH-induced structural changes in cell wall xyloglucans in *Vigna angularis* epicotyl segments. Plant Sci Lett 28:87–94

Nishitani K, Masuda Y (1983) Auxin-induced changes in the cell wall xyloglucans: effects of auxin on the two different subfractions of xyloglucans in the epicotyl cell wall of *Vigna angularis*. Plant Cell Physiol 24:345–355

Nishitani K, Tominaga R (1991) In vitro molecular weight increase in xyloglucans by an apoplastic enzyme preparation from epicotyls of *Vigna angularis*. Plant Physiol 82:490–497

Nishitani K, Tominaga R (1992) Endo-xyloglucan transferase, a novel class of glycosyltransferase that catalyzes transfer of a segment of xyloglucan molecule to another xyloglucan molecule. J Biol Chem 267:21058–21064

Obel N, Neumetzler L, Pauly M (2007) Hemicelluloses and cell expansion. In: Verbelen JP, Vissenberg K (eds) The expanding cell. Plant Cell Monographs, vol 5. Springer, Berlin Heidelberg New York (in press)

Oh MH, Romanow WG, Smith RC, Zamski E, Sasse J, Clouse SD (1998) Brassinosteroids promote soybean epicotyl elongation and regulate expression of BRU1, a gene with homology to xyloglucan endotransglycosylases (XETs). Plant Cell Physiol 39:124–130

Ohmiya Y, TakedaT, Nakamura S, Sakai F, Hayashi T (1995) Purification and properties of a wall-bound endo-1,4-β-glucanase from suspension-cultured poplar cells. Plant Cell Physiol 36:607–614

Okazawa K, Sato Y, Nakagawa T, Asada K, Kato I, Tomita E, Nishitani K (1993) Molecular cloning and cDNA sequencing of endoxyloglucan transferase, a novel class of glycosyltransferase that mediates molecular grafting between matrix polysaccharides in plant cell walls. J Biol Chem 268:25364–25368

Okushima Y, Overvoorde PJ, Arima K, Alonso JM, Chan A, Chang C, Ecker JR, Hughes B, Lui A, Nguyen D, Onodera C, Quach H, Smith A, Yu G, Theologis T (2005) Functional genomic analysis of the AUXIN RESPONSE FACTOR gene family members in *Arabidopsis thaliana*: Unique and overlapping functions of ARF7 and ARF19. Plant Cell 17:444–463

Oogawara R, Satoh S, Yoshioka T, Ishizawa K (2005) Expression of alpha-expansin and xyloglucan endotransglucosylase/hydrolase gene associated with shoot elongation enhanced by anoxia, ethylene and carbon dioxide in arrowhead (*Sagittaria pygmaea* Mig.) tuber. Ann Bot 96:693–702

Osato Y, Yokoyama R, Nishitani K (2006) A principal role for AtXTH18 in *Arabidopsis thaliana* root growth – a functional analysis using RNAi plants. J Plant Res 119:153–162

Palmer SJ, Davies WJ (1996) An analysis of relative elemental growth rate, epidermal cell size and xyloglucan endotransglycosylase activity through the growing zone of ageing maize leaves. J Exp Bot 47:339–347

Peschke VM, Sachs MM (1994) Characterization and expression of transcripts induced by oxygen deprivation in maize (*Zea mays* L). Plant Physiol 104:387–394

Potter I, Fry SC (1993) Xyloglucan endotransglycosylase activity in Pea internodes: effects of applied gibberellic acid. Plant Physiol 103:235–241

Pritchard J, Hetherington PR, Fry SC, Tomos AD (1993) Xyloglucan endotransglycosylase activity, microfibril orientation and the profiles of cell wall properties along growing regions of maize roots. J Exp Bot 44:1281–1289

Purugganan MM, Braam J, Fry SC (1997) The Arabidopsis TCH4 xyloglucan endotransglycosylase – substrate specificity, pH optimum, and cold tolerance. Plant Physiol 115:181–190

Rose JKC, Braam J, Fry SC, Nishitani K (2002) The XTH family of enzymes involved in xyloglucan endotransglucosylation and endohydrolysis: current perspectives and a new unifying nomenclature. Plant Cell Physiol 43:1421–1435

Saab IN, Sachs MM (1995) Complete cDNA and genomic sequence encoding a flooding-responsive gene from maize (*Zea mays* L.) homologous to xyloglucan endotransglycosylase. Plant Physiol 108:439–440

Saab IN, Sachs MM (1996) A flooding-induced xyloglucan endotransglycosylase homolog in maize is responsive to ethylene and associated with aerenchyma. Plant Physiol 112:385–391

Saura-Valls M, Faure R, Rags S, Piens K, Brumer H, Teeri TT, Cottaz S, Driguez H, Planas A (2006) Kinetic analysis using low molecular mass xyloglucan oligosaccharides defines the catalytic mechanism of a Populus xyloglucan endotransglycosylase. Biochem J 395:99–106

Schröder R, Atkinson RG, Langenkämper G, Redgwell RJ (1998) Biochemical and molecular characterisation of xyloglucan endotransglycosylase from ripe kiwifruit. Planta 204:242–251

Steele NM, Sulová Z, Campbell P, Braam J, Farkaš V, Fry SC (2001) Ten isoenzymes of xyloglucan endotransglycosylase from plant cell walls select and cleave the donor substrate stochastically. Biochem J 355:671–679

Strohmeier M, Hrmova M, Fischerl M, Harvey AJ, Fincher GB, Pleiss J (2004) Molecular modeling of family GH16 glycoside hydrolases: Potential roles for xyloglucan transglucosylases/hydrolases in cell wall modification in the poaceae. Protein Sci 13:3200–3232

Sulová Z, Farkaš V (1999) Purification of xyloglucan endotransglycosylase based on affinity sorption of the active glycosyl-enzyme intermediate complex to cellulose. Protein Express Purif 16:231–235

Sulová Z, Baran R, Farkaš V (2003) Divergent modes of action on xyloglucan of two isoenzymes of xyloglucan endotransglycosylase from Tropaeolum majus. Plant Physiol Biochem 41:431–437

Sulová Z, Lednicka M, Farkas V (1995) A colorimetric assay for xyloglucan-endotransglycosylase from germinating seeds. Anal Biochem 229:80–85

Sulová Z, Takacova M, Steele NM, Fry SC, Farkaš V (1998) Xyloglucan endotransglycosylase: evidence for the existence of a relatively stable glycosyl-enzyme intermediate. Biochem J 330:1475–1480

Thompson JE, Fry SC (2001) Restructuring of wall-bound xyloglucan by transglycosylation in living plant cells. Plant J 26:23–34

Thompson JE, Smith RC, Fry SC (1997) Xyloglucan undergoes inter-polymeric transglycosylation during binding to the plant cell wall in vivo: evidence from $^{13}C/^{3}H$ dual labeling and isopycnic centrifugation in caesium trifluoroacetate. Biochem J 327:699–708

Ulmasov T, Hagen G, Guilfoyle TJ (1997) ARF1, a transcription factor that binds to auxin response elements. Science 276:1865–1868

Uozu S, Tanaka-Ueguchi M, Kitano H, Hattori K, Matsuoka M (2000) Characterization of XET-related genes of rice. Plant Physiol 122:853–860

Valent BS, Albersheim P (1974) The structure of plant cell walls V: on the binding of xyloglucan to cellulose fibres. Plant Physiol 54:105–108

Van Sandt VST, Guisez Y, Verbelen J-P, Vissenberg K (2006) Analysis of a xyloglucan endotransglycosylase/hydrolase (XTH) from the lycopodiophyte Selaginella kraussiana suggests that XTH sequence characteristics and function are highly conserved during the evolution of vascular plants. J Exp Bot (in press) doi:10.1093/jxb/erl064

Verma DPS, Maclachlan GA, Byrne H, Ewing D (1975) Regulation and in vitro translation of messenger ribonucleic acid for cellulase from auxin-treated pea epicotyls. J Biol Chem 250:1019–1026

Vissenberg K, Martinez-Vilchez IM, Verbelen J-P, Miller JG, Fry SC (2000) In vivo colocalization of xyloglucan endotransglycosylase activity and its donor substrate in the elongation zone of Arabidopsis roots. Plant Cell 12:1229–1238

Vissenberg K, Verbelen J-P, Fry SC (2001) Root hair initiation is coupled to a highly localized increase of xyloglucan endotransglycosylase action in Arabidopsis roots. Plant Physiol 127:1125–1135

Vissenberg K, Van Sandt V, Fry SC, Verbelen J-P (2003) Xyloglucan endotransglucosylase action is high in the root elongation zone and in the trichoblasts of all vascular plants from *Selaginella* to *Zea mays*. J Exp Bot 54:335–344

Vissenberg K, Oyama M, Osato Y, Yokoyama R, Verbelen J-P, Nishitani K (2005a) Differential expression of AtXTH17, -18, -19 and -20 genes in Arabidopsis roots. Physiological roles in specification in cell wall construction. Plant Cell Physiol 46:192–200

Vissenberg K, Fry SC, Pauly M, Höfte H, Verbelen J-P (2005b) XTH acts at the microfibril-matrix interface during cell elongation. J Exp Bot 56:673–683

Whitney SEC, Gothard MGE, Mitchell JT, Gidley MJ (1999) Roles of cellulose and xyloglucan in determining the mechanical properties of primary plant cell walls. Plant Physiol 121:657–663

Wu YJ, Spollen WG, Sharp RE, Hetherington PR, Fry SC (1994) Root growth maintenance at low water potentials - Increased activity of xyloglucan endotransglycosylase and its possible regulation by abscisic acid. Plant Physiol 106:607–615

Xu W, Purugganan MM, Polensky DH, Antosiewicz DM, Fry SC, Braam J (1995) Arabidopsis TCH4, regulated by hormones and the environment, encodes a xyloglucan endotransglycosylase. Plant Cell 7:1555–1567

Xu W, Campbell P, Vargheese AK, Braam J (1996) The Arabidopsis XET-related gene family: environmental and hormonal regulation of expression. Plant J 9:879–889

Yokoyama R, Nishitani K (2000) Functional diversity of xyloglucan-related proteins and its implications in the cell wall dynamics in plants. Plant Biol 2:598–604

Yokoyama R, Nishitani K (2001a) Endoxyloglucan transferase is localized both in the cell plate and in the secretary pathway destined for the apoplast in tobacco cells. Plant Cell Physiol 42:292–300

Yokoyama R, Nishitani K (2001b) A comprehensive expression analysis of all members of a gene family encoding cell-wall enzymes allowed us to predict cis-regulatory regions involved in cell wall construction in specific organs of Arabidopsis. Plant Cell Physiol 42:1025–1033

Yokoyama R, Nishitani K (2004) Genomic basis for cell-wall diversity in plants: a comparative approach to gene families in rice and Arabidopsis. Plant Cell Physiol 45:1111–1121

Yokoyama R, Nishitani K (2006) Identification and characterization of Arabidopsis thaliana genes involved in xylem secondary cell walls. J Plant Res 3:189–194

Yokoyama R, Rose JKC, Nishitani K (2004) A surprising diversity and abundance of XTHs (xyloglucan endotransglucosylase/hydrolases) in rice: classification and expression analysis. Plant Physiol 134:1088–1099

Zurek DM, Clouse SD (1994) Molecular cloning and characterization of a brassinosteroid-regulated gene from elongating soybean (Glycine max L.) epicotyls. Plant Physiol 104:161–170

Zurek DM, Rayle DL, Mcmorris TC, Clouse SD (1994) Investigation of gene expression, growth kinetics, and wall extensibility during brassinosteroid-regulated stem elongation. Plant Physiol 104:505–513

Plant Cell Monogr (5)
J.-P. Verbelen · K. Vissenberg: The Expanding Cell
DOI 10.1007/7089_2006_073/Published online: 7 October 2006
© Springer-Verlag Berlin Heidelberg 2006

Expansins

Simon McQueen-Mason (✉) · Ngoc Tuan Le · David Brocklehurst

CNAP, Biology Department, University of York, University Road, Heslington YO10 5DD, UK
sjmm1@york.ac.uk

Abstract The plant cell wall is a complex composite material that has evolved to meet the seemingly paradoxical requirements of providing sufficient strength to withstand cell turgor whilst accommodating rapid cell expansion during growth. Expansins are small proteins that act as key modulators of cell expansion by catalyzing the process of acid-induced cell wall extension. Expansins induce rapid wall extension by disrupting non-covalent interactions (most likely hydrogen bonds) between cellulose microfibrils and the hemicellulose polymers that coat them in the wall. As well as cell expansion, expansins are involved in a variety of other processes in plants such as vascular differentiation, fruit ripening and organ abscission. Expansins are encoded by a substantial gene family that may have arisen specifically in land plants.

1 Introduction

The cell wall serves numerous functions in plants; the most obvious of which is to provide skeletal support, allowing trees to attain heights exceeding 100 metres. Strength is an important feature of plant cell walls, not only in the context of providing support to a plant but also at the cellular level in containing the high hydrostatic pressures (turgor) generated in plant cells. Most plant cells maintain pressure in the region of 0.5 MPa (or 5 atmospheres) but pressure can exceed 10 times this value in specialized examples such as stomatal guard cells. Cell turgor is contained and balanced by a relatively thin cell wall requiring considerable strength in the material of the wall in order to sustain the stresses imposed by turgor pressure. Enclosing a cell in such a strong material might potentially serve to physically constrain the growth of cells. However, this is not the case as most plant cells undergo substantial and rapid increases in size during development. To accommodate this expansion, the plant cell wall has to be sufficiently plastic whilst maintaining sufficient integrity to avoid cell rupture under the potentially explosive force of cell turgor.

Not only have plants produced a material that accommodates both strength and flexibility, but also the ability of the cell wall to expand is under very precise dynamic control. Indeed, changes in plant growth rates are generally effected through changes in plant cell wall extensibility, and modulations in wall extensibility have been measured to occur extremely rapidly. For example, during blue light inhibition of hypocotyl elongation, growth was inhibited within 45 s of seedling irradiation, and this inhibition was preceded by an abrupt change in cell wall extensibility (Spalding and Cosgrove 1989). Thus, cell wall extensibility appears to be subject to fine tuning in both time and space, allowing plant cells to adopt their distinctive sizes and shapes. The rapid changes in cell wall extensibility that have been observed led to the notion that there must be effector molecules in the matrix of the cell wall able to rapidly modulate the mechanical properties of the wall in response to signals from the protoplast.

2 Expansin Discovery and Mode of Action

During the 1970s a number of groups applied the use of direct mechanical measurements to develop an understanding of plant cell wall extension during growth. In particular, Robert Cleland's work using constant load extensometry led to the formation of the acid growth hypothesis. Cleland's experiments showed that the cell walls of growing plant tissues showed very distinctive mechanical properties. At pH values of neutrality or above, growing cell walls showed little or no extension when placed under a constant load. In contrast, at lower pH values (between pH 5.5 and 4.0) cell walls from growing plant tissues showed much more rapid and sustained long-term extension (Rayle et al. 1970). It was later shown by Cosgrove (1989) that this pH-dependent wall extension could be eliminated by heat treatment or proteolysis and was inhibited by various heavy metals suggesting an enzyme-catalyzed process. These observations helped to establish a scientific search for the biochemical basis of plant cell wall extension during growth and for the enzymes catalyzing these events.

In the early 1990s McQueen-Mason et al. (1992) showed that wall extensibility could be reconstituted in cell walls that had been inactivated by heat or proteolysis, by adding back proteins extracted from growing plant cell walls. They further showed that acid-induced wall extension could be fully restored to inactivated cell walls by adding a single purified protein of a class that was subsequently named expansins. To date, expansins are the only proteins shown capable of catalyzing long-term acid-induced cell wall extension in a manner consistent with a role in cell expansion.

It should be noted that at least two other protein groups have now been shown to have a direct effect on cell wall mechanical properties. Yieldins have so far only been characterized from one plant species and have rather subtle effects, apparently modifying the yield threshold (critical point of applied force after which a material begins to extend) but have received little recent attention (Okamoto-Nakazato et al. 2000a,b). Lipid transfer proteins (LTPs) were recently shown to induce cell wall extension in vitro as assayed with an extensometer, however, the effect was independent of pH and the biological relevance of this observation has yet to be determined (Nieuwland et al. 2005).

It is also important to note that a wide range of other enzymes undoubtedly contribute at different levels to overall cell wall mechanical properties. At the front end of this are enzymes involved in cell wall biosynthesis as the composition, interaction and orientation of the various wall components undoubtedly exert profound effects on wall properties (Hématy and Höfte 2007, in this volume). At the next level will be those enzymes of the wall that modify the structure and crosslinking of the matrix, including those involved in hydrolysis, transglycolysis and free radical chemistry to name a few examples (Nishitani and Vissenberg 2007; Lindsay and Fry 2007; in this volume). However, none of these types of enzymes have been shown to directly bring about measurable effects on wall extensibility in the manner of expansins but instead may be thought of as secondary wall loosing enzymes producing the overall material context of the wall in which expansins or other primary wall loosening factors operate (Cosgrove 1999).

Although expansins have profound and direct effects on cell wall extension, the biochemistry underlying their effects remains hypothetical, albeit supported by several key observations. When expansins were first isolated there was a general expectation that they would prove to be cell wall hydrolases that would loosen the wall matrix by reducing the sizes of polysaccharides or perhaps operate as endotransglycosylases that would relieve wall stress by cleaving load-bearing polysaccharides prior to re-joining polymers in new positions. Neither of these expectations was fulfilled. Exhaustive attempts were made to examine the first expansins isolated for hydrolytic activity. These included examining the ability of expansins to hydrolyze a range of commercially available polysaccharides as well as matrix polysaccharides from cell walls, but no such activity was detected (McQueen-Mason and Cosgrove 1994, 1995). In addition, when expansins were incubated with plant cell walls over protracted periods no sugars or oligosaccharides were detected to be released (McQueen-Mason et al. 1992). Expansins were also assayed for xyloglucan endotransglycosylase (XET) activity, but again with negative results (McQueen-Mason et al. 1993).

An interesting observation regarding expansin effects was that a small quantity of protein was able to induce substantial wall elongation over periods of hours. During these measurements extension rates stayed rather linear or slowly decreased in rate, contrary to what might be expected of a hydrolytic enzyme (McQueen-Mason and Cosgrove 1995). Indeed, comparisons between expansin-induced extension and the effects of hydrolases in extension are revealing. When cell wall extension is measured in the presence of cellulases, there is initially little or no detectable effect of the enzyme. However, with time, accelerating rates of extension are seen leading rapidly to tissue breakage as the material is weakened by the hydrolase. In contrast, expansin effects on extension are apparent in less than a minute and, after an initial acceleration, the walls soon settle into a steady rate of extension that slowly declines (presumably due to protein inactivation as extension can be stabilized with reducing agents). This tends to suggest that expansin action does not lead to any cumulative weakening of the cell wall—in other words, that wall integrity is maintained during expansin action.

This suggestion is strongly supported by experiments examining the effects of expansins on the stress-relaxation properties of cell walls (McQueen-Mason and Cosgrove 1995). In these experiments, cell walls are stretched until a predetermined amount of force has been generated by the stretching. At this point, the cell walls are held at a constant size and the decay of stress in the material (as the wall polymers rearrange themselves) is monitored. Stress relaxation is, in part, a function of polymer sizes—bigger molecular movements take more time. McQueen-Mason and Cosgrove (1995) showed that expansin fully restores the stress-relaxation properties of cell walls that have been inactivated by heat treatment in much the similar way that they restore the extension properties. Furthermore, it was shown that the effect of expansins on the stress relaxation profile was always the same, independent of how long the walls had been incubated with the proteins. This data strongly suggests that during prolonged incubation with expansins there is no reduction (or increase) in polymer sizes in the cell walls and supports the idea that expansins most likely function to dissociate non-covalent interactions in the wall.

The hypothesis that expansins act by disrupting non-covalent interactions was given support by experiments looking at the effects of these proteins on substrates other than plant cell walls. McQueen-Mason and Cosgrove (1994) showed that expansins can enhance the extension and breakage of pieces of cellulosic paper in a constant load extensometer, and subsequent work showed that expansin treatment substantially weakens these papers (Bolam et al. 1998). These weakening effects of expansins are not accompanied by any detectable hydrolysis of cellulose in the paper indicating that they most probably disrupt the non-covalent interactions (largely H-bonds) between the fibres of the paper. In addition, the extension-inducing activity of expansins in cell walls was shown to be enhanced by adding urea up to 3 M, compatible with an H-bond disrupting process, and expansin-induced extension was also shown to be slowed down in D_2O compared to in H_2O also supporting the case for H-bond disruption (McQueen-Mason and Cosgrove 1994).

Experimental evidence, therefore, indicates that expansins operate by disrupting H-bonding between cellulose microfibrils and the hemicellulose polymers that coat them in the wall (Obel, Neumetzler and Pauly 2007, in this volume) Binding studies showed that expansins did not bind detectably to any of the matrix polysaccharides of the cell wall in isolation (McQueen-Mason and Cosgrove 1995). The proteins did bind to cellulose microfibrils, but only weakly to crystalline cellulose. In contrast, binding to cellulose was greatly enhanced when the microfibrils were coated with hemicellulose polymers and also partially enhanced by disordering the crystallinity of glucan chains in cellulose microfibrils (McQueen-Mason and Cosgrove 1995). These data indicate that expansins bind at junctions between hemicellulose polymers (or disordered glucan chains) and cellulose microfibril surfaces. This suggests that expansins potentially induce wall loosening by weakening the interactions between microfibrils and the hemicelluloses that potentially tether them to one another in the wall.

This hypothetical mode of action was tested by work examining the effects of expansins on the mechanical properties of synthetic composite materials made by bacterial cultures. *Acetobacter xylinus* has been used extensively to produce high-performance papers for various applications such as loud speaker cones. When grown in the presence of abundant glucose, these bacteria individually secrete long ribbon-like cellulose microfibrils into the culture medium which become interwoven and form a cohesive cellulosic network. It was found that if cell wall matrix polysaccharides were included in the culture medium during bacterial growth then these were incorporated into the cellulose network to form a composite material that reflected several characteristics of plant cell walls (Whitney et al. 1995). In particular, hemicelluloses such as xyloglucans and glucomannans appeared to coat the microfibrils and to form cross bridges between them when examined by transmission electron microscopy (TEM) (Whitney et al. 1995, 1998). The inclusion of matrix polysaccharides gave rise to far more extensible materials compared to cellulose only networks and this, in part, seemed to be due to the microfibrils being more ordered in the material. It was found that cucumber expansin 1 induced very rapid extension when added to such a composite that contained xyloglucans in a similar manner to its effects on plant cell walls. In contrast to the effects on xyloglucan-containing composites, no effects of expansin addition could be detected on a composite containing glucomannan in spite of the fact that this material shared many properties with that made with xyloglucan (Whitney et al. 2000). This demonstrates that the action of this expansin is quite specific for xyloglucan/cellulose networks, and many more expansins have also been shown to be active on this material suggesting that cellulose/xyloglucan interfaces may be a common substrate for these proteins (e.g. Jones et al. 2004; Belfield et al. 2005). Interestingly, in the experiments with bacterial composites, expansin effects were found to be significantly bigger on composites made with long xyloglucans than with shorter versions, and TEM analysis indicated less extensive interfibrillar cross-linking by these shorter polysaccharides, perhaps confirming that tethering xyloglucans are the targets for expansin effects.

Taken together, the preceding observations indicate a general mechanism where expansins induce cell wall extension by weakening the bonds between xyloglucan tethers and cellulose microfibrils. However, as will become clear, there are instances where expansins are correlated with events where such a mechanism does not make immediate sense. In addition, there is considerable sequence diversity among expansins and it is not yet clear to what extent these differences in protein sequences may indicate differences in substrate specificity or perhaps even mechanism of action. One of the major barriers yet to be overcome in expansin research is the difficulty in isolating the proteins in a purified form to allow critical biochemical characterization. Expansins are encoded by a large gene family and most tissues appear to express several different expansin genes simultaneously making purification from plants difficult. This is exacerbated by our current inability to produce active recombinant expansins from a non-plant expression system. There are however, a number of pertinent observations that can be made in the context of expansin action and substrate specificity.

As will be discussed shortly, the majority of plant expansins fall into two distinct groups called α- and β-expansins. The preceding discussion has focused on the α-expansins as they remain the better-characterized group. β-expansins share roughly 25% sequence identity with α-expansins and are predicted to have a similar three-dimensional structure. Until now, the activity of only one β-expansin protein has been characterized and this is Zea m1 (ZmEXPB1). This particular protein was first described as a pollen allergen and is highly abundant in maize pollen grains and released in large quantities on imbibition (Cosgrove et al. 1997). It has been shown that this β-expansin can induce extension in cell walls derived from maize silks and in maize coleoptiles, but unlike α-expansins has little extension-inducing activity in dicot tissues such as cucumber hypocotyls (Cosgrove et al. 1997). This indicates that the substrate specificity of this β-expansin is probably different to that of α-expansins. Similarly, perhaps, α-expansins have been found in association with non-growing tissues such as softening fruit and Rose et al. (1997) published data suggesting that these expansins may be less effective at inducing wall extension than α-expansins from growing tissues, possibly indicating functional differences in these proteins. Whilst there are a number of areas in need of clarification, it is probably safe to say that in relation to cell expansion, the principle mechanism of expansin action involves disruption of hydrogen bonds between cellulose microfibrils and xyloglucans.

3
Structural Features

Expansin proteins are defined by a common structure with two distinct domains (I and II) that are preceded by a signal peptide (Fig. 1) which directs the protein into the secretory pathway and is cleaved off during the formation of the mature protein (Shcherban et al. 1995). The N-terminal domain I contains an HFD motif that is distantly related to the catalytic domain of glycoside hydrolase family 45 (GH45), the majority of which are fungal β-1,4-

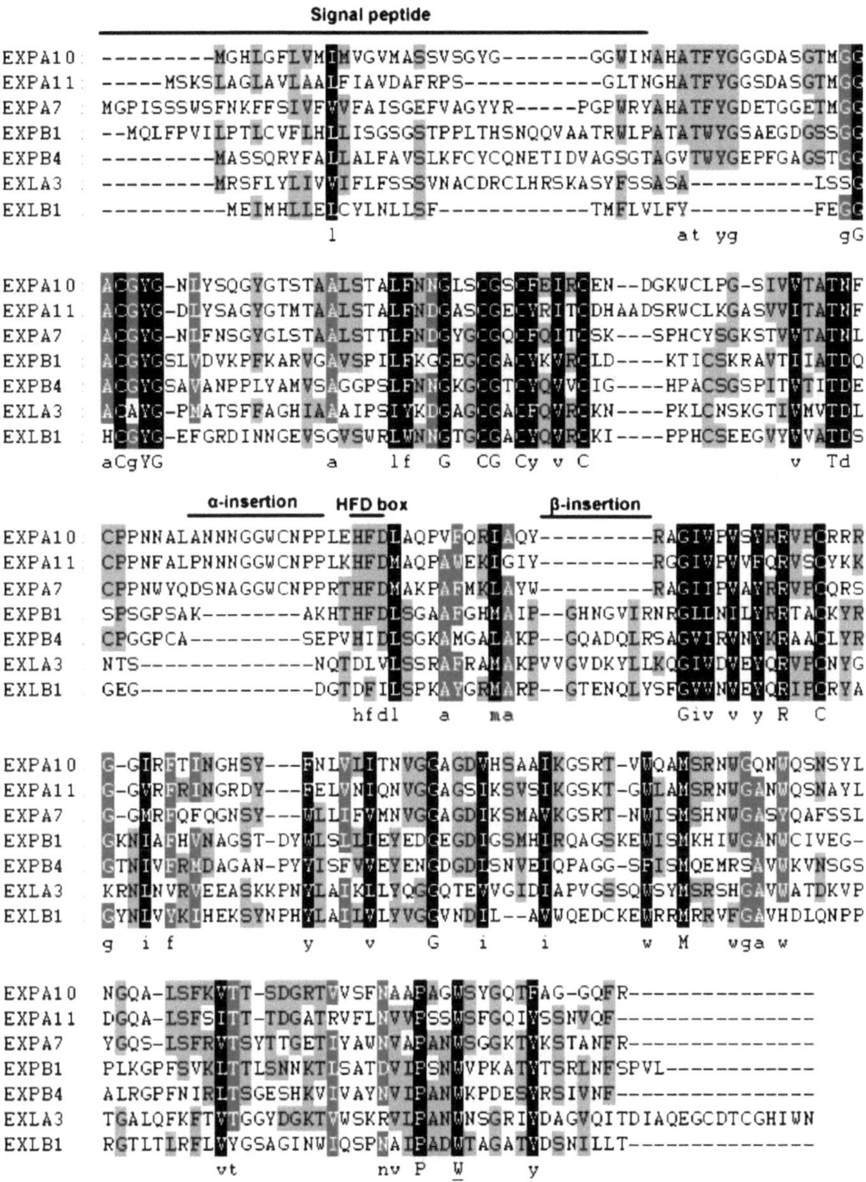

Fig. 1 Alignment of expansins from the *α*-, *β*-, EXLA, and EXLB families in Arabidopsis. Conserved amino acids are highlighted in bold. The level of shading at each position indicates the degree of similarity among the sequences. The consensus sequence is given below the alignment. The HFD box, and insertions characteristic of *α*- and *β*-expansins are indicated by bars. Conserved Trp residues within the *C*-terminal region are underlined

D-endoglucanases (Coutinho and Henrissat 1999). Domain I also contains a number of other conserved amino acids found in the fungal enzymes, including several Cys residues that may be involved in the formation of disulphide bonds, and is thus predicted to fold like GH45s (Li et al. 2002; Sampedro and Cosgrove 2005). However, despite the presence of these conserved GH45 motifs, no hydrolytic activity has been detected for either EXPA or EXPB proteins (McQueen-Mason et al. 1992; McQueen-Mason and Cosgrove 1995). The C-terminal domain II is distantly related to group-2 grass pollen allergens which appear to have arisen from recent duplications of β-expansin genes (Li et al. 2003). Domain II also contains several non-contiguous conserved Trp residues (Li et al. 2002), the spacing between which resembles that found between the same residues in the cellulose-binding domain of some cellulases (Gilkes et al. 1991). Thus, it has been hypothesized, but not yet demonstrated, that domain II may be responsible for expansin binding to cellulose and other related cell wall glycans (Shcherban et al. 1995).

4 Classification of Expansins

Under current nomenclature, expansins most similar to the first expansins discovered by McQueen-Mason et al. (1992) are classified as the EXPA or α-expansin family (Kende et al. 2004). The second family, EXPB or β-expansins, were first identified as major allergens of grass pollen and were later shown, like the α-expansins, to possess cell wall-loosening activity (Cosgrove et al. 1997). Two smaller families, the expansin-like A (EXLA) and the expansin-like B (EXLB), have been identified from genomic analyses, and these were previously classified as β2- and β3-expansins, respectively (Li et al. 2002). However, no biological or biochemical function (including cell wall-loosening activity) has yet been established for any member of either the EXLA or EXLB family (Kende et al. 2004). EXLA and EXLB proteins lack the HFD motif, suggesting that their mode of action may differ from that of the EXPA and EXPB expansins (Sampedro and Cosgrove 2005).

Current nomenclature only recognizes proteins that contain both domains I and II as expansins (Kende et al. 2004). Expansins have been found in all land plants ranging from pteridophytes including mosses (Schipper et al. 2002; Richard et al. 2005), to monocots and dicots. However, genes that encode proteins with homology to either expansin domain I or II have also been found in mussels, slime mould (*Dictyostelium*), fungi, nematodes, protists and bacteria (Li et al. 2002; Darley et al. 2003). Such proteins are classified as expansin-like family X (EXLX), a polyphyletic group containing all non-plant expansins (Kende et al. 2004). The EXLX genes found in bacteria, fungi and animals are restricted to organisms involved in plant pathogenesis or plant cell wall digestion, while those in *Dictyostelium* might lubricate the movement of the cellulose microfibrils during cell growth and wall extension and/or maintain the fluid state of cell walls in the multicellular stalk structure (Darley et al. 2003). The evolutionary relationship between the EXLX proteins and "true" expansins is unclear. The EXLX genes may have existed before the origin of land plants and subsequently diverged or they may have been acquired later through horizontal gene transfer of a plant expansin gene (Sampedro and Cosgrove 2005).

5
Phylogeny and Gene Diversity

It is unclear when expansins first appeared in evolution. No expansin sequences have yet been identified in algal species suggesting that these genes have arisen in land plants. In this context, it is interesting to note that xyloglucans also appear to be peculiar to terrestrial plants and have not been found in algae (Popper and Fry 2003), suggesting that expansins may have arisen in parallel with their polysaccharide substrate. Although the origin of expansins remains uncertain, the EXPA and EXPB genes were already present when vascular plants and mosses diverged (Li et al. 2002; Schipper et al. 2002), while the EXLA and EXLB families can be traced back only to the last ancestor of angiosperms and gymnosperms (Sampedro and Cosgrove 2005). Analysis of expansin sequences from a gymnosperm, *Pinus taeda*, suggests that the last ancestor of angiosperms and gymnosperms had at least 6 EXPA genes and one from each of the EXPB, EXLA and EXLB subfamilies (Sampedro et al. 2006). The number of genes within each expansin family has almost doubled in the last common ancestor of eudicots and monocots, to a total of 17 independent gene lineages or clades. The number of expansin clades were derived from a comparison of the divergence of expansin sequences in the *Arabidopsis* and rice genomes, and also from an examination of the conservation of gene order and orientation in duplicated gene segments, usually referred to as gene colinearity or microsynteny (Sampedro and Cosgrove 2005). The existence of 17 expansin clades is strongly supported by a recent study of the expansin superfamily in poplar, *Populus trichocarpa*, a second eudicot species to have its genome sequenced (almost complete at time of writing) (Sampedro et al. 2006).

The complexity of the expansin gene family appears to have increased significantly during the evolution of land plants. The total number of expansin genes has doubled in *Arabidopsis* (eudicot) and more than tripled in rice (monocot) since the last common ancestor between the two species. The main sources of increase in the expansin superfamily are segmental and tandem duplications, with translocations being relatively rare (Sampedro et al. 2005). Although rice contains almost the same number of EXPA, EXLA, and EXLB genes as *Arabidopsis*, it has three times the number of EXPB genes (Li et al.

2003). The rise in the number of β-expansins in rice is due to the greater number of segmental or tandem duplications in its genome, compared to *Arabidopsis*, since the two species diverged approximately 150 million years ago (Sampedro and Cosgrove 2005). The functional significance of this increase may be that these β-expansin genes are required to act on unique cell wall polysaccharides, such as mixed-linked β-1,3 : 1,4-glucan or glucuronoarabinoxylans, that are found in rice (and other grasses) but not in other angiosperms (Wu et al. 2001).

Expansin proteins from different families share only 20–40% identity with each other, with the highest degree of conservation within domain I (Li et al. 2002). There is approximately 25% identity between the EXPA and EXPB families. Within each family, α-expansins show a greater degree of conservation than β-expansins. Sequence conservation within the *Arabidopsis* α-expansin family ranges from 52% to 99% identity at the protein level (Li et al. 2002), while the rice β-expansins, of which there are 18 compared to only 6 in *Arabidopsis*, show between 28% to 76% identity (Wu et al. 2001). These numbers illustrate that, although there is significant conservation of expansin gene families across the different plant species, there is also large gene diversity within each expansin gene family.

6
Functional Specialization

Several studies have suggested some degree of functional specialization among the different expansin clades before the separation of the angiosperms and gymnosperms (Sampedro et al. 2006). For example, α-expansins in clade IV (EXPA-IV) have been shown to play a role in xylem development in both aspen (Gray-Mitsumune et al. 2004) and zinnia (Im et al. 2000), while the *Arabidopsis* EXPA-X genes are specifically expressed in trichoblasts (Cho and Cosgrove, 2002). To test the hypothesis of functional specialization within individual expansin clades, we undertook a phylogenetic analysis of several expansin genes which have been shown to be up-regulated during fruit ripening. These expansins have been isolated from such diverse fruits as apricot, tomato, and banana (see Sect. 8 for references). Despite the obvious differences in fruit size and structure among the various plants, we found that all of the expansins with increased levels of expression during ripening were distributed in three clades (EXPA-I, EXPA-III, and EXPA-IV). To simplify matters, we have only shown the relationship of the ripening-regulated expansins with the *Arabidopsis* α-expansins from clades EXPA-I to EXPA-VI (Fig. 2). Several interesting observations should be noted. Firstly, apart from xylem development, clade EXPA-IV expansins also appear to play a role during fruit ripening. Secondly, expansins from diverse fruits such as banana, strawberry and apricot are grouped together in clade EXPA-III. This suggests

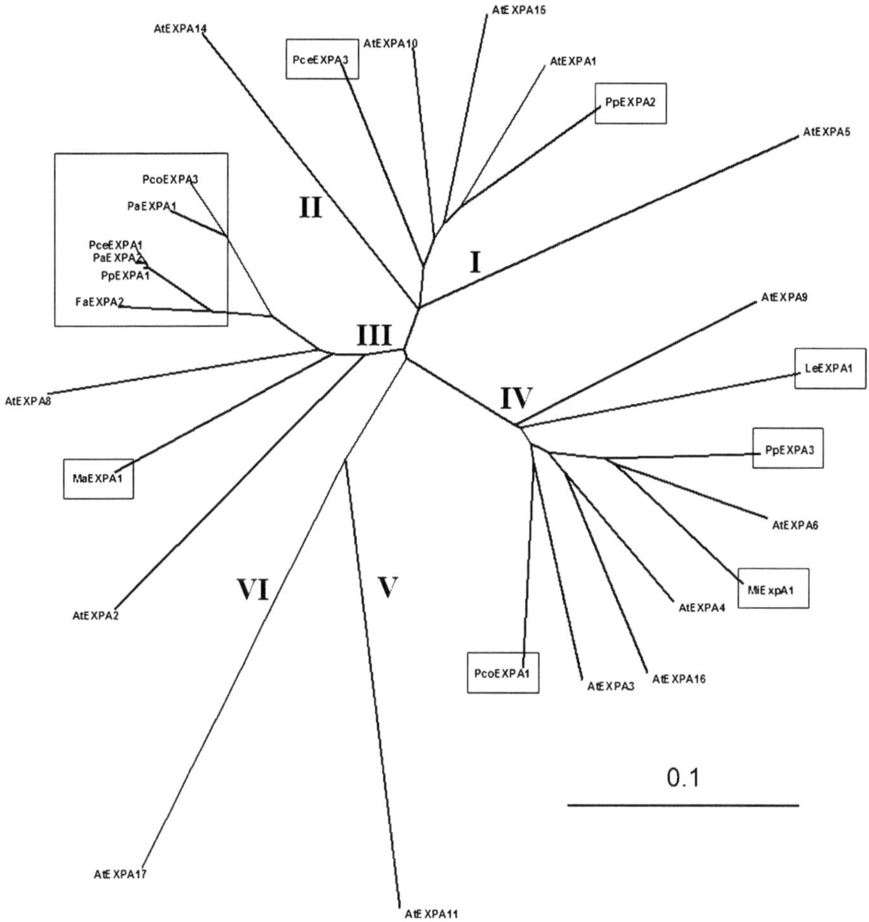

Fig. 2 Phylogenetic tree of Arabidopsis clade EXPA-I to EXPA-VI genes and expansin genes up-regulated during fruit ripening from other species (*boxed*). Full-length protein sequences were aligned with CLUSTALX (Thompson et al. 1997) and a bootstrap (5000 replicates) neighbour-joining tree was constructed with modified parameters (exclusion of positions with gaps and corrections for multiple substitutions). Clades are indicated at their roots and the phylogenetic distance is indicated by a bar (*bottom right* 0.1 substitutions per 1000 residues). Protein sequences include those from mango (*Mangifera indica*), pear (*Pyrus communis*), peach (*Prunus persica*), sour cherry (*Prunus cerasus*), apricot (*Prunus armeniaca*), strawberry (*Fragaria x ananassa*), tomato (*Lycopersicon esculentum*), and banana (*Musa acuminata*). Accession numbers are AF548376 (LeEXPA1), AY600964 (MiEXPA1), AX392017 (PcoEXPA1), AB093030 (PcoEXPA3), AY083168 (MaEXPA1), AB029083 (PpEXPA1), AB047518 (PpEXPA2), AB047519 (PpEXPA3), AF350936 (PceEXPA1), AF350938 (PceEXPA3), PAU93167 (PaEXPA1), and AF159563 (FaEXPA2)

that, although fruit structure in these species has diverged, the role of these expansins in ripening has been conserved. Thirdly, ripening-regulated expansins from peach (*Prunus persica*) are distributed in three different clades: PpEXPA1 (EXPA-III), PpEXPA2 (EXPA-I), and PpEXPA3 (EXPA-IV). This is also the case for sour cherry (*Prunus cerasus*) when all ripening-regulated expansins from this species are included in the phylogenetic analysis (data not shown). These results suggest that the activity of several different classes of expansins may be required during the ripening process in *Prunus* species and perhaps in other fruit-bearing plants.

7
Expansins in Growth and Development

Approaches based on immunolocalization and gene expression analysis have demonstrated a clear spatio-temporal link between expansin activity and wall loosening during growth. Since their discovery in actively growing regions of cucumber hypocotyls (McQueen-Mason et al. 1992), both α- and β-expansins have been detected in a broad range of tissues including elongation regions of maize, soybean and pine roots (Zhang and Hasenstein 2000; Lee et al. 2003; Hutchinson et al. 1999), growing internodes of wheat and deepwater rice (Lin et al. 2005; Cho and Kende 1997; Lee and Kende 2001), in germinating seed and developing fruit of tomato (Chen et al. 2001; Catalá et al. 2000; Rose et al. 2000) and in elongating cotton fibres (Orford and Timmis 1998; Harmer et al. 2002).

Studies using transgenic plants with enhanced levels of expansin expression have provided further, although not unequivocal, support for a role in growth. Overexpression of a soybean α-expansin in tobacco resulted in larger plants with a bushy phenotype and faster growing roots (Lee et al. 2003) and *Arabidopsis* plants overexpressing an endogenous α-expansin (*AtEXPA10*) exhibited increased growth of leaves and petioles (Cho and Cosgrove 2000). Other studies have reported the opposite correlation however; over-expression of an endogenous α-expansin in tomato, for example, resulted in fruits half the size of controls (Brummel et al. 1999). Similarly, Rochange et al. (2001) reported stunted growth in tomato plants overexpressing a cucumber α-expansin. Isolated cell walls from these plants were found to be less extensible and less sensitive to expansin activity than non-transgenic controls suggesting that the wall structure had been modified to compensate for the increased expansin activity. Choi et al. (2003) also reported stunted growth in rice plants overexpressing an α-expansin gene (*Os-EXP4*) at high levels but obtained taller plants at intermediate levels of transgene expression. As the majority of transgenic plants developed additional leaves and occasionally tillers, the authors proposed that the stunting could be due to morphological changes at the shoot apical meristem restricting internodal growth.

No obvious phenotypic defects in *Arabidopsis* insertion-mutants of various expansin genes have yet been reported (Cosgrove et al. 2002; Li et al. 2003) and the knockout of four expansin genes in the moss, *Physcomitrella patens* similarly had no observable effect (Schipper et al. 2002). These results point to functional redundancy within the expansin family although antisense suppression experiments in *Arabidopsis* (Cho and Cosgrove 2000) and rice seedlings (Choi et al. 2003) have shown a consequent reduction in growth. Zenoni et al. (2004) have also reported that silencing of a petunia α-expansin (*PhEXP1*) resulted in reduced petal limb size and alterations in the morphology of epidermal cells. No observable effects in other organs were noted however. It therefore appears likely that some expansins have specific in vivo activities which cannot be substituted for by the activity of other family members.

The direct application of partially purified expansins to growing cells has also been used to study the effects of expansin activity on growth. The addition of cucumber expansins to tobacco bright yellow 2 cell cultures induced a 3-fold increase in growth rate (Link and Cosgrove 1998) and the elongation of excised *Arabidopsis* hypocotyls was found to be stimulated by treatment with exogenous expansins, comparable to the effect of 1 μM auxin (Cosgrove et al. 2002). It was also shown that the application of expansins to cucumber root hairs caused swelling and at higher concentrations, the bursting of tips (Cosgrove et al. 2002). Differences in the orientation of cellulose microfibrils—anisotropic in hypocotyl cell walls but isotropic in root hair tips—was proposed to account for the contrasting effects of expansin on these tissues.

Fleming et al. (1997) showed that application of expansin-soaked beads to vegetative meristems of tomato could induce the formation of aberrant primordia, suggesting that the control of cell enlargement is important in regulating developmental pathways. This finding was strengthened by Reinhardt et al. (1998) who demonstrated that, in the tomato meristem, localized expression of an α-expansin gene (*LeEXP18*) was predictive for the emergence of the leaf primordium. It was later shown, using an inducible promoter system, that expansin expression in the tobacco apical meristem was sufficient to initiate leaf development at the site of induction and could alter leaf shape when expression was induced on the flanks of developing primordia (Pien et al. 2001). These findings suggest that the precise spatial distribution and timing of expansin activity may be required for the appropriate initiation of developmental and morphogenic pathways.

Analogous to their role at the shoot apical meristem, expansins have also been found to accumulate at the bulged domains from which root hair development is initiated in maize (Baluška et al. 2000) and *Arabidopsis* (Cho and Cosgrove 2002). The expression of specific α-expansin genes is also upregulated in the roots of the parasitic plant *Striga asiatica* during development of haustoria (O'Malley and Lynn 2000). This process involves swelling of cells just distal to the root tip and is achieved by switching from a longitudinal to a radial direction of cell expansion. Intriguingly, the root knot nematode *Meloidogyne javanica*, is able to induce the expression of a tomato α-expansin, *LeEXP5*, in roots during the establishment of feeding sites which are composed of giant cells up to 100 times larger than normal (Bar-Or et al. 2005; Gal et al. 2006). Silencing of *LeEXP5* expression in roots was found to reduce the ability of the nematodes to establish parasitism (Gal et al. 2006).

8 Expansins in Wall Weakening and Disassembly

Apart from the role of expansins in wall-expansion during growth, they are also implicated in processes which require weakening and breakdown of the wall such as ripening, abscission and certain developmental processes including pollen-tube growth and xylem formation. There are now numerous reports of high levels of α-expansin expression during fruit ripening. These include tomato (Rose et al. 1997; Brummell et al. 1999), apple (Wakasa et al. 2003), pear (Hiwasa et al. 2003; Fonseca et al. 2005), strawberry (Harrison et al. 2001), peach (Hayama et al. 2000, 2003; Obenland et al. 2003), mango (Sane et al. 2005), fig (Owino et al. 2004), apricot (Mbéguié-A-Mbéguié et al. 2002), banana (Trivedi and Nath 2004), olive (Ferrante et al. 2004) and sour cherry (Yoo et al. 2003). Using antibodies raised to a ripening-specific tomato expansin, LeEXP1, Rose et al. (2000) have also detected expansin expression in melon, avocado, persimmon, kiwi, pineapple and pepper. In the same study, it was found that antibodies to CsEXP1 cross-reacted with expansins expressed in early stages of fruit development but not during ripening stages, providing evidence that distinct expansin isoforms are involved in growth and ripening processes.

The function of expansin in facilitating ripening has been most extensively investigated in tomato by Brummel and co-workers. Antisense-inhibition of *LeEXP1* expression was not found to influence fruit size but did cause significant reductions in fruit softening (Brummell et al. 1999). The depolymerization of hemicellulose was not affected in antisense-lines, suggesting that LeEXP1 functions primarily to loosen walls during ripening rather than to modify the activity of hydrolases. Delayed ripening was, however, associated with reduced pectin breakdown and a strong down-regulation in polygalacturonase (PG) expression (Brummell and Harpster 2001). The authors suggested that reduced access of PG to its substrate in *LeEXP1*-suppressed fruit caused accumulation of the protein in abnormal locations and led to reduced expression, possibly through a feedback regulation mechanism. In contrast, overexpression of *LeEXP1* substantially increased fruit softening and caused an increased breakdown of hemicelluloses while pectin depolymerization was unaffected. These findings suggest that in addition to a direct effect on wall loosening during ripening, expansins may also function indirectly by exposing substrate sites to the action of wall hydrolases at the appropriate times in ripening.

Cosgrove et al. (1997) showed that the β-expansin, Zea m1 from maize pollen was able to induce rapid extension of maize silks (the stigmas and styles of the maize flower) suggesting that these highly abundant pollen allergens function in aiding penetration of the pollen tube through maternal tissues. Interestingly, Wang et al. (2004) have shown that a reduced level of Zea m1 is associated with the sterile phenotype of the maize gaMS-2 mutant. Another β-expansin, PPAL from tobacco, was found to accumulate in the stigmatic exudate during pollination (Pezzotti et al. 2002) but it was later shown that a purified form of this protein was not able to induce extension in wall specimens from cucumber hypocotyls or wheat coleoptiles and the wall-loosening activity was attributed to a lipid transfer protein (Nieuwland et al. 2005).

In another example of expansin-mediated intrusive growth, three α-expansins were associated with the growth of procambium cells during the differentiation of tracheary elements in zinnia stems (Im et al. 2000). In situ hybridization showed that two of these expansins were localized to the apical end and one to the basal end of putative cambial cells. Similar findings were reported for an α-expansin from hybrid aspen which was found to accumulate at the tips of intrusively growing xylem fibres (Gray-Mitsumune et al. 2004). Using a cDNA-AFLP screening technique, Milioni et al. (2001) identified two α-expansins and three β-expansins in zinnia mesophyl cell cultures which were more abundant at later stages of differentiation, during or after the deposition of secondary wall thickenings.

Expansins have also been implicated in the process of abscission which involves the coordinated breakdown of cell walls in discrete zones. Belfield et al. (2005) cloned two α-expansin cDNAs from *Sambucus nigra* which accumulated specifically in petiole abscission zones in response to ethylene treatment. This followed an earlier study where fusion of the *Arabidopsis AtEXP10* promoter region to the GUS reporter gene resulted in GUS staining at the base of petioles (Cho and Cosgrove 2000). Another α-expansin from tomato, LeEXP4, may play a similar role in wall breakdown of seed tissue during germination as it is specifically localized to the micropylar endosperm cap and is thought to cause weakening of this region to facilitate radicle emergence (Chen and Bradford 2000). Mella et al. (2004) also showed that an expansin which was present in the same region of *Datura ferox* seeds was upregulated in response to red light and that this effect was reversible by far-red light.

9 Adaptive Responses

The expression of some expansin genes is also regulated by environmental signals and expansin activity may aid in the adaptation of plants to conditions of abiotic stress. At low water potentials, expansins are associated with the maintenance of root elongation in maize (Wu et al. 1996, 2001) and rice

(Yang et al. 2004), an important response that helps to reestablish plant water supply. Expansins may fulfil a similar function in allowing the continued expansion of vegetative tissues during periods of decreased water availability. Sabirzhanova et al. (2005) showed an increase in the expression of a maize α-expansin prior to the resumption in growth of leaves after PEG-induced water stress and, in microarray experiments, Buchanan et al. (2005) found a greater than 100-fold increase in the expression of a β-expansin in sorghum shoots in response to ABA treatment and osmotic stress.

The leaves of the desiccation-tolerant plant, *Craterostigma plantagineum*, undergo significant shrinkage due to extensive folding of cell walls during drying. This is thought to be an important mechanism of maintaining connections between the plasma membrane and cell wall. Jones and McQueen-Mason (2004) showed that wall extensibility increased significantly during drying and rehydration and that this was correlated with increased expansin activity and expression of three α-expansins.

During flooding, several species of semi-aquatic plants are able to maintain their foliage above the water level by extremely rapid shoot elongation. In deepwater rice, an accelerated growth rate of up to 5 mm h^{-1} has been reported (Stünzi and Kende 1989) and is associated with increased expression of several expansin genes in internode regions, most of which are also induced by gibberellin (Cho and Kende 1997; Lee and Kende 2002). Likewise, the dicot *Rumex palustris*, responds to complete submergence by a strong increase in expansin activity within 4–6 h resulting in rapid elongation of petiole cells (Vriezen et al. 2000; Vreeburg et al. 2005). This is thought to be triggered by ethylene which accumulates in submerged tissue. A similar response to flooding has been described for arrowhead (*Sagittaria pygmaea*), a common weed present in rice fields. Two expansins in the tubers of this species were upregulated by anoxia but only one of these was responsive to ethylene (Ookawara et al. 2005). Additionally, expression of an α-expansin was found to increase in the rachis of a semi-aquatic fern *Regnellidium diphyllum* during submergence (Kim et al. 2000).

Expansins have also been implicated in the establishment of symbioses with arbuscular mycorrizal fungi and nitrogen-fixing bacteria. During colonization of cucumber roots by the fungus *Glomus versiforme*, Balestrini et al. (2005) observed an increase in the size and thickness of infected cortical cells. Using antibodies against two α-expansins, it was shown that cross-reacting proteins were mostly associated with the interface region, suggesting that a wall-loosening activity is required for penetration of the fungal hyphae through the cell wall. Similarly, during nodule formation induced by *Sinorhizobium*, Giordano and Hirsch (2004) identified an α-expansin gene in the legume, *Melilotus alba* which was upregulated in roots and showed that a protein which cross-reacted with a cucumber expansin antibody accumulated in expanding nodules.

10
Future Prospects

An important role for expansins in growth and wall breakdown is now well supported by several lines of evidence but it is not yet clear how these seemingly diverse processes are mediated by apparently similar proteins. In addition, gene-suppression studies suggest that at least some expansins have tissue-specific functions which are not complemented by the activity of others. Addressing these observations will require a better understanding of expansin activity in the context of differing cell wall architectures and compositions. Progress in understanding the mechanism of action and substrate specificity of expansins is currently hampered by our inability to produce them in an heterologous system. Another remaining challenge lies in elucidating the signalling pathways which regulate expansin expression during plant development and in response to external stimuli.

References

Balestrini R, Cosgrove DJ, Bonfante P (2005) Differential location of α-expansin proteins during the accommodation of root cells to an arbuscular mycorrhizal fungus. Planta 220:889–899

Baluška F, Salaj J, Mathur J, Braun M, Jasper F, Samaj J, Chua NH, Barlow PW, Volkmann D (2000) Root hair formation: F-actin-dependent tip growth is initiated by local assembly of profilin-supported F-actin meshworks accumulated within expansin-enriched bulges. Dev Biol 227:618–632

Bar-Or C, Kapulnik Y, Koltai H (2005) A broad characterization of the transcriptional profile of the compatible tomato response to the plant parasitic root knot nematode *Meloidogyne javanica*. Eur J Plant Pathol 111:181–192

Belfield EJ, Ruperti B, Roberts JA, McQueen-Mason S (2005) Changes in expansin activity and gene expression during ethylene-promoted leaflet abscission in *Sambucus nigra*. J Exp Bot 56:817–823

Bolam DN, Ciruela A, McQueen-Mason S, Simpson P, Williamson MP, Rixon JE, Boraston A, Hazlewood GP, Gilbert HJ (1998) Pseudomonas cellulose-binding domains mediate their effects by increasing enzyme substrate proximity. Biochem J 331:775–781

Brummell DA, Harpster MH (2001) Cell wall metabolism in fruit softening and quality and its manipulation in transgenic plants. Plant Mol Biol 47:311–340

Brummell DA, Harpster MH, Civello PM, Palys JM, Bennett AB, Dunsmuir P (1999) Modification of expansin protein abundance in tomato fruit alters softening and cell wall polymer metabolism during ripening. Plant Cell 11:2203–2216

Buchanan CD, Lim SY, Salzman RA, Kagiampakis L, Morishige DT, Weers BD, Klein RR, Pratt LH, Cordonnier-Pratt MM, Klein PE, Mullet JE (2005) *Sorghum bicolor's* transcriptome response to dehydration, high salinity and ABA. Plant Mol Biol 58:699–720

Catalá C, Rose JKC, Bennett AB (2000) Auxin-regulated genes encoding cell wall-modifying proteins are expressed during early tomato fruit growth. Plant Physiol 122:527–534

Chen F, Bradford KJ (2000) Expression of an expansin is associated with endosperm weakening during tomato seed germination. Plant Physiol 124:1265–1274

Chen F, Dahal P, Bradford KJ (2001) Two tomato expansin genes show divergent expression and localization in embryos during seed development and germination. Plant Physiol 127:928–936

Cho HT, Cosgrove DJ (2000) Altered expression of expansin modulates leaf growth and pedicel abscission in *Arabidopsis thaliana*. Proc Natl Acad Sci USA 97:9783–9788

Cho HT, Cosgrove DJ (2002) Regulation of root hair initiation and expansin gene expression in Arabidopsis. Plant Cell 14:3237–3253

Cho HT, Kende H (1997) Expansins and internodal growth of deepwater rice. Plant Physiol 113:1145–1151

Choi DS, Lee Y, Cho HT, Kende H (2003) Regulation of expansin gene expression affects growth and development in transgenic rice plants. Plant Cell 15:1386–1398

Cosgrove DJ (1989) Characterization of long-term extension of isolated cell-walls from growing cucumber hypocotyls. Planta 177:121–130

Cosgrove DJ (1997) Assembly and enlargement of the primary cell wall in plants. Annu Rev Cell Dev Biol 13:171–201

Cosgrove DJ (1999) Enzymes and other agents that enhance cell wall extensibility. Annu Rev Plant Physiol Plant Mol Biol 50:391–417

Cosgrove DJ, Bedinger P, Durachko DM (1997) Group I allergens of grass pollen as cell wall-loosening agents. Proc Natl Acad Sci USA 94:6559–6564

Cosgrove DJ, Li LC, Cho HT, Hoffmann-Benning S, Moore RC, Blecker D (2002) The growing world of expansins. Plant Cell Physiol 43:1436–1444

Coutinho PM, Henrissat B (1999) Carbohydrate-Active Enzymes.
URL: http://afmb.cnrs-mrs.fr/CAZY/

Coutinho PM, Henrissat B (1999) Carbohydrate-active enzymes: an integrated database approach. In: Gilbert HJ, Davies G, Henrissat B, Svensson B (eds) Recent Advances in Carbohydrate Bioengineering. The Royal Society of Chemistry, Cambridge, p 3–12

Darley CP, Li Y, Schaap P, McQueen-Mason SJ (2003) Expression of a family of expansin-like proteins during the development of *Dictyostelium discoideum*. FEBS Lett 546:416–418

Ferrante A, Hunter DA, Reid MS (2004) Towards a molecular strategy for improving harvesting of olives (*Olea europaea* L.). Postharvest Biol Tec 31:111–117

Fleming AJ, McQueen-Mason S, Mandel T, Kuhlemeier C (1997) Induction of leaf primordia by the cell wall protein expansin. Science 276:1415–1418

Fonseca S, Monteiro L, Barreiro MG, Pais MS (2005) Expression of genes encoding cell wall modifying enzymes is induced by cold storage and reflects changes in pear fruit texture. J Exp Bot 56:2029–2036

Gal T, Aussenberg E, Burdman S, Kapulnik Y, Koltai H (2006) Expression of a plant expansin is involved in the establishment of root knot nematode parasitism in tomato. Planta 224:155–162

Gilkes NR, Henrissat B, Kilburn DG, Miller RC, Warren RAJ (1991) Domains in microbial beta-1,4-glycanases—sequence conservation, function, and enzyme families. Microbiol Rev 55:303–315

Giordano W, Hirsch AM (2004) The expression of MaEXP1, a *Melilotus alba* expansin gene, is upregulated during the sweetclover-*Sinorhizobium meliloti* interaction. Mol Plant Microbe Interact 17:613–622

Gray-Mitsumune M, Mellerowicz EJ, Abe H, Schrader J, Winzell A, Sterky F, Blomqvist K, McQueen-Mason S, Teeri TT, Sundberg B (2004) Expansins abundant in secondary xylem belong to subgroup a of the alpha-expansin gene family. Plant Physiol 135:1552–1564

Harmer SE, Orford SJ, Timmis JN (2002) Characterisation of six alpha-expansin genes in *Gossypium hirsutum* (upland cotton). Mol Genet Genomics 268:1–9

Harrison EP, McQueen-Mason SJ, Manning K (2001) Expression of six expansin genes in relation to extension activity in developing strawberry fruit. J Exp Bot 52:1437–1446

Hayama H, Ito A, Moriguchi T, Kashimura Y (2003) Identification of a new expansin gene closely associated with peach fruit softening. Postharvest Biol Tec 29:1–10

Hayama H, Shimada T, Haji T, Ito A, Kashimura Y, Yoshioka H (2000) Molecular cloning of a ripening-related expansin cDNA in peach: evidence for no relationship between expansin accumulation and change in fruit firmness during storage. J Plant Physiol 157:567–573

Hiwasa K, Rose JKC, Nakano R, Inaba A, Kubo Y (2003) Differential expression of seven alpha-expansin genes during growth and ripening of pear fruit. Physiol Plant 117:564–572

Hutchison KW, Singer PB, McInnis S, Diaz-Salaz C, Greenwood MS (1999) Expansins are conserved in conifers and expressed in hypocotyls in response to exogenous auxin. Plant Physiol 120:827–831

Im KH, Cosgrove DT, Jones AM (2000) Subcellular localization of expansin mRNA in xylem cells. Plant Physiol 123:463–470

Jones L, McQueen-Mason S (2004) A role for expansins in dehydration and rehydration of the resurrection plant *Craterostigma plantagineum*. FEBS Lett 559:61–65

Kende H, Bradford KJ, Brummell DA, Cho HT, Cosgrove DJ, Fleming AJ, Gehring C, Lee Y, McQueen-Mason S, Rose JKC, Voesenek L (2004) Nomenclature for members of the expansin superfamily of genes and proteins. Plant Mol Biol 55:311–314

Kim JH, Cho HT, Kende H (2000) alpha-Expansins in the semiaquatic ferns *Marsilea quadrifolia* and *Regnellidium diphyllum*: evolutionary aspects and physiological role in rachis elongation. Planta 212:85–92

Lee DK, Ahn JH, Song SK, Choi YD, Lee JS (2003) Expression of an expansin gene is correlated with root elongation in soybean. Plant Physiol 131:985–997

Lee Y, Kende H (2001) Expression of beta-expansins is correlated with internodal elongation in deepwater rice. Plant Physiol 127:645–654

Li Y, Darley CP, Ongaro V, Fleming A, Schipper O, Baldauf SL, McQueen-Mason SJ (2002) Plant expansins are a complex multigene family with an ancient evolutionary origin. Plant Physiol 128:854–864

Li Y, Jones L, McQueen-Mason S (2003) Expansins and cell growth. Curr Opin Plant Biol 6:603–610

Lin Z, Ni ZF, Zhang Y, Yao YY, Wu HY, Sun QX (2005) Isolation and characterization of 18 genes encoding alpha- and beta-expansins in wheat (*Triticum aestivum* L.). Mol Genet Genomics 274:548–556

Link BM, Cosgrove DJ (1998) Acid-growth response and alpha-expansins in suspension cultures of bright yellow 2 tobacco. Plant Physiol 118:907–916

Mbéguié-A-Mbéguié D, Gouble B, Gomez RM, Audergon JM, Albagnac G, Fils-Lycaon B (2002) Two expansin cDNAs from *Prunus armeniaca* expressed during fruit ripening are differently regulated by ethylene. Plant Physiol Biochem 40:445–452

McQueen-Mason SJ, Cosgrove DJ (1994) Disruption of hydrogen-bonding between plant-cell wall polymers by proteins that induce wall extension. Proc Natl Acad Sci USA 91:6574–6578

McQueen-Mason SJ, Cosgrove DJ (1995) Expansin mode of action on cell-walls—analysis of wall hydrolysis, stress-relaxation, and binding. Plant Physiol 107:87–100

McQueen-Mason SJ, Durachko DM, Cosgrove DJ (1992) Two endogenous proteins that induce cell-wall extension in plants. Plant Cell 4:1425–1433

McQueen-Mason SJ, Fry SC, Durachko DM, Cosgrove DJ (1993) The relationship between xyloglucan endotransglycosylase and in vitro cell wall extension in cucumber hypocotyls. Planta 190:327–331

Mella RA, Burgin MJ, Sanchez RA (2004) Expansin gene expression in *Datura ferox* L. seeds is regulated by the low-fluence response, but not by the high-irradiance response, of phytochromes. Seed Sci Res 14:61–71

Milioni D, Sado PE, Stacey NJ, Domingo C, Roberts K, McCann MC (2001) Differential expression of cell-wall-related genes during the formation of tracheary elements in the *Zinnia* mesophyll cell system. Plant Mol Biol 47:221–238

Nieuwland J, Feron R, Huisman BAH, Fasolino A, Hilbers CW, Derksen J, Mariani C (2005) Lipid transfer proteins enhance cell wall extension in tobacco. Plant Cell 17:2009–2019

Obenland DM, Crisosto CH, Rose JKC (2003) Expansin protein levels decline with the development of mealiness in peaches. Postharvest Biol Tec 29:11–18

Okamoto-Nakazato A, Nakamura T, Okamoto H (2000) The isolation of wall-bound proteins regulating yield threshold tension in glycerinated hollow cylinders of cowpea hypocotyl. Plant Cell Environ 23:145–154

Okamoto-Nakazato A, Takahashi K, Kido N, Owaribe K, Katou K (2000) Molecular cloning of yieldins regulating the yield threshold of cowpea cell walls: cDNA cloning and characterization of recombinant yieldin. Plant Cell Environ 23:155–164

O'Malley RC, Lynn DG (2000) Expansin message regulation in parasitic angiosperms: Marking time in development. Plant Cell 12:1455–1465

Ookawara R, Satoh S, Yoshioka T, Ishizawa K (2005) Expression of alpha-expansin and xyloglucan endotransglucosylase/hydrolase genes associated with shoot elongation enhanced by anoxia, ethylene and carbon dioxide in arrowhead (*Sagittaria pygmaea* Miq.) tubers. Ann Bot 96:693–702

Orford SJ, Timmis JN (1998) Specific expression of an expansin gene during elongation of cotton fibres. Biochim Biophys Acta, Gene Struct Expression 1398:342–346

Owino WO, Nakano R, Kubo Y, Inaba A (2004) Coordinated expression patterns of genes encoding cell wall modifying enzymes during ripening in distinct anatomical tissue regions of the fig (*Ficus carica* L.) fruit. Postharvest Biol Tec 32:253–261

Pezzotti M, Feron R, Mariani C (2002) Pollination modulates expression of the PPAL gene, a pistil-specific beta-expansin. Plant Mol Biol 49:187–197

Pien S, Wyrzykowska J, McQueen-Mason S, Smart C, Fleming A (2001) Local expression of expansin induces the entire process of leaf development and modifies leaf shape. Proc Natl Acad Sci USA 98:11812–11817

Popper ZA, Fry SC (2003) Primary cell wall composition of bryophytes and charophytes. Ann Bot 91:1–12

Rayle DL, Haughton PM, Cleland R (1970) An in-vitro system that simulates plant cell extension growth. Proc Natl Acad Sci USA 67:1814–1817

Reinhardt D, Wittwer F, Mandel T, Kuhlemeier C (1998) Localized upregulation of a new expansin gene predicts the site of leaf formation in the tomato meristem. Plant Cell 10:1427–1437

Richard O, Paquet N, Haudecoeur E, Charrier B (2005) Organization and expression of the GSK3/Shaggy kinase gene family in the moss Physcomitrella patens suggest early gene multiplication in land plants and an ancestral response to osmotic stress. J Mol Evol 61:99–113

Rochange SF, Wenzel CL, McQueen-Mason SJ (2001) Impaired growth in transgenic plants over-expressing an expansin isoform. Plant Mol Biol 46:581–589

Rose JKC, Cosgrove DJ, Albersheim P, Darvill AG, Bennett AB (2000) Detection of expansin proteins and activity during tomato fruit ontogeny. Plant Physiol 123:1583–1592

Rose JKC, Lee HH, Bennett AB (1997) Expression of a divergent expansin gene is fruit-specific and ripening-regulated. Proc Natl Acad Sci USA 94:5955–5960

Sabirzhanova IB, Sabirzhanov BE, Chemeris AV, Veselov DS, Kudoyarova GR (2005) Fast changes in expression of expansin gene and leaf extensibility in osmotically stressed maize plants. Plant Physiol Biochem 43:419–422

Sampedro J, Carey R, Cosgrove D (2006) Genome histories clarify evolution of the expansin superfamily: new insights from the poplar genome and pine ESTs. J Plant Res 119:11–21

Sampedro J, Cosgrove DJ (2005) The expansin superfamily. Genome Biol 6:242

Sampedro J, Lee Y, Carey RE, dePamphilis C, Cosgrove DJ (2005) Use of genomic history to improve phylogeny and understanding of births and deaths in a gene family. Plant J 44:409–419

Sane VA, Chourasia A, Nath P (2005) Softening in mango (*Mangifera indica* cv. Dashehari) is correlated with the expression of an early ethylene responsive, ripening related expansin gene, *MiExpA1*. Postharvest Biol Tec 38:223–230

Schipper O, Schaefer D, Reski R, Fleming A (2002) Expansins in the bryophyte *Physcomitrella patens*. Plant Mol Biol 50:789–802

Shcherban TY, Shi J, Durachko DM, Guiltinan MJ, McQueen-Mason SJ, Shieh M, Cosgrove DJ (1995) Molecular cloning and sequence analysis of expansins—a highly conserved, multigene family of proteins that mediate cell-wall extension in plants. Proc Natl Acad Sci USA 92:9245–9249

Spalding EP, Cosgrove DJ (1989) Large plasma-membrane depolarization precedes rapid blue-light-induced growth inhibition in cucumber. Planta 178:407–410

Stünzi JT, Kende H (1989) Light-dependent short-term modulations of elongation in rice plants. Plant Cell Physiol 30:415–422

Thompson JD, Gibson TJ, Plewniak F, Jeanmougin F, Higgins DG (1997) The ClustalX windows interface: flexible strategies for multiple sequence alignment aided by quality analysis tools. Nucleic Acids Res 25:4876–4882

Trivedi PK, Nath P (2004) *MaExp1*, an ethylene-induced expansin from ripening banana fruit. Plant Sci 167:1351–1358

Vreeburg RAM, Benschop JJ, Peeters AJM, Colmer TD, Ammerlaan AHM, Staal M, Elzenga TM, Staals RHJ, Darley CP, McQueen-Mason SJ, Voesenek LACJ (2005) Ethylene regulates fast apoplastic acidification and expansin A transcription during submergence-induced petiole elongation in *Rumex palustris*. Plant J 43:597–610

Vriezen WH, De Graaf B, Mariani C, Voesenek LACJ (2000) Submergence induces expansin gene expression in flooding- tolerant *Rumex palustris* and not in flooding-intolerant *R- acetosa*. Planta 210:956–963

Wakasa Y, Hatsuyama Y, Takahashi A, Sato T, Niizeki M, Harada T (2003) Divergent expression of six expansin genes during apple fruit ontogeny. Eur J Hortic Sci 68:253–259

Wang W, Scali M, Vignani R, Milanesi C, Petersen A, Sari-Gorla M, Cresti M (2004) Male-sterile mutation alters Zea m 1 (beta-expansin 1) accumulation in a maize mutant. Sex Plant Reprod 17:41–47

Whitney SEC, Brigham JE, Darke AH, Reid JSG, Gidley MJ (1995) In-vitro assembly of cellulose/xyloglucan networks—ultrastructural and molecular aspects. Plant J 8:491–504

Whitney SEC, Brigham JE, Darke AH, Reid JSG, Gidley MJ (1998) Structural aspects of the interaction of mannan-based polysaccharides with bacterial cellulose. Carbohydr Res 307:299–309

Whitney SEC, Gidley MJ, McQueen-Mason SJ (2000) Probing expansin action using cellulose/hemicellulose composites. Plant J 22:327–334

Wu YJ, Cosgrove DJ (2000) Adaptation of roots to low water potentials by changes in cell wall extensibility and cell wall proteins. J Exp Bot 51:1543–1553

Wu YJ, Meeley RB, Cosgrove DJ (2001) Analysis and expression of the alpha-expansin and beta-expansin gene families in maize. Plant Physiol 126:222–232

Wu YJ, Sharp RE, Durachko DM, Cosgrove DJ (1996) Growth maintenance of the maize primary root at low water potentials involves increases in cell-wall extension properties, expansin activity, and wall susceptibility to expansins. Plant Physiol 111:765–772

Yang L, Zheng B, Mao C, Qi X, Liu F, Wu P (2004) Analysis of transcripts that are differentially expressed in three sectors of the rice root system under water deficit. Mol Genet Genomics 272:433–442

Yoo SD, Gao ZF, Cantini C, Loescher WH, van Nocker S (2003) Fruit ripening in sour cherry: Changes in expression of genes encoding expansins and other cell-wall-modifying enzymes. J Am Soc Hortic Sci 128:16–22

Zenoni S, Reale L, Tornielli GB, Lanfaloni L, Porceddu A, Ferrarini A, Moretti C, Zamboni A, Speghini S, Ferranti F, Pezzotti M (2004) Downregulation of the *Petunia hybrida* alpha-expansin gene *PhEXP1* reduces the amount of crystalline cellulose in cell walls and leads to phenotypic changes in petal limbs. Plant Cell 16:295–308

Zhang NG, Hasenstein KH (2000) Distribution of expansins in graviresponding maize roots. Plant Cell Physiol 41:1305–1312

Plant Cell Monogr (5)
J.-P. Verbelen · K. Vissenberg: The Expanding Cell
DOI 10.1007/7089_2006_074/Published online: 30 September 2006
© Springer-Verlag Berlin Heidelberg 2006

Pectic Polysaccharides and Expanding Cell Walls

Yves Verhertbruggen · J. Paul Knox (✉)

Centre for Plant Sciences, University of Leeds, Leeds LS2 9JT, UK
j.p.knox@leeds.ac.uk

Abstract Pectic polysaccharides are major components of extendable primary cell walls and are amongst the most complex macromolecules in nature. The three major pectic polymers are homogalacturonan, rhamnogalacturonan I and rhamnogalacturonan II. Here we review the current understanding of how these structurally distinct polymers are integrated into primary cell wall pectic networks and how they function in cell wall matrices. We also review how the structure and properties of pectic networks can be modified in muro and how they can contribute to environments for the controlled slippage of the cellulose and cross-linking glycan network during cell expansion processes.

1
Introduction

Remodelling of the appropriate cell walls of extending cells to allow organ extension in terms of both direction and extent is a major feature of the growth of plants. In addition to any alterations to the existing sets of cell wall polymers, a major aspect of this cell wall remodelling is the continued synthesis and assembly of all cell wall components to ensure that cell walls maintain the appropriate tensile strength of their surfaces. This continuous formation of cell wall material requires the extensive Golgi apparatus-based synthesis of matrix polysaccharides, their transport to, and deposition at, the cell surface and the integration with the synthesis of cellulose microfibrils and cell wall assembly at the plasma membrane (Cosgrove 2005). Of the three major sets of polysaccharides that comprise extendable primary cell walls – cellulose microfibrils (see Hématy and Höfte, in this volume), cross-linking glycans (CLGs) or hemicelluloses (see Obel et al., in this volume) and pectins – it is pectins that are often not present in non-extendable secondary cell walls. It is a striking point that it is pectins for which we have the least clear understanding of their role in cell enlargement. This limited understanding in part arises from the greater structural complexity of the pectic polymers, over cellulose and CLGs, but also from their likely extended multifunctionality and contribution to factors such as cell wall architecture, mechanical properties, porosity, ionic status, cell adhesion and plant defence responses, in addition to any functions that are directly required for cell expansion.

Pectin-rich primary cell walls resist tensile forces, generate turgor pressure and accommodate turgor-driven expansion, whereas secondary cell walls (composites of cellulose and CLGs that are often lignified) are deposited in fully extended cells to resist compressive forces. Thus, in simple terms, the pectic polymers can be viewed as providing a three-dimensional hydrated matrix in which the controlled stretching of the oriented CLG-tethered cellulose microfibril network can take place. However, in addition to this role of providing an environment for the cellulose–CLG network during cell wall stretching, it is worth indicating the possible role of pectic polymers in the assembly of new primary cell wall material during the expansion process. The presence of pectic polymers may in fact be a requirement for cell wall assembly at a rapidly expanding plasma membrane surface. This is in contrast to the situation of static or decreasing surface area of the plasma membrane during the deposition of secondary cell walls. In this possible role, pectic polysaccharides may impact upon the formation or spacing of cellulose microfibrils and this may underpin extensibility. Such a role for pectin has been suggested by a study of its presence during the in vitro formation of *Acetobacter*-derived cellulose composites (Chanliaud and Gidley 1999).

In current models of primary cell walls, the pectic network is a collection of distinct polysaccharides that have a capacity for a variety of covalent and non-covalent linkages that contribute to a dynamic three-dimensional matrix that surrounds and embeds the cellulose–CLG network (Cosgrove 2005). However, evidence is emerging of direct links between specific polymers of the pectic matrix and both cellulose microfibrils and CLGs (Sect. 3). The putative roles for pectic polymers in maintaining the coherence and properties of the cellulose microfibril–CLG layers and networks, the adherence of adjacent cell walls, and cell wall associations with the plasma membrane are far from clear. At the heart of the many remaining questions concerning pectin and plant growth is the structure of the pectin network, its capacity to be modified in muro and its ability to influence cell wall properties.

2 Pectin Structure: Domains, Networks, Matrices, Models and Megamacromolecules

Pectins are defined as polysaccharides rich in galacturonic acid and they appear to be generally restricted to land plants. Recent reviews have collected information on the structure and biology of pectic polymers (Mohnen 1999; Ridley et al. 2001; Willats et al. 2001a; Vincken et al. 2003a,b). Currently, it is thought that there are three major classes of pectic polysaccharide that are to be found to some extent in all primary cell walls. These polysaccharides are homogalacturonan and rhamnogalacturonans I and II. Other related polysaccharides, such as xylogalacturonan, have been characterized and appear to have more limited distributions (Ridley et al. 2001; Willats et al. 2001a, 2004). As no evidence currently links the occurrence of these minor or less characterized pectic polymers with cell extension processes they will not be considered further here.

Homogalacturonan (HG) is the most abundant pectic polymer and it is thought to account for about 50–70% of primary cell wall pectin (Mohnen 1999). HG is an α-1,4-linked galacturonan that can be substituted with methyl-esters and/or it can be acetylated. It is the pectic polymer that is often directly equated with "pectin" in the literature. HG is synthesized in a largely methyl-esterified form and subject to de-methylesterification by cell wall-based enzymes known as pectin methyl esterases (PMEs) (Ridley et al. 2001). The action of PMEs generates negatively charged carboxyl groups, and the presence of contiguous de-esterified galacturonic acid residues allows cross-linking of HG chains by ionic interaction with calcium ions. Such interactions can have a significant impact upon cell wall structure and properties, as shown schematically in Fig. 1 and as discussed in Sect. 4.

Potential for covalent links between HG chains in muro is generated by the presence of the polymer rhamnogalacturonan II (RGII). RGII is an extreme heteropolymer based around an HG backbone and is experimentally released from cell walls by polygalacturonase action, indicating attachment to HG chains (O'Neill et al. 2004). The structure of RGII, which appears to be conserved in land plants, has four distinct oligosaccharide chains that occur on contiguous or near galacturonic acid residues and accounts for about 10% of pectin in tracheophytes and about 0.1% of pectin in bryophytes (O'Neill et al. 2004). The presence of RGII domains within an HG polymer can directly influence the interaction of HG domains by allowing a novel type of interaction. Apiose residues in the side chain of separate RGII polymers can form a covalent cross-link in the form of a borate diester. This results in the potentially stable cross-linking of two HG chains that can contribute to cell wall properties and that can also influence cell wall porosity (Fig. 1; Fleischer et al. 1999). The extreme heteropolymeric nature of RGII appears to promote stability within the cell wall environment and no plant enzyme has been identified that can disrupt its structure (O'Neill et al. 2004).

Rhamnogalacturonan I (RGI) is a distinct, highly variable set of pectic polymers that can also be isolated by the polygalacturonase treatment of cell wall preparations (indicating attachment by glycosidic links to HG polysaccharides). However, RGI polysaccharides are based on a rhamnogalacturonan backbone (RG) in which rhamnose residues alternate with those of galacturonic acid. The rhamnose residues can act as points of attachment for a range of polymers that are generally rich in galactose and arabinose residues. The structural variation within these galactans, arabinans and arabinogalactans is large (Ridley et al. 2001; Hinz et al. 2005). There are two broad classes of arabinogalactans, type I and type II. Type I is characterized by a β-1,4-galactan core and type II by a β-1,3-β-1,6-galactan core. Type II arabinogalactans are also associated with protein components in arabinogalactan-proteins (AGPs) that are associated with plasma membranes and cell walls (Showalter 2001).

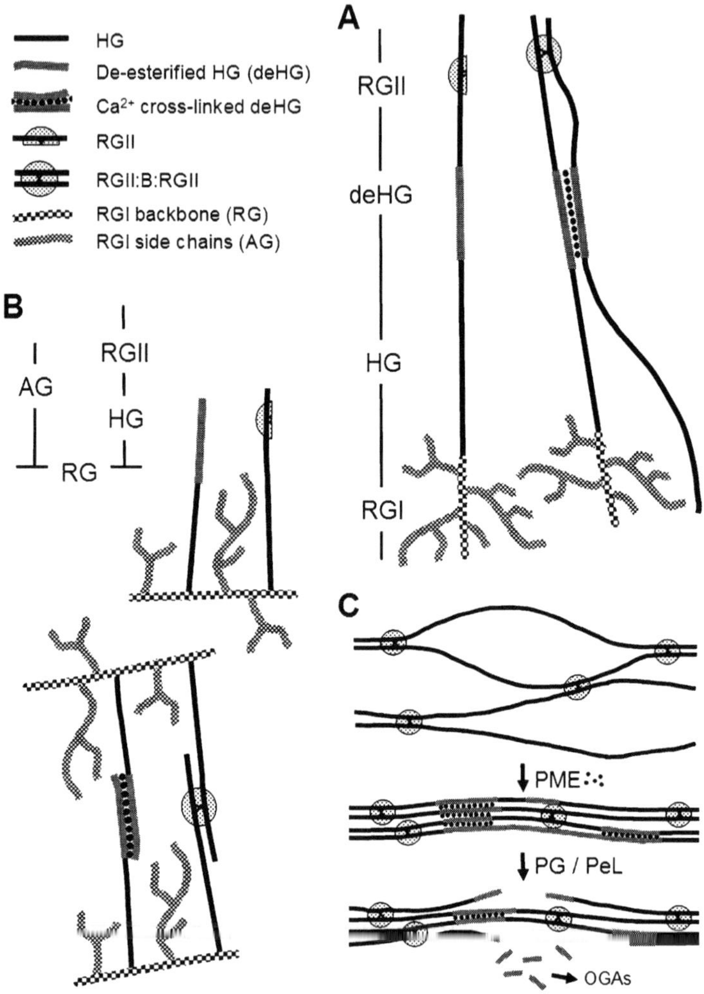

Fig. 1 Making a pectic network. Highly schematic representations of the major pectic polysaccharides and possible models for their interrelations as domains within larger macromolecules of the pectic network. **a** Conventional linear galacturonan model for the linkage between homogalacturonan (HG), rhamnogalacturonan II (RGII) and rhamnogalacturonan I (RGI) domains. **b** Vincken model for the linkage of HG (with RGII) along with arabinan, galactan and arabinogalactan (AG) domains as side chains of RGI backbones (RG). **c** Diagram indicating the impact of pectin methylesterase enzyme (PME) and polygalacturonase (PG) or pectate lyase (PeL) enzymes on the properties of an HG-rich matrix stabilized by RGII dimers. The combined action of cell wall enzymes may result in the release of bioactive oligogalacturonides (OGAs), which are shown as monomers but can act in the form of Ca^{2+} cross-linked dimers. For clarity the linear model is used and RGI domains are omitted although they may contribute to the generation of microenvironments within an HG matrix

The nature of biosynthetic or functional links between type II AG domains of RGI and AGPs are far from clear. The RGI sets of polymer appear to display heterogeneity within isolated samples and structural features of RGI polymers display considerable variations in their occurrence between taxons, organs, tissues, and even within the wall domains of individual cells (Knox 2002; Willats et al. 2003). The taxonomic aspects of pectic polymer structures, which have relevance for both our understanding of the evolution of land plants and the evolution of pectic polymer functions, are a promising line of enquiry for future studies (Popper and Fry 2003). It is the diverse RGI set of pectic polymers that is likely to contain the most taxonomically significant information in this regard.

In general terms, the analysis of pectin components indicates that HG, RGI and RGII comprise approximately 60, 30 and 10%, respectively, of pectic polymers in primary cell walls of dicotyledons (Mohnen 1999), although this cannot take into account local variations associated with diverse cell types or distinct cell wall regions. To understand how these polymers are integrated into functional networks that act in cell expansion and other cell processes it is important to know their occurrence in relation to individual cell wall architectures. For example, does a polymer or a particular form resulting from in muro modification occur evenly across primary cell walls and intercellular matrices, or is it restricted to middle lamellae, or to primary cell walls and is absent from middle lamellae? Other possibilities include a spatial restriction to inner cell wall regions close to the plasma membrane, or an even greater restriction to cell wall regions of longitudinal or transverse cell walls, or to cell walls around pit fields/plasmodesmata or at intercellular spaces. One of the most effective techniques to ascertain such details of pectin and cell wall microstructure is the use of defined antibodies in immunocytochemical procedures (Willats et al. 2001a). For an overview of the status of this technology, see Techniques box 1. Most of the above patterns of occurrences have been documented for epitopes of pectic polymers (Willats et al. 2001a, 2003; Knox 2002). In an elongating organ, cell walls transverse to the direction of growth are not subject to the extensive stretching undergone by the longitudinal cell walls. In the case of such cell walls, extensive stretching will be accompanied by the synthesis of new cell wall material and it is the inner region of the cell wall that is most recently assembled and that therefore may have a distinct microstructure. Moreover, the middle lamella adhesion zone between longitudinal cell walls in an elongating organ will also be subject to significant stretching. How the cell-to-cell links are maintained at these cell junctions and if any specific components are maintained is far from clear.

A key facet of the pectic matrix that impinges significantly on any understanding of the microstructure of primary cell walls outlined above is how the HG, RGI and RGII polysaccharides are linked together: How are they connected and integrated within a pectic network of a particular cell wall? How do they function as domains of a matrix? How do they influence or create microenvironments? The conventional model suggests that all three polymers are covalently linked along linear galacturonic acid-rich backbones, which results in distinct RGI and RGII domains, as shown in Fig. 1a. However, a recent proposal has put forward a model that suggests that the RGI backbone forms a core domain, to which HG (with RGII) and neutral side chains with diverse arabinans, galactans and arabinogalactans (AGs) and all other pectic domins are attached – much like hairs on a brush (Vincken et al. 2003a,b) as shown in Fig. 1b. This new model is suggested to account for observations of the branching of HG backbones, as imaged by atomic force microscopy, and of aspects of cell wall microstructure (Vincken et al. 2003a,b).

The Vincken model of the relationship of domains provides a system for the insertion of varied multi-domain pectins into primary cell walls and aims to account for aspects of cell wall assembly and the integration of the formation of primary cell walls, along with the formation and maintenance of middle lamellae and the control of cell wall thickness. In effect, it proposes a series of molecular brushes that are assembled in a cell wall in a highly oriented, stacked fashion with each brush having appropriate pectic hairs – attachments rich in HG, HG/RGII or AGs. The model proposes that these pectic hairs are radial to the plasma membrane surface spanning cell wall layers. This raises questions of how oriented deposition is achieved. Evidence for this model is not compelling at this stage; in striving to account for middle lamella formation it overlooks the fact that much cell wall assembly does not occur at a cell plate in transverse walls but in longitudinally orientated walls of rapidly expanding cells. Moreover, the appropriate cross-link between a galacturonic acid residue of HG and a rhamnose of RGI has yet to be found.

Both models probe and stimulate important questions concerning the arrangement of HG, RG and AG domains within covalently linked polymers. Improved molecular tools arising from the identification and functional dissection of genes underpinning pectic polymer synthesis and methods for imaging specific domains and their arrangements in isolated polysaccharides should answer some of these important questions.

What seems to be clear is that in cell walls HG, RGI and RGII polymers are covalently linked through glycosyl links to generate large pectic polymers with HG, RGI and RGII as distinct structural regions or domains. As indicated above, the form of these chains and the arrangements, patterns and frequencies of specific domains within these multi-domain chains are unknown. An additional factor contributing to the structure and properties of the pectic network is the extent to which these chains are covalently linked through borate diesters and/or through calcium ion cross-links. Additionally, in the Amaranthaceae, a family of plants that includes spinach and sugar beet, feruloylation of AG side chains allows an additional potential mechanism for cross-linking of pectic chains through the oxidative coupling of AG side chains of RGI (Clausen et al. 2004). Clearly, pectic components could be interlinked as part of macromolecular structures that ramify over many cells, an organ or even a plant. Whether such pectic megamacromolecules exist and whether they have any functional significance above local remodelling to suit requirements of growth and physiology, awaits further study.

3 Pectin Synthesis, Cell Wall Assembly and Turnover, and Interactions with Other Polymer Systems

An increased understanding of pectin synthesis could answer some of the above questions concerning the structure and architecture of the pectic network. All pectic polymers are thought to be synthesized in the Golgi apparatus. Evidence for this has been obtained for HG (Sterling et al. 2001). It is estimated that over 50 glycosyltransferases (GTs) are required for the direct synthesis of pectic polymers (Mohnen 1999). This is a challenging area of research, which is currently seeing progress and is providing important tools for pectic structure–function analyses (Schieble and Pauly 2004). A galacturonosyltransferase involved in the synthesis of HG and a related superfamily of GT sequences has recently been characterized (Sterling et al. 2006). A GT has been identified that catalyses a glucuronosyl linkage of RGII (Iwai et al. 2002) and a putative glycosyltransferase has been identified for pectic arabinan (Harholt et al. 2006). The continued identification of GTs should greatly facilitate the functional analyses of pectic domains.

Almost nothing is known about the molecular aspects of the regulation of the transport of pectic polymers in vesicles from the Golgi apparatus to the plasma membrane, their unloading and the regulation of their involvement (alongside CLGs and cellulose microfibrils) into assembling cell wall materials. The generation of synthetic cellulose-based composites by *Acetobacter xylinus* has indicated that the presence of pectin can influence composite properties including extensibility (Chanliaud and Gidley 1999). In these studies it was clear that pectin exerted its effect during the formation of the cellulose networks and also that the extent of HG methyl-esterification influenced composite properties (Chanliaud and Gidley 1999).

Another factor that could influence cell wall properties is the possible turnover and internalization of cell wall components from the cell wall. Analysis has indicated that RGII and some HG and RGI domains are endocytosed in meristematic cells (Baluška et al. 2002; Yu et al. 2002). Such a process has the potential to generate internal stores of cell wall material that are used for the construction of cell plates during cytokinesis (Baluška et al. 2005; Dhonukshe et al. 2006). These are important observations raising interesting sets of questions: Is the turnover and cycling of cell wall components a major activity in all cells? Does it have a role or is it down-regulated in cells undergoing extensive elongation? Is the passage of a pectic polymer through the cell wall environment and its modification by cell wall enzymes important for its subsequent function?

The nature of the interactions of domains of the pectic matrix with the other major polymer systems (cellulose microfibrils and CLGs) is also of considerable importance for understanding cell wall properties and functions. The idea that the pectic network merely embeds and is co-extensive with the cellulose-CLG polymers and that it can be solubilized by calcium chelators is an over-simplification, although it may apply to certain cell walls of certain species. Pectic epitopes are readily detected in cellulose residues of isolated cell walls (Orfila et al. 2002) and evidence for links between pectin and xyloglucan are accumulating (Abdel-Massih et al. 2003; Cumming et al. 2005). A study demonstrating that RGI side chains and particularly de-branched arabinan have the capacity to adhere to the surface of cellulose microfibrils in a manner similar to CLGs indicates that these domains may directly function in cell wall assembly or extensibility (Zykwinska et al. 2005).

4 HG: A Versatile Domain of the Pectic Matrix

HG is the most abundant polymer of the pectic matrix and has been most often implicated in having a role in cell expansion processes. In broad terms, HG can be viewed as regulating both the porosity or size-exclusion limit (SEL) of the matrix and its mechanical properties that impact upon stiffness and flexibility. Two major factors stabilize cross-links of HG and impinge on these properties: calcium ion cross-linking of de-esterified HG domains and borate diester linked RGII dimers, as shown in Fig. 1c. Current evidence indicates that 95% of RGII is cross-linked with boron and that this is a relatively stable linkage in pectic matrices (O'Neill et al. 2004). RGII links are clearly important for matrix integrity and cell wall porosity. Disruption of RGII dimers by boron depletion or by mutations that affect RGII structure indicate that RGII-mediated HG links are required for growth and that cell expansion seems to be the process that is disrupted (O'Neill et al. 2001, 2004). Loss of glucuronosyltransferase activity involved in construction of RGII resulted in a severe disruption of meristematic cells, so any direct specific effects on cell expansion processes could not be ascertained (Iwai et al. 2002). The level of RGII is estimated to be in the order of 100-fold less in occurrence in bryophyte cell walls (O'Neill et al. 2004). Comparative studies in bryophyte and tracheophyte systems may provide insight into the relative roles of HG cross-linking mechanisms in cell expansion processes.

The non-RGII-mediated interaction of HG chains and the modification of HG structure and its impact on cell wall properties are highly complex and would seem to relate directly and indirectly to cell wall extensibility. Several sets of cell wall-located enzymes have the capacity to influence the HG domains and thus control SELs and cell wall properties. HG is synthesized in a highly methyl-esterified form with 70–80% of galacturonic acid residues carrying methyl esters (Mohnen 1999). HG can be acted upon in the cell wall by PMEs that remove methyl-esters, resulting in the controlled formation of unesterified galacturonic acid residues. Blocks of de-esterified HG can be cross-linked by calcium ions. This cross-linking can influence local properties with a tendency to tighten the cell wall matrix. This may directly contribute to a reduction of extensibility and cell wall porosity that could restrict access of wall-loosening factors, such as expansins, to sites of action (Cosgrove 2005). The pattern as well as the extent of methyl-esterification is also likely to be important for influencing the properties of the pectic matrix (Willats et al. 2001b). After the action of PMEs results in the production of unesterified HG (or polygalacturonic acid) regions in the cell wall matrix, these can be cleaved by the action of other cell wall-based enzymes, such as polygalacturonases (PGs) and pectate lyases (PeLs), thus loosening the pectic network (Fig. 1c). The interplay of the diverse actions of these enzymes could clearly have contradictory influences on the coherence and interactions of HG polymers. For example, a PME may increase the acidic pectin content, promoting calcium ion cross-linking that would strengthen a cell wall. However, PME action may also allow subsequent PG or PeL action that cleaves HG chains and thus loosen the pectic network. These pectin-acting enzymes (PMEs, PGs and PeLs) occur in relatively large gene families in *Arabidopsis*. They have not been fully characterized in terms of developmental roles (Torki et al. 2000; Micheli 2001; Willats et al. 2001a; Bosch and Hepler 2005) but could clearly have significant involvement in cell expansion processes.

Another factor that relates to the role of HG metabolism in growth processes and that can impact upon expansion is the release of oligogalacturonides from combined PME and PG action (Ridley et al. 2001). Oligogalacturonides (OGAs) are products of enzyme action of de-esterified HG and are known signals in a range of plant processes. A report indicates that they can counteract auxin promotion of pea stem extension (Branca et al. 1988), but how OGAs act in such a system is unknown. They could have a signalling role in a receptor-based mechanism or they may bind to other HG domains and directly modify the properties of the pectic matrix (Ridley et al. 2001). Reactive oxygen species and, specifically, hydroxyl radicals are implicated in cell elongation processes (Vreeburg and Fry 2005, see Lindsay and Fry (2007), in this volume). De-esterified HG chains may be particularly susceptible to the action of hydroxyl radicals, although direct evidence for scission of pectic polymers within cell walls is currently lacking (Dumville and Fry 2003; Rose et al. 2004; Vreeburg and Fry 2005).

Calcium ion cross-linking of HG chains also provides the extensive capacity of primary cell walls to retain calcium ions, which appears to be an important factor for cell growth (Ezaki et al. 2005). The hydration status of an HG network seems to directly influence its mechanical properties. Studies of the properties of concentrated pectin gels that closely mimic the situation in muro, have indicated that swelling (or hydration status) and stiffness are greatly influenced by the ionic environment (Zsivanovits et al. 2004). The concentrated gel system indicates that cell wall hydration is a balance between the osmotic activity of a cell exerting stress on the wall, which restricts swelling, and the capacity of a network to attract water, which counteracts this (Zsivanovits et al. 2004). The properties of these systems also depend upon the source of the pectin, presumably reflecting distinct structural features in HG and/or RGI components (Zsivanovits et al. 2004). Cations other than calcium ions have not been ruled out from having roles in the stabilization of or the influencing of pectic networks, nor have polyamines or basic cell wall proteins such as extensins (Morris et al. 2003).

Genetic approaches to explore HG function in growth and cell expansion processes have not yet provided any clear answer as to how HG modulation may impact upon expansive growth – other than that it appears to be important. Manipulations to result in the knockout or disruption of glycosyltransferase (GT) genes or in the expression of hydrolases (with the aim to reduce particular components of the pectic matrix), often have pleiotropic effects that can be difficult to interpret in terms of direct effects on cell elongation. Proteins whose action could directly impact upon HG structure are GTs, PMEs, PGs and PeLs. Mutations of putative galacturonosyltransferases (part of a large superfamily related to a functionally characterized galacturonosyltransferase; Sterling et al. 2006) can result in dwarf plants, but also in serious disruptions of cell adhesion and other aspects of growth (Bouton et al. 2002; Lao et al. 2003; Shao et al. 2004), making dissection of the specific roles of HG difficult. Similarly, over-expression of PGs can result in a wide range of effects including anatomical changes and reduced plant stature (Atkinson et al. 2002; Capodicasa et al. 2003). A study of genes that influence the resistance to plant pathogens has led to important observations on the relationship between pectin and cell expansion (Vogel et al. 2004). This study identified *pmr5* and *pmr6* mutants that have altered cell wall compositions with elevated levels of pectins (with possibly reduced methyl-esterification) and a considerably reduced cell size (Vogel et al. 2004). Analysis of a double mutant indicated that the increase in pectin may be responsible for the restriction of cell expansion. The *pmr6* mutant was found to be defective in a novel glycosylphosphatidylinositol (GPI)-anchored pectate lyase-like gene (Vogel et al. 2004).

Analysis of the extension of pollen tubes has provided strong evidence that methyl-esterification status and HG modification are key elements in the regulation of the stiffness of three-dimensional pectic networks that can influence cell expansion. Extending pollen tubes are good systems for such studies as the growing tips are pectin-rich without significant cellulose or CLG components. Moreover, in these single cell systems there is reduced multifunctionality of the pectic matrix as there is no cell adhesion. Current understanding indicates that highly methyl-esterified HG is secreted at extending tips and PME action is required to change its properties and to control the yielding characteristics of the cell wall. Analysis of VANGUARD1, a PME, has indicated its role in maintaining the rapid growth rates of pollen tubes in *Arabidopsis* (Jiang et al. 2005). Recent work on pollen tubes from *Solanum chacoense* has indicated that pectin structure can change in response to the stiffness of the growth medium and that a gradient in HG de-esterification, away from the extending pollen tube tip, results in decreased viscoelasticity and increased rigidity (Parre and Geitmann 2005). Several studies have shown gradients in HG de-esterification using the monoclonal antibodies JIM5 and JIM7 (Bosch and Hepler 2005; Parre and Geitmann 2005).

There are indications that levels of HG methyl-esterification are higher in cell walls during diffuse growth in multicellar organs and that the ester level is reduced after growth has ceased (Liberman et al. 1999). However, immunocytochemical analysis of the methyl-esterification status of HG in elongating organs with antibody probes has not indicated any clear correlation of methyl-esterification status of HG and cell expansion, or with load-bearing layers within an expanding organ (Liberman et al. 1999; Willats et al. 2001a). This is likely due to lack of a distinct spatial focus of cell wall yielding in cells showing diffuse growth, to HG multifunctionality and its contribution to processes such as cell adhesion, and to the contribution of other polymer systems to cell wall properties.

5
RGI and Cell Expansion Processes: Diverse Polymers with Unknown Functions

As discussed above, the set of polymers designated as RGI is likely to be highly varied, diverse and heterogeneous within an elongating organ. RGI is a pectic domain that carries extensive and highly varied AG side chains (Ridley et al. 2001). These side chains are highly mobile and are proposed to interact with water (Ha et al. 2005). They may therefore have the capacity to create microenvironments in the cell wall matrix, with specific cell wall properties relating to water-holding capacity or porosity (Ulvskov et al. 2005). The demonstration of a key role for pectic arabinans in maintaining the flexibility of guard cell walls, required to mediate stomatal opening and closure, indicates a possible role in the spacing of cell wall components including HG (Jones et al. 2003). Such a capacity of RGI domains to interfere with HG chain interaction and gel formation may simply be steric hindrance and could prevent both calcium ion cross-linking or RGII dimerization. However, the indication that RGI side chains can attach to cellulose microfibrils also suggests a possible involvement in the cross-linking of cell wall components (Zykwinska et al. 2005).

Immunocytochemical analyses focussed on RGI have largely centred upon the occurrence of epitopes for β-1,4-galactan and α-1,5-arabinan, features of RGI side chains. These studies have indicated that these two epitopes are developmentally highly regulated and that they can often occur in distinct regions of cell walls and distinct stages of cell development (Willats et al. 2001a, 2003). The strongest evidence for a role in cell expansion processes is associated with the LM5 β-1,4-galactan epitope (McCartney et al. 2003). This epitope appears in cell walls at some stage during cell development in many angiosperm systems. In the *Arabidopsis* root, the developmental appearance of the LM5 galactan epitope reflects cell elongation (McCartney et al. 2003). LM5 has a restricted occurrence at the root surface and also in underlying endodermal, cortical and epidermal cell walls that marks the onset of the acceleration of cell elongation (McCartney et al. 2003). The epitope occurs in cell walls of the outer cell layers of the *Arabidopsis* root when cell lengths are in the region of 10–50 μm but cannot be detected in cells of greater lengths, up to the final length of about 200 μm (McCartney et al. 2003). LM5 epitope abundance during this transient occurrence is modulated by hormonal and genetic factors that regulate cell extension and root growth. The LM6 arabinan epitope is also modulated but shows distinct patterns of occurrence, as shown in Fig. 2. These observations give rise to several key questions: What properties does a particular RGI structure impart to a cell wall? How do these

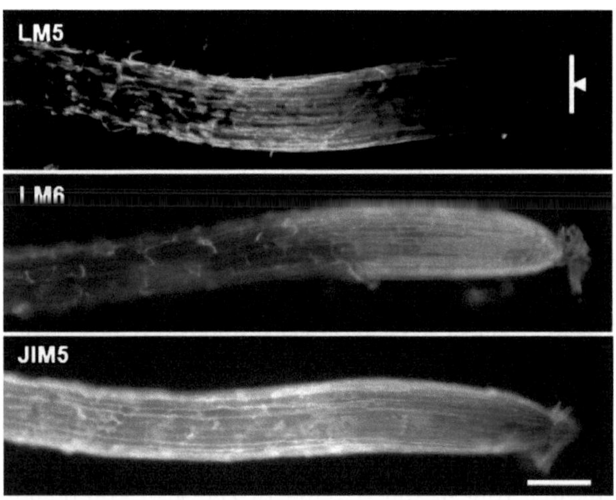

Fig. 2 Tracking pectic epitopes in relation to cell development. Equivalent indirect immunofluorescence labelling of the surface of intact *Arabidopsis thaliana* seedling roots with monoclonal antibodies LM5 (β-1,4-galactan), LM6 (α-1,5-arabinan) and JIM5 (HG). The epitopes show different patterns of occurrence at the surface of the root apex with the LM5 epitope marking the region of cells that are beginning rapid cell elongation. *Arrowhead* indicates the position of the root tip. *Bar* = 100 μm. For further details see McCartney et al. (2003)

modulatable RGI structures function in the physiological framework of factors that regulate cell elongation in this organ?

Attempts to explore RGI function by specific removal of RGI components (β-1,4-galactan, α-1,5-arabinan and/or the RG backbone) by expression of microbial hydrolytic enzymes or disruption of GTs have caused both limited and severe effects on development, but have made nothing clear in terms of specific roles in cell expansion (Sørensen et al. 2000; Oomen et al. 2002; Skjøt et al. 2002; Ulvskov et al. 2005; Harholt et al. 2006). Any specific adaptations of RGI or other pectin domains in response to these genetic impacts to maintain cell wall properties are not known. As indicated above, AGPs share structural glycan features with RGI polymers. It is of interest that AGPs are implicated in plant cell expansion processes; these factors are discussed briefly in Sect. 6.

6
Pectin-Binding Factors and Pectin-Plasma Membrane Interactions

An important feature of any cell wall, and particularly of an extending cell wall, are the links between the cell wall, the plasma membrane and the cell's interior. Pectin is a major component of the primary cell wall and has been implicated in providing functional links to the plasma membrane. This may be important in the regulation of cell expansion and growth processes. Two sets of plasma membrane-associated molecules that are implicated in these links are AGPs and wall-associated kinases (WAKs).

As discussed above, AGPs are a highly complex set of proteoglycans that are part of a large group of hydroxyproline-containing cell wall proteins (Showalter 2001). AGPs are strongly implicated in cell expansion processes and in tip growth (McCartney et al. 2003; Lee et al. 2005). A significant proportion of AGPs are attached to the plasma membrane by means of glycosylphosphatidylinositol anchors (Showalter 2001). Pickard and co-workers postulate a plasma membrane array of AGPs that interacts with the cell wall and call it the "plasmalemmal reticulum" (Pickard and Fujiki 2005). Recent work with salt-adapted suspension-cultured cells has led to the idea that AGPs released from the plasma membrane into the cell wall could act as cell wall plasticizers and could promote growth, possibly by modifying the pectin network (Lamport et al. 2006; Knox 2006). There is some evidence that AGPs can bind to pectins (Ridley et al. 2001), but this needs to be studied systematically with specific AGPs and specific pectin domains. It is possible that AGPs become attached to the pectic network during growth, accounting for AGP glycan association with isolated RGI polymers. The potential links between these polymers need to be explored in more detail.

The *Arabidopsis* genome encodes five wall-associated kinases (WAKs) that have putative kinase domains in the cytoplasm and have domains outside the plasma membrane that include EGF motifs, which interact with the cell wall (Kohorn 2001). WAKs associate tightly with cell wall components and can be released by pectinase treatment. The released WAKs have been found to co-migrate with the JIM5 and JIM7 HG epitopes in gels (Wagner and Korhorn 2001). The cell wall domain of WAK1 has been shown to interact in vitro with a glycine-rich protein *At*GRP-3 (Park et al. 2001). Antisense analyses of WAK2 and WAK4 have resulted in the reduction of all WAKs, which in turn resulted in a serious disruption of cell expansion processes (Wagner and Kohorn 2001; Lally et al. 2001). Recent study of in vitro binding of the WAK1 domain outside the plasma membrane to pectins has indicated that this domain requires un-esterified calcium cross-linked pectin for recognition (Decreux and Messiaen 2005). It remains to be seen if there is a covalent aspect to this binding. What is not known is the significance of WAK–pectin binding nor what is being signalled to the cell's interior (Decreux and Messiaen 2005). Are WAKs sensing the presence of particular calcium-sensitive HG domains within cell wall matrices?

A class of proteins that also have putative calcium-pectate binding domains is cell wall-located peroxidases. The mutation of two such genes in *Arabidopsis* have resulted in a significant reduction of cell elongation in the root (Passardi et al. 2006).

7
Future Prospects

Pectic polymers are ubiquitous and their properties are of industrial relevance, yet our understanding of their structure–function relationships is fragmentary. We have a broad understanding of how the regulation of HG structure and its cross-links could influence extensibility. However, the dissection of the specific functions of the domains of the pectic matrix in plant cell expansion processes continues to be a challenge. Genetic approaches to this dissection are fraught by the difficulties in assessment of phenotypes, the genetic redundancy that may make the analysis of single gene disruption difficult, and the fact that cell walls display considerable plasticity of composition and structure (e.g. His et al. 2001; Roudier et al. 2005).

We are just starting to understand how the precise structural forms of components and their arrangements within pectic networks are responsive to developmental needs and physiological changes. How the links and integrity of a pectic network and/or how pectic polymer links to cellulose and CLGs maintain an environment for cell wall stretching is not clear. Disruption of one component or linkage within a pectic network may be compensated for by the up-regulation of others – just as cells can cope with disruption of cellulose formation by an increase in the abundance of pectic polymers (Manfield et al. 2004).

The identification of all the GTs required for the construction of pectic polymers will establish a powerful set of tools that should enable new insights into the organization and the functions of pectic domains. In addition, what is also required is a comprehensive approach to understand the genetic and physiological factors that act and interact to construct pectic networks, and to modify them in response to local needs. We also need appropriate methods or systems to study how the roles of pectin in cell expansion processes are integrated with the other cellular roles of this complex and intriguing set of polymers.

Techniques Box

Tracking Polysaccharides in Plant Cell Walls: Antibodies and Cell Wall Microstructures

The complexity of pectic polymers requires appropriate techniques for the determination of their precise structure in relation to cell wall architecture and cell expansion. Currently, the best way to do this is by the use of monoclonal antibodies to defined structural features or epitopes of pectic polysaccharides in conjunction with immunocytochemistry procedures (Table 1).

Antibody generation. For the generation of probes to pectic epitopes the most effective strategy is to generate a neoglycoprotein with a defined oligosaccharide of interest attached to a protein. This neoglycoprotein is then used in immunization procedures or used for the screening of phage display antibody libraries. However, this approach requires the availability of appropriate amounts of isolated pectic oligosaccharides for coupling and for

Table 1 Monoclonal antibodies to pectic epitopes

Antibody	Antigen/Epitope	Refs.
JIM5	HG/partially methyl-esterified/no ester	Clausen et al. 2003
JIM7	HG/partially methyl-esterified	Clausen et al. 2003
LM7	HG/partially methyl-esterified	Clausen et al. 2003
PAM1	HG/unesterified blocks	Manfield et al. 2005
2F4	HG/unesterified Ca^{2+} cross-linked	Liners et al. 1992
LM5	RGI/β-1,4-galactan	Willats et al. 2001
LM6	RGI/α-1,5-arabinan	Willats et al. 2001
LM9	RGI/feruloylated β-1,4-galactan	Clausen et al. 2004
CCRCM7	RGI/arabinosylated β-1,6-galactan	Steffan et al. 1995
6D8	RGI/arabinan	Baluška et al. 2005
LM8	Xylogalacturonan	Willats et al. 2004

screening procedures. An alternative is to immunize an animal, or screen a library, with a complex polysaccharide, and then select useful antibodies by their capacity to bind to panels of isolated polymers and/or appropriate cell walls.

Use of antibodies. Initial immunocytochemical studies with antibodies on sections of plant material should always be at the light microscope level as the immediate examination with an electron microscope may lead to aspects of developmental regulation being overlooked. In addition to the immunolabelling of sectioned materials, whole mount immunolabelling can be an efficient and effective way to ascertain patterns of cell wall polymers in relation to developmental processes. Examples of the immunolabelling of intact roots of *Arabidopsis* seedlings are shown in Fig. 2.

Further references: Willats et al. 2001a, 2003; Willats and Knox 2003.

References

Abdel-Massih RM, Baydoun EA, Brett CT (2003) In vitro biosynthesis of 1,4-β-galactan attached to a pectin-xyloglucan complex in pea. Planta 216:502–511

Atkinson RG, Schroder R, Hallett IC, Cohen D, MacRae EA (2002) Overexpression of polygalacturonase in transgenic apple trees leads to a range of novel phenotypes involving changes in cell adhesion. Plant Physiol 129:122–133

Baluška F, Hlavacka A, Samaj J, Palme K, Robinson DG, Matoh T, McCurdy DW, Menzel D, Volkmann D (2002) F-actin-dependent endocytosis of cell wall pectins in meristematic root cells: insights from brefeldin A-induced compartments. Plant Physiol 130:422–431

Baluška F, Liners F, Hlavaèka A, Schlicht M, Van Cutsem P, McCurdy DW, Menzel D (2005) Cell wall pectins and xyloglucans are internalized into dividing root cells and accumulate within cell plates during cytokinesis. Protoplasma 225:141–155

Bosch M, Hepler PK (2005) Pectin methylesterases and pectin dynamics in pollen tubes. Plant Cell 17:3219–3226

Bouton S, Leboeuf E, Mouille G, Leydecker MT, Talbotec J, Granier F, Lahaye M, Hofte H, Truong HN (2002) QUASIMODO1 encodes a putative membrane-bound glycosyltransferase required for normal pectin synthesis and cell adhesion in Arabidopsis. Plant Cell 14:2577–2590

Branca C, DeLorenzo G, Cervone F (1988) Competitive inhibition of the auxin-induced elongation by α-D-oligogalacturonides in pea stem segments. Physiol Plant 72:499–504

Capodicasa C, Vairo D, Zabotina O, McCartney L, Caprari C, Mattei B, Manfredini C, Aracri B, Benen J, Knox JP, De Lorenzo G, Cervone F (2003) Targeted modification of homogalacturonan by transgenic expression of a fungal polygalacturonase alters plant growth. Plant Physiol 135:1294–1304

Chanliaud E, Gidley MJ (1999) In vitro synthesis and properties of pectin/*Acetobacter xylinus* cellulose composites. Plant J 20:25–35

Clausen MH, Willats WGT, Knox JP (2003) Synthetic methyl hexagalacturonate hapten inhibitors of anti-homogalacturonan monoclonal antibodies LM7, JIM5 and JIM7. Carbohydr Res 338:1797–1800

Clausen MH, Ralet M-C, Willats WGT, McCartney L, Marcus SE, Thibault J-F, Knox JP (2004) A monoclonal antibody to feruloylated-(1→4)-β-D-galactan. Planta 219:1036–1041

Cosgrove DJ (2005) Growth of the plant cell wall. Nature Rev Mol Cell Biol 6:850–861

Cumming CM, Rizkallah HD, McKendrick KA, Abdel-Massih RM, Baydoun EA, Brett CT (2005) Biosynthesis and cell-wall deposition of a pectin-xyloglucan complex in pea. Planta 222:546–555

Decreux A, Messiaen J (2005) Wall-associated kinase WAK1 interacts with cell wall pectins in a calcium-induced conformation. Plant Cell Physiol 46:268–278

Dhonukshe P, Baluška F, Schlicht M, Hlavacka A, Šamaj J, Friml J, Gadella TW Jr (2006) Endocytosis of cell surface material mediates cell plate formation during plant cytokinesis. Dev Cell 10:137–150

Dumville JC, Fry SC (2003) Solubilisation of tomato fruit pectins by ascorbate: a possible non-enzymic mechanism of fruit softening. Planta 217:951–961

Ezaki N, Kido N, Takahashi K, Katou K (2005) The role of wall Ca^{2+} in the regulation of wall extensibility during the acid-induced extension of soybean hypocotyl cell walls. Plant Cell Physiol 46:1831–1838

Fleischer A, O'Neill MA, Ehwald R (1999) The pore-size of non-graminaceous plant cell walls is rapidly decreased by borate ester cross-linking of the pecitc polysaccharide rhamnogalacturonan II. Plant Physiol 121:829–838

Ha MA, Viëtor RJ, Jardine GD, Apperley DC, Jarvis MC (2005) Conformation and mobility of the arabinan and galactan side chains of pectin. Phytochemistry 66:1817–1824

Harholt J, Jensen JK, Sørensen SO, Orfila C, Pauly M, Scheller HV (2006) ARABINAN DEFICIENT 1 is a putative arabinosyltransferase involved in biosynthesis of pectic arabinan in Arabidopsis. Plant Physiol 140:49–58

Hématy K, Höfte H (2007) Cellulose and cell elongation. In: Verbelen JP, Vissenberg K (eds) The expanding cell. Plant Cell Monographs, vol 5. Springer, Berlin Heidelberg New York (in press)

Hinz SWA, Verhoef R, Schols HA, Vincken J-P, Voragen AGJ (2005) Type I arabinogalactan contains β-D-Gal*p*-(1→3)-β-D-Gal*p* structural elements. Carbohydr Res 340:2135–2143

His I, Driouich A, Nicol F, Jauneau A, Hofte H (2001) Altered pectin composition in primary cell walls of korrigan, a dwarf mutant of *Arabidopsis* deficient in a membrane-bound endo-1,4-β-glucanase. Planta 212:348–358

Iwai H, Masaoka N, Ishii T, Satoh S (2002) A pectin glucuronyltransferase gene is essential for intercellular attachment in the plant meristem. Proc Natl Acad Sci USA 99:16319–16324

Jiang LX, Yang SL, Xie LF, Puah CS, Zhang XQ, Yang WC, Sundaresan V, Ye D (2005) VANGUARD1 encodes a pectin methylesterase that enhances pollen tube growth in the Arabidopsis style and transmitting tract. Plant Cell 17:584–596

Jones L, Milne J, Ashford D, McQueen-Mason SJ (2003) Cell wall arabinan is essential for guard cell function. Proc Natl Acad Sci USA 100:11783–11788

Knox JP (2002) Cell and developmental biology of pectins. In: Seymour GB, Knox JP (eds) Pectins and their manipulation. Blackwell/CRC, Oxford, pp 131–149

Knox JP (2006) Up against the wall: arabinogalactan-protein dynamics at cell surfaces. New Phytol 169:443–445

Kohorn BD (2001) WAKs; cell wall associated kinases. Curr Opin Cell Biol 13:529–533

Lally D, Ingmire P, Tong H-Y, He Z-H (2001) Antisense expression of a cell wall-associated protein kinase, WAK4, inhibits cell elongation and alters morphology. Plant Cell 13:1317–1331

Lamport DTA, Kieliszewski MJ, Showalter AM (2006) Salt-stress upregulates arabinogalactan-proteins: using salt-stress to analyse AGP function. New Phytol 169:479–492

Lao NT, Long D, Kiang S, Coupland G, Shoue DA, Carpita NC, Kavanagh TA (2003) Mutation of a family 8 glycosyltransferase gene alters cell wall carbohydrate composition and causes a humidity-sensitive semi-sterile dwarf phenotype in *Arabidopsis*. Plant Mol Biol 53:647–661

Lee KJD, Sakata Y, Mau S-L, Pettolino F, Bacic A, Quatrano RS, Knight CD, Knox JP (2005) Arabinogalactan-proteins are required for apical cell extension in the moss *Physcomitrella patens*. Plant Cell 17:3051–3065

Liberman M, Mutaftschiev S, Jauneau A, Vian B, Catesson AM, Goldberg R (1999) Mung bean hypocotyls homogalacturonan: localization, organization and origin. Ann Bot 84:225–233

Lindsay SE, Fry SC (2007) Redox and wall-restructuring. In: Verbelen JP, Vissenberg K (eds) The expanding cell. Plant Cell Monographs, vol 5. Springer, Berlin Heidelberg New York (in press)

Liners F, Thibault J-F, van Cutsem P (1992) Influence of the degree of polymerization of oligogalactturonates and of esterification pattern on pectin on their recognition by monoclonal antibodies. Plant Physiol 99:1099–1104

Manfield IW, Orfila C, McCartney L, Harholt J, Bernal AJ, Scheller HV, Gilmartin PM, Mikkelsen JD, Knox JP, Willats WGT (2004) Novel cell wall architecture of isoxaben-habituated *Arabidopsis* suspension-cultured cells: global transcript profiling and cellular analysis. Plant J 40:260–275

Manfield IW, Bernal AJ, Møller I, McCartney L, Riess NP, Knox JP, Willats WGT (2005) Re-engineering of the PAM1 phage display monoclonal antibody to produce a soluble, versatile anti-homogalacturonan scFv. Plant Sci 169:1090–1095

McCartney L, Steele-King CG, Jordan E, Knox JP (2003) Cell wall pectic (1→4)-β-D-galactan marks the acceleration of cell elongation in the *Arabidopsis* seedling root meristem. Plant J 33:447–454

Micheli F (2001) Pectin methylesterases: cell wall enzymes with important roles in plant physiology. Trends Plant Sci 6:414–419

Mohnen D (1999) Biosynthesis of pectins and galactomannans. In: Barton D, Nakanishi K, Meth-Cohn O (eds) Comprehensive natural products chemistry, vol 3. Elsevier, Amsterdam, pp 497–527

Morris VJ, Ring SG, MacDougall AJ, Wilson RH (2003) Biophysical characterization of plant cell walls. In: Rose JKC (ed) The plant cell wall. Blackwell/CRC, Oxford, pp 55–91

Obel N, Neumetzler L, Pauly M (2007) Hemicelluloses and cell expansion. In: Verbelen JP, Vissenberg K (eds) The expanding cell. Plant Cell Monographs, vol 5. Springer, Berlin Heidelberg New York (in press)

O'Neill MA, Eberhard S, Albersheim P, Darvill AG (2001) Requirement of borate cross-linking of cell wall rhamnogalacturonan II for *Arabidopsis* growth. Science 294:846–849

O'Neill MA, Ishii T, Albersheim P, Darvill AG (2004) Rhamnogalacturonan II: structure and function of a borate cross-linked cell wall pectic polysaccharide. Annu Rev Plant Biol 55:109–139

Oomen R, Doeswijk-Voragen C, Bush M, Vincken J-P, Borkhardt B, Van den Broek L, Corsar J, Ulvskov P, Voragen A, McCann MC, Visser R (2002) In muro fragmentation of the rhamnogalacturonan I backbone in potato (*Solanum tuberosum* L.) results in a reduction and altered location of the galactan and arabinan side-chains and abnormal periderm development. Plant J 30:403–413

Orfila C, Huisman MMH, Willats WGT, van Alebeek GJWM, Schols HA, Seymour GB, Knox JP (2002) Altered cell wall disassembly during ripening of *Cnr* tomato fruit: implications for cell adhesion and fruit softening. Planta 215:440–447

Park AR, Cho SK, Yun UJ, Jin MY, Lee SH, Schetto-Martins G, Park OK (2001) Interaction of the *Arabidopsis* receptor protein kinase Wak1 with a glycine-rich protein, AtGRP-3. J Biol Chem 276:26688–26693

Parre E, Geitmann A (2005) Pectin and the role of the physical properties of the cell wall in pollen tube growth of *Solanum chacoense*. Planta 220:582–592

Passardi F, Tognolli M, De Meyer M, Penel C, Dunand C (2006) Two cell wall associated peroxidases from *Arabidopsis* influence root elongation. Planta 223:965–974

Pickard BG, Fujiki M (2005) Ca^{2+} pulsation in BY-2 cells and evidence for the control of mechanosensory Ca^{2+}-selective channels by the plasmalemmal reticulum. Funct Plant Biol 32:863–879

Popper ZA, Fry SC (2003) Primary cell wall composition of bryophytes and charophytes. Ann Bot 91:1–12

Ridley BL, O'Neill MA, Mohnen D (2001) Pectins: structure, biosynthesis, and oligogalacturonide-related signaling. Phytochemistry 57:929–967

Rose JKC, Saladié M, Catalá C (2004) The plot thickens: new perspectives of primary cell wall modification. Curr Opin Plant Biol 7:296–301

Roudier F, Fernandez AG, Fujita M, Himmelspach R, Borner GH, Schindelman G, Song S, Baskin TI, Dupree P, Wasteneys GO, Benfey PN (2005) COBRA, an *Arabidopsis* extracellular glycosyl-phosphatidyl inositol-anchored protein, specifically controls highly anisotropic expansion through its involvement in cellulose microfibril orientation. Plant Cell 17:1749–1763

Scheible W-R, Pauly M (2004) Glycosyltransferases and cell wall biosynthesis: novel players and insights. Curr Opin Plant Biol 7:285–295

Shao M, Zheng H, Hu Y, Liu D, Jang JC, Ma H, Huang H (2004) The GAOLAOZHUANGREN1 gene encodes a putative glycosyltransferase that is critical for normal development and carbohydrate metabolism. Plant Cell Physiol 45:1453–1460

Showalter AM (2001) Arabinogalactan-proteins: structure, expression and function. Cell Mol Life Sci 58:1399–1417

Skjøt M, Pauly M, Bush M, Borkhardt B, McCann MC, Ulvskov P (2002) Direct interference with rhamnogalacturonan I biosynthesis in Golgi vesicles. Plant Physiol 129:95–102

Sørensen SO, Pauly M, Bush M, Skjøt M, McCann MC, Borkhardt B, Ulvskov P (2000) Pectin engineering: modification of potato pectin by in vivo expression of an endo-1,4-β-D-galactanase. Proc Natl Acad Sci USA 97:7639–7644

Steffan W, Kovàc P, Albersheim P, Darvill AG, Hahn MG (1995) Characterization of a monoclonal antibody that recognizes an arabinosylated (1$\rightarrow$6)-β-D-galactan epitope in plant complex carbohydrates. Carbohydr Res 275:295–307

Sterling JD, Quigley HF, Orellana A, Mohnen D (2001) The catalytic site of the pectin biosynthetic enzyme alpha-1,4-galacturonosyltransferase is located in the lumen of the Golgi. Plant Physiol 127:360–371

Sterling JD, Atmodjo MA, Inwood SE, Kolli VSK, Quigley HF, Hahn MG, Mohnen D (2006) Functional identification of an Arabidopsis pectin biosynthetic homogalacturonan galacturonosyltransferase. Proc Natl Acad Sci USA 103:5236–5241

Torki M, Mandaron P, Mache R, Falconet D (2000) Characterization of a ubiquitous expressed gene family encoding polygalacturonase in *Arabidopsis thaliana*. Gene 242:427–436

Ulvskov P, Wium H, Bruce D, Jørgensen B, Qvist KB, Skjøt M, Hepworth D, Borkhardt B, Sørensen SO (2005) Biophysical consequences of remodelling the neutral side chains of rhamnogalacturonan I in tubers of transgenic potatoes. Planta 220:609–620

Vincken JP, Schols HA, Oomen RJFJ, Beldman G, Visser RGF, Voragen AGJ (2003a) Pectin – the hairy thing. Evidence that homogalacturonan is a side chain of rhamnogalacturonan I. In: Voragen F, Schols H, Visser R (eds) Advances in pectin and pectinase research. Kluwer, Dordrecht, pp 47–59

Vincken JP, Schols HA, Oomen RJ, McCann MC, Ulvskov P, Voragen AG, Visser RG (2003b) If homogalacturonan were a side chain of rhamnogalacturonan I: implications for cell wall architecture. Plant Physiol 132:1781–1789

Vogel JP, Raab TK, Somerville CR, Somerville SC (2004) Mutations in *PMR5* result in powdery mildew resistance and altered cell wall composition. Plant J 40:968–978

Vreeburg RAM, Fry SC (2005) Reactive oxygen species in cell walls. In: Smirnoff N (ed) Antioxidants and reactive oxygen species in plants. Blackwell, Oxford, pp 215–249

Wagner TA, Kohorn BD (2001) Wall-associated kinases are expressed throughout plant development and are required fro cell expansion. Plant Cell 13:303–318

Willats WGT, McCartney L, Mackie W, Knox JP (2001a) Pectin: cell biology and prospects for functional analysis. Plant Mol Biol 47:9–27

Willats WGT, Orfila C, Limberg G, Buchholt HC, van Alebeek G-JWM, Voragen AGJ, Marcus SE, Christensen TMIE, Mikkelsen JD, Murray BS, Knox JP (2001b) Modulation of the degree and pattern of methyl-esterification of pectic homogalacturonan in plant cell walls: implications for pectin methyl esterase action, matrix properties and cell adhesion. J Biol Chem 276:19404–19413

Willats WGT, Knox JP (2003) Molecules in context: probes for cell wall analysis. In: Rose JKC (ed) The plant cell wall. Blackwell/CRC, Oxford, pp 92–110

Willats WGT, McCartney L, Knox JP (2003) Pectin cell biology: complexity in context. In: Voragen F, Schols H, Visser R (eds) Advances in pectin and pectinase research. Kluwer, Dordrecht, pp 147–157

Willats WGT, McCartney L, Steele-King CG, Marcus SE, Mort A, Huisman M, van Alebeek G-J, Schols HA, Voragen AGJ, Le Goff A, Bonnin E, Thibault J-F, Knox JP (2004) A xylogalacturonan epitope is specifically associated with plant cell detachment. Planta 218:673–681

Yu Q, Hlavacka A, Matoh T, Volkmann D, Menzel D, Goldbach HE, Baluška F (2002) Short-term boron deprivation inhibits endocytosis of cell wall pectins in meristematic cells of maize and wheat root apices. Plant Physiol 130:415–421

Zsivanovits G, MacDougall AJ, Smith AC, Ring SG (2004) Material properties of concentrated pectin networks. Carbohydr Res 339:1317–1322

Zykwinska AW, Ralet M-CJ, Garnier CD, Thibault J-FJ (2005) Evidence for in vitro binding of pectin side chains to cellulose. Plant Physiol 139:397–407

Plant Cell Monogr (5)
J.-P. Verbelen · K. Vissenberg: The Expanding Cell
DOI 10.1007/7089_2006_075/Published online: 28 October 2006
© Springer-Verlag Berlin Heidelberg 2006

Redox and Wall-Restructuring

Shona E. Lindsay · Stephen C. Fry (✉)

The Edinburgh Cell Wall Group, Institute of Molecular Plant Sciences,
School of Biological Sciences, The University of Edinburgh, Daniel Rutherford Building,
The King's Buildings, Mayfield Road, Edinburgh EH9 3JH, UK
S.Fry@ed.ac.uk

Abstract Diverse mechanisms contribute to primary cell wall re-structuring, causing wall loosening and tightening (increasing and decreasing extensibility, respectively). Wall loosening can occur by enzymic hydrolysis and possibly also elimination-degradation of polysaccharides; by enzymic transglycosylation of xyloglucan; by expansin-mediated rupture of hemicellulose–cellulose tethers; and by non-protein-mediated scission of polysaccharides through hydroxyl radical attack. Tightening can occur by enzymic de-esterification of pectin enabling Ca^{2+}-bridge formation; and by peroxidase-catalysed coupling of phenol–polysaccharide complexes and of tyrosine-containing glycoproteins. Several loosening and tightening mechanisms involve redox reactions; low-molecular-weight oxidants and anti-oxidants in the apoplast can therefore control wall extensibility. Apoplastic ascorbate is unusual in potentially being either an anti-oxidant or a pro-oxidant (the latter via Fenton reaction-mediated production of hydroxyl radicals). Many wall-localised reactions are known only from model experiments in vitro: an important future challenge is to explore the relative contributions of postulated reactions in the walls of *living* plant cells. To this end, a clear distinction is required between enzyme activity (assayed in vitro) and enzyme action (occurring in vivo).

1 Wall Restructuring and How It Can Be Detected In Vivo

1.1 Expanding and Non-Expanding Cell Walls Compared

A plant cell wall layer that is (or was) capable of irreversibly expanding in area (i.e., "growing", as defined here) is described as a primary wall. The drive for cell expansion is provided by turgor pressure (Fricke and Chaumont, in this volume). Cell growth is often regulated by changes in the primary wall's susceptibility to turgor-driven irreversible expansion.

A wall acquires the ability to undergo rapid, turgor-driven, irreversible expansion if wall-loosening agents [such as XTHs (Nishitani and Vissenberg, in this volume), expansins (McQueen-Mason et al., in this volume), H^+ ions (Rayle and Cleland 1992) or possibly certain reactive oxygen species (ROS; present chapter)] are able to act within the wall matrix. It is a familiar observation that some very small cells possessing only thin, primary walls nevertheless fail to undergo rapid cell expansion even if they have a high turgor pressure. Such cells include those of apical meristems, and their lack of rapid expansion may be due to the absence (or inaction) of a suitable combination of wall-loosening factors. Acquisition of the ability to succumb to turgor-driven expansion is described here as wall "loosening".

Conversely, an expanding primary wall can lose its growth ability in three principal ways: it can have additional substances such as lignin deposited within itself, or the structural polymers already present in the primary wall can become cross-linked, or the cell can acquire a secondary wall (between the primary wall and the plasma membrane) that is inherently resistant to expansion. Here, we discuss the first two mechanisms. Loss of the ability to succumb to turgor-driven expansion is described here as wall "tightening"; the terms stiffening and rigidification are not favoured because stiffness and rigidity conjure up an image of inflexibility (inability to bend) rather than inextensibility (inability to stretch). As an everyday illustration, string is almost inextensible without being stiff or rigid.

An important challenge, and a recurrent theme of the present work, is how we can observe and quantify the changes that take place in the primary cell wall in vivo during wall-loosening and -tightening. This issue centres on the distinction between enzyme "activity" and enzyme "action" (Fry 2004). An *active* enzyme (its activity assayed in vitro) does not necessarily exhibit *action* (in vivo). However, the current work goes beyond this question: we also consider non-enzymic reactions that may occur in the cell wall in vivo.

1.2 Deposition of New Polymers

In principle, a cell could enhance its ability to grow by depositing into the primary wall additional polysaccharides or glycoproteins that favour wall expansion, for example, those that are targets for wall-loosening factors. Such "lubricating" polymers could potentially include certain pectins (Verhertbruggen and Knox, in this volume), arabinogalactan-proteins, and low-M_r hemicelluloses (Obel et al., in this volume; Talbott and Ray 1992).

The deposition of new polysaccharides and glycoproteins can be looked for by analysis of samples taken at intervals. Such analysis can be by chemical methods, yielding exact structural information but a relatively vague picture of the spatial distribution of the polymers. Alternatively, the analysis can be microscopy-based, yielding precise information on the whereabouts of the polymers but less detailed structural information and less quantitative data. However, both these approaches can suffer from the limitation that they give information only about wall composition *at a given moment* in time, without indicating whether a changing wall composition is due to deposition of new molecules different from those already present, or alternatively to the modification of polymers that were already present. For example, pectic polysaccharides are widely thought to be secreted in a highly methylesterified form and then to lose some of their methyl ester groups, but this scenario is difficult to distinguish from one in which young cells deposit high-methylester pectins and older ones deposit low-methylester pectins.

The ambiguity can be resolved by kinetic studies involving in vivo isotopic labelling, for example, with radioactive (^{3}H-, ^{14}C- or ^{35}S-labelled; Fry 2000) precursors or with "heavy" (labelled with stable isotopes such as ^{2}H, ^{13}C or ^{15}N; Morrison et al. 1993; Alonso et al. 2005) precursors.

Conversely, cell expansion can potentially be stopped by the deposition within the primary wall of the phenolic polymer, lignin. Since lignin is readily distinguished from the polysaccharides and glycoproteins that constitute the expanding primary wall, and lignin cannot be manufactured by modification of polysaccharides and glycoproteins, it can be assumed that it is acceptable to monitor lignin deposition by simple measurements (or microscopical observations) of the lignin content of wall samples taken at timed intervals.

1.3 Oxidation of Existing Wall Polymers

Some plant cell wall matrix polysaccharides carry phenolic groups, especially ferulate, as side-chains. These phenolic groups can be oxidatively coupled to form dimers (dehydrodiferulates) and larger oligomers (Sect. 4.2), a process that may tighten the cell wall, suppressing cell expansion. Old cells tend to have a higher dehydrodiferulate : ferulate ratio in their wall polysaccharides than do young ones (Tan et al. 1991; Wakabayashi et al. 1997), compatible with a role of phenolic cross-linking in wall tightening. However, it is never clear from simple measurements of dehydrodiferulate and ferulate in differently aged cells whether this change in ratio is due to *in-muro* oxidative coupling of previously deposited feruloyl-polysaccharides: there remains the alternative possibility that older cells deposit dehydrodiferuloyl-polysaccharides rather than feruloyl-polysaccharides so that by the time old cells are analysed, the dehydrodiferulate : ferulate ratio is higher despite no oxidative coupling occurring in the cell wall itself.

More rigorous evidence for oxidative coupling of previously polysaccharide-bound phenolic residues can be sought by two main strategies: (a) detecting the changes that occur with time in the oxidation-state of previously radiolabelled phenolics, and (b) demonstrating the occurrence of [antioxidant-inhibitable] increases in M_r of wall polysaccharides.

For the first strategy, the feruloyl and dehydrodiferuloyl residues of polysaccharides can be radio-labelled if cells are fed [^{14}C]cinnamate. This precursor is not itself oxidised in the apoplast (owing to its lack of a phenolic hydroxyl group it is not a peroxidase substrate), but it is converted intracellularly to precursors of feruloylation reactions, probably mainly

[^{14}C]feruloyl-CoA (Sect. 4.1). Thus, in pulse–chase experiments, it is possible to follow the timing of the conversion of polysaccharide-bound ferulate to polysaccharide-bound dehydrodiferulate. The results, with cultured maize cells, showed that at least a proportion of the dimerisation occurs very rapidly after the feruloyl group has been attached to the polysaccharide chain (Fry et al. 2000). A further proportion of the dimerisation occurs several hours later, after the molecules have been secreted into the wall/apoplast. Thus, the oxidative coupling of feruloyl-polysaccharides takes place in part intraprotoplasmically (probably still within the Golgi system) and partly *in muro* after secretion. Much earlier work had made the false assumption that feruloyl dimerisation occurs exclusively in the apoplast after polysaccharide secretion.

For the second strategy, plant cells can be fed any suitable radio-labelled precursor of polysaccharides, and the subsequent changes in M_r of the polysaccharide monitored. In this way, the "careers" of xylan or xyloglucan molecules can be followed. In cultured maize cells, the [^{3}H]hemicelluloses undergo increases in M_r at three stages of their careers: (a) inside the Golgi vesicular system, (b) during or immediately after deposition in the cell wall, and (c) after several days' residence in the apoplast (Kerr and Fry 2003). This last phase was of particular interest in the present context because it could be blocked by the presence of anti-oxidants, especially competing phenolic substances, suggesting that it is due to oxidative cross-linking (Kerr and Fry 2004).

1.4 Hydrolysis of Existing Wall Polymers

Certain polysaccharides undergo partial hydrolysis after their initial deposition in the growing plant cell wall. The evidence is clearer in the case of certain *non-growing* cell walls such as those of ripening fruits, lysing abscission zones, and storage cells in cotyledons after germination—this will not be discussed further in this chapter.

In grasses and cereals, one particular wall polysaccharide, "mixed-linkage β-(1→3),(1→4)-D-glucan" (MLG), has frequently been reported to decrease in amount per leaf or per coleoptile after cell expansion stops (Roulin et al. 2002). It is unclear whether this polysaccharide degradation is causally involved in growth deceleration; indeed the precise role of MLG in wall architecture is not known. It appears very possible that the free glucose obtained by MLG hydrolysis is used as a carbon and energy source (Roulin et al. 2002). A definite decrease in a particular wall matrix polysaccharide *per plant organ* is clear evidence for the in vivo re-structuring of the cell wall. In contrast, a decrease of MLG *per gram of total cell wall* (Gibeaut et al. 2005) could represent either MLG breakdown or enhanced synthesis of non-MLG polysaccharides.

Radio-labelling experiments of "pulse–chase" design can give incontrovertible evidence for the degradation of a wall polysaccharide (even in the face of continued biosynthesis of identical polysaccharide molecules). An early example was provided by Franz (1972), who showed that in a given zone of a mung-bean hypocotyl, pulse-labelled β-[^{14}C]glucan (possibly callose) disappeared from the zone during a "chase" (= no further feeding of ^{14}C) period. Another example of possible post-synthetic wall re-structuring in vivo may be provided by the loss of arabinose side-chains from the xylan backbones of arabinoxylans—a major class of hemicelluloses in the primary cell wall, especially in grasses and cereals (Darvill et al. 1978; Carpita 1984).

There are reports of partial xyloglucan hydrolysis in the stems of legume seedlings. One way in which this can be detected is by demonstration of the disappearance of a proportion of the previously deposited, ^{14}C-labelled wall polysaccharides from a defined zone of the stem (Labavitch and Ray 1974). Since wall polysaccharides are not mobile within the plant, their disappearance must be interpreted as degradation to products so small that they are lost during dialysis. Another line of evidence in favour of wall polysaccharide degradation (possibly hydrolysis) during growth comes from the observation of xyloglucan "sloughing": i.e., some of the previously radiolabelled wall polysaccharides later become water-extractable (not firmly wall-bound), unlike their initial condition in which xyloglucans are tightly hydrogen-bonded to the wall's cellulosic microfibrils. Similar results were obtained in growing cell-suspension cultures of *Rosa* (Edelmann and Fry 1992). In addition, there are reports of xyloglucans rapidly decreasing in mean M_r during episodes of growth, this decrease being enhanced in tissues treated with growth-promoting agents such as auxins and H^+, and occurring so rapidly that an explanation based solely on the subsequent deposition of newly synthesised lower-M_r xyloglucans is implausible (Nishitani and Masuda 1982).

Pre-labelling with ^{3}H has enabled the post-synthetic changes in M_r to be monitored in a carefully defined cohort of xyloglucan molecules. It was found that in growing cultured cells of maize a sub-population of the [^{3}H]xyloglucan molecules underwent a gradual decrease in M_r (Kerr and Fry 2003). An even smaller proportion was degraded to oligosaccharides such as the nonasaccharide XXFG (McDougall and Fry 1991). However, the majority of the [^{3}H]xyloglucan remained of high M_r.

The disappearance, solubilisation and decrease in M_r of xyloglucans during growth have often been tacitly assumed to be due to hydrolysis, presumably catalysed by endo-β-(1$\rightarrow$4) glucanases (cellulases) or XTHs acting in XEH mode (Nishitani and Vissenberg, in this volume). However, there is no direct evidence for this assumption; alternative possibilities include the release of xyloglucan fragments during XTH action in XET mode, the solubilisation of xyloglucans by expansin-mediated "unzipping" of xyloglucan–cellulose hydrogen-bonds, or the ROS-mediated scission of xyloglucan chains (Sect. 5).

1.5
Isomerisation of Existing Wall Polymers

There is at least equally strong evidence that certain polysaccharides undergo *increases* in M_r with time after their deposition in the primary cell wall. Such changes have been observed in legume stem segments by use of bulk polysaccharide analysis at timed intervals after a growth-decelerating increase in pH (Nishitani and Masuda 1982) and by use of ^{14}C-labelling in vivo (Talbott and Ray 1992). In the ^{3}H-labelling study of maize cells (Kerr and Fry 2003), part of the increase in M_r of ^{3}H-polysaccharides, especially in the case of (arabino)xylans, could be by oxidative coupling of phenolic side-chains (Sect. 1.3). However, another likely explanation, especially in the case of xyloglucans, is that transglycosylation was responsible: i.e., an isomerisation reaction.

The in vivo action of XTHs (in XET mode) has been demonstrated, by dual isotopic labelling, in two significantly different physiological situations in cultured *Rosa* cells. The first involved the integration of newly secreted xyloglucans into the cell wall, a process shown to be accompanied, and probably caused, by polysaccharide-to-polysaccharide transglycosylation (Thompson et al. 1997). In this situation, a newly secreted xyloglucan molecule became molecularly "grafted" onto a pre-existing wall-bound xyloglucan molecule. The second situation involved inter-polysaccharide transglycosylation between two xyloglucan molecules, both of which had already been in the wall for some time (Thompson and Fry 2001). This process, termed "re-structuring" transglycosylation, is proposed to play a role in loosening the wall to facilitate cell expansion. Both types of transglycosylation, integrational and re-structuring, were demonstrated in vivo in experiments involving the dual labelling of xyloglucan with ^{13}C (which is dense but non-radioactive) and, at a different time, ^{3}H (which did not measurably increase the polysaccharides' density at the quantity supplied, but did make them radioactive). The subsequent transfer of radioactivity from polysaccharides of one density to those of the other density provided an experimental demonstration of polysaccharide-to-polysaccharide transglycosylation within the walls of living plant cells.

Additional evidence for the action of endogenous XTHs (in XET mode) in the cell wall has come from studies in which exogenous, fluorescently labelled oligosaccharides (which act as acceptor substrates) were integrated into the cell wall (Ito and Nishitani 1999; Vissenberg et al. 2000, 2005). Endogenous, high-M_r, wall-bound xyloglucans acted as the donor substrates in this case, demonstrating that the XTHs were in contact with, and able to act upon, the wall's xyloglucans.

In conclusion, there is now strong evidence that XTHs, acting in XET mode, contribute to both the construction and the re-structuring of the primary cell wall in expanding plant cells. These in vivo demonstrations of enzyme *action* lend credibility to hypotheses drawn from transcriptional analyses and enzyme *activity* studies that have suggested roles for XTHs in wall loosening and thus cell expansion.

2
Low-M_r Oxidising Agents in the Apoplast

Five small, inorganic, oxygen-based molecules occur in the cell wall and are relevant to this discussion. These molecules differ in oxidation state, as indicated formally in Scheme 1, each theoretical "step" consisting of the addition of one hydrogen atom:

$$H^+ + O_2^{\bullet -} \text{ (superoxide radical)}$$

$$\downarrow\uparrow (pK_a \approx 4.8)$$

$$\mathbf{O_2} \text{ (oxygen)} \rightarrow \mathbf{HO_2^{\bullet}} \text{ (hydroperoxyl radical)} \rightarrow \mathbf{H_2O_2} \text{ (hydrogen peroxide)} \rightarrow \mathbf{H_2O + {}^{\bullet}OH} \text{ (water + hydroxyl radical)} \rightarrow \mathbf{2\,H_2O} \text{ (water)}$$

Scheme 1

Although H_2O is the most highly reduced simple molecule containing oxygen, it does not normally act as a reducing reagent (the exception being in the light-dependent "water-splitting" step of photosystem II). Oxidative reactions occurring in the cell wall require a "small" substrate, usually O_2, H_2O_2, $O_2^{\bullet -}$ (or $HO_2^{\bullet}$), or $OH^{\bullet}$. In the following sections, we discuss the presence and source of these oxidising agents.

2.1
Molecular Oxygen (O_2)

All plants require O_2 for the completion of their life cycles, even if for short periods they can go into "anaerobic retreat" (Crawford 1989) and survive temporarily in the absence of O_2. Therefore, O_2 is normally present dissolved in the apoplastic fluid, which permeates the primary cell wall. O_2 can participate in non-enzymic reactions, for example with ascorbate, and also in oxidase-catalysed reactions. The requirement of plants for O_2 may well be partly due to the roles of this oxidant in cell wall biochemistry (Fujii 1978; Arwell et al. 1982).

2.2 Hydrogen Peroxide and Superoxide

H_2O_2 is often present in the cell wall and its synthesis is initiated by a range of factors including developmental stage and fungal infection (Vreeburg and Fry 2005). There is controversy as to the major source of H_2O_2 production in the cell wall. One possibility is the contribution of oxidases, which catalyse reactions of the type:

$$RH_2 + O_2 \rightarrow R + H_2O_2 \,. \tag{1}$$

Substrates (RH_2) to which this scheme may apply include oxalate (oxalate oxidase is known to occur in many gramineous cell walls; Lane et al. 1993), polyamines (Federico and Angelini 1986); and possibly dehydroascorbate (Sect. 3.1).

Another reaction by which H_2O_2 can be formed in vivo is by the dismutation of superoxide:

$$O_2^{\bullet -} + O_2^{\bullet -} + 2H^+ \rightarrow H_2O_2 + O_2 \tag{2}$$

(Wojtaszek 1997). This reaction can take place non-enzymically or it can be catalysed by superoxide dismutase (SOD), which occurs in the cell wall amongst other cellular locations (Schinkel et al. 1998).

The mechanism of $O_2^{\bullet -}$ and H_2O_2 production has mainly been studied in relation to the oxidative burst caused by pathogenic attack (e.g. Shetty et al. 2003). During an oxidative burst, ROS are thought to be synthesised via two main routes involving NADPH oxidase and peroxidase, respectively.

In the NADPH oxidase system, an elicitor molecule from a potential pathogen triggers synthesis of $O_2^{\bullet -}$ by a plasma membrane-bound NADPH oxidase, as detected for example in potato tubers (Doke and Miura 1995) and rose cell-suspension cultures (Auh and Murphy 1995).

In the peroxidase system, the signalling pathway leads to a temporary increase in the pH of the apoplast, as found in tobacco cell-suspension cultures (Keppler et al. 1989). The rise in pH may activate wall peroxidases which, in the presence of a suitable reductant, can generate H_2O_2. One such reductant is NADPH (reviewed by Vreeburg and Fry 2005), although it is unlikely that this compound is routinely present in the apoplast.

As well as in response to elicitation, plants also produce apoplastic H_2O_2 in significant amounts as a matter of course (Bindschedler et al. 2001). This H_2O_2 is available as a substrate for the oxidative coupling of polymer-esterified phenolics (Sect. 4.2) and tyrosine residues (Sect. 4.4). Indeed the mechanism for normal, non-stressed production of H_2O_2 may be in part similar to that of the oxidative burst since increases in peroxidase activity prior to growth cessation have often been recorded (Sect. 4.1).

2.3 Hydroxyl Radicals ($^{\bullet}OH$) and Their "Biosynthesis"

OH is the most reactive known molecule. Few studies have been made of its ability to cause oxidative cross-linking (Miller and Fry 2004); but it seems clear that polysaccharide scission greatly exceeds any cross-linking that OH may cause. OH can be formed by the one-electron reduction of H_2O_2 (Scheme 1), for example by the reduced form of a transition metal ion, Cu^+ being particularly effective. This is a "Fenton reaction", and it occurs non-enzymically at room temperature and physiological pH:

$$H_2O_2 + Cu^+ \rightarrow {}^{\bullet}OH^- + OH + Cu^{2+} . \qquad (3)$$

or

$$H_2O_2 + Fe^{2+} \rightarrow {}^{\bullet}OH^- + OH + Fe^{3+} . \qquad (4)$$

The two reactants (e.g. Cu^+ and H_2O_2) can be produced from well-established apoplastic components: O_2, Cu^{2+} and ascorbate. Apoplastic ascorbate (Sect. 3.1) will reduce Cu^{2+} to Cu^+, and O_2 to H_2O_2, both again being non-enzymic reactions occurring readily at room temperature and physiological pH. It is interesting that "Nature's favourite reducing agent" (ascorbate) can in this way drive the generation of the most reactive known oxidising agent, $^{\bullet}OH$.

An alternative proposed mechanism for $^{\bullet}OH$ production involves the Haber–Weiss reaction:

$$H_2O_2 + O_2^{\bullet-} \rightarrow {}^{\bullet}OH + OH^- + O_2 \qquad (5)$$

"catalysed" by peroxidase-bound iron (Chen and Schopfer 1999). This proposal may be supported by the ability of peroxidases to generate H_2O_2 and $O_2^{\bullet-}$ in vitro when given O_2 and a suitable reducing agent such as NADH. However, it is not expected that peroxidase should be involved in the production of a wall-loosening factor (OH) since there are numerous examples of negative correlations between peroxidase activity and cell expansion rate (Sect. 4.1).

3 Low-M_r Reducing Agents in the Apoplast

Other low-M_r solutes occurring in the apoplast and potentially able to influence redox-based wall-restructuring reactions are reducing agents.

3.1 Ascorbate and Its Metabolites

L-Ascorbate (vitamin C) is a very important protoplasmic anti-oxidant in plants and animals (Asard et al. 2004). However, a significant minority of the plant cell's ascorbate is located in the apoplast, where it could potentially influence both the loosening and tightening of the cell wall. Although chemically a reducing agent, it is intriguing that apoplastic ascorbate can act as both an anti-oxidant and as a "pro-oxidant".

Treatment of onion roots with an ascorbate/dehydroascorbate mixture caused membrane hyperpolarisation, promoted the uptake of sugars and nitrate (González-Reyes et al. 1994), and (as a cause or a consequence?) promoted root cell expansion (Hidalgo et al. 1991).

Ascorbate oxidase causes apoplastic ascorbate to react with O_2, forming dehydroascorbate + water. Ascorbate also undergoes non-enzymic oxidation, reacting with O_2 to form dehydroascorbate + H_2O_2 instead of water—a very significant functional difference. About 50% of the [^{14}C]ascorbate added to the medium of cultured rose cells underwent enzymic oxidation, the other half non-enzymic (Green and Fry 2005a).

Some of the dehydroascorbate formed can be taken up again via a dehydroascorbate transporter into the protoplast, where it can be re-reduced to ascorbate. Thus, at any moment there can be a steady-state concentration of ascorbate in the apoplast, although the ascorbate:dehydroascorbate ratio there is lower than in the cytoplasm (Takahama 1993).

The ascorbate (AH_2) in the apoplast can react not only with O_2 but also with the phenolic free radicals ($\Phi^{\bullet}$) that are produced transiently during oxidative phenolic coupling reactions (Sect. 4):

$$\Phi^{\bullet} + AH_2 \rightarrow \Phi H + AH^{\bullet} . \tag{6}$$

In this way, the ascorbate is oxidised (to semidehydroascorbate, $AH^{\bullet}$, also known as monodehydroascorbate), while the phenolic radical is reduced back to the original phenol (ΦH) (Takahama and Oniki 1994). In this way, apoplastic ascorbate can potentially block the oxidative cross-linking of polymer-bound phenols in the primary cell wall, and thereby prevent wall tightening. The semidehydroascorbate does not accumulate to high concentrations, but undergoes non-enzymic disproportionation to ascorbate (AH_2) and dehydroascorbate (A):

$$2AH^{\bullet} \leftrightarrow AH_2 + A . \tag{7}$$

The kinetics of this reaction are such that a mixture containing 0.5 mM ascorbate and 0.5 mM dehydroascorbate will at equilibrium also contain approximately 0.1 μM semidehydroascorbate (González-Reyes et al. 1994).

The dehydroascorbate formed from apoplastic ascorbate by any of the three mechanisms discussed (with $\Phi H^{\bullet}$; or enzymically or non-enzymically with O_2) has several possible fates. It can be reabsorbed into the protoplast and re-cycled. Alternatively, it can remain in the apoplast, where it can follow at least two different pathways:

i hydrolysis to diketogulonate followed by an ill-defined (probably oxidative) reaction to form two interconverting compounds designated C and E; or

ii oxidation to cyclic and non-cyclic oxalyl esters of threonate, eventually yielding free L-threonate plus oxalate (Green and Fry 2005a).

In those cells (mainly of grasses and cereals) that have wall-bound oxalate oxidase (Lane et al. 1993), dehydroascorbate metabolism can therefore lead to the generation of hydrogen peroxide:

$$(COOH)_2 + O_2 \rightarrow 2CO_2 + H_2O_2 \,. \tag{8}$$

There are also several other steps in the pathways of dehydroascorbate catabolism at which H_2O_2 could potentially be generated (Green and Fry 2005b), and experiments both in vitro and in living spruce cell-cultures have demonstrated the production of H_2O_2 after addition of ascorbate, dehydroascorbate or diketogulonate (Kärkönen and Fry 2006). In this way, ascorbate (and its metabolites) can be regarded as a "pro-oxidant".

Curiously, the H_2O_2 generated by ascorbate as a "pro-oxidant" could contribute to either the loosening or the tightening of the plant cell wall: loosening by serving as a precursor of hydroxyl radicals, which can cause polysaccharide scission (Sect. 5); and tightening by serving as a substrate for peroxidases, which can catalyse the cross-linking of wall phenolics (Sect. 4). The balance between loosening and tightening may be governed by the activity of peroxidase: in the presence of high peroxidase activity, phenolic cross-linking would prevail, while in its absence $OH^{\cdot}$ would formed.

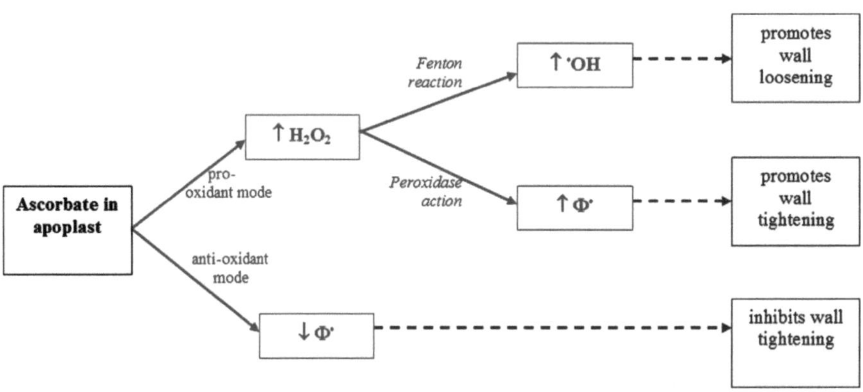

Fig. 1 The "split personality" of vitamin C. A summary of possible cell-expansion-related consequences of apoplastic ascorbate. *Key*: ↑, ↓ = increase and decrease in concentration, H_2O_2 = hydrogen peroxide, $^{\cdot}OH$ = hydroxyl radicals, $\Phi^{\cdot}$ = phenolic free-radicals

As an anti-oxidant (reducing $\Phi^{\cdot}$ back to ΦH; see above), ascorbate appears likely to act to prevent wall tightening. The various opposing possibilities for the action of apoplastic ascorbate on wall extensibility are summarised in Fig. 1.

3.2
Sulphydryl Reagents

Major intraprotoplasmic reductants possessing an – SH group include cysteine and glutathione. There are a few reports of these substances also in the apoplast. For example, cultured tobacco cells accumulated up to 700 μM glutathione in the extracellular medium when grown in the light, but less than 40 μM in the dark (Bergmann and Rennenberg 1978). However, these concentrations appear to be exceptional. The apoplast of spruce needles contained ∼ 6 μM glutathione and 0.6 μM cysteine (Polle et al. 1990). In oat leaves about 2% of the total glutathione was located in the apoplast (Vanacker et al. 1999). There is scope for further studies of sulphydryl compounds as significant anti-oxidants in the apoplast.

3.3
Transfer of Reducing Equivalents Across the Plasma Membrane

There is evidence that the protoplast can effect the reduction of apoplastic semidehydroascorbate back to ascorbate, by means of a trans-plasmalemmar reduction system involving a *b*-type cytochrome (Asard et al. 1992). In addition, the ability of plasmalemmar NAD(P)H oxidase to reduce O_2 to superoxide (Sect. 2.2) suggests the existence of a system for transferring reducing equivalents from cytosolic NAD(P)H to an apoplastic electron acceptor —in this case O_2. The trans-plasmalemmar transfer of reducing equivalents can also be monitored as the ability of living cells to reduce exogenous (apoplastic) ferricyanide to ferrocyanide (Morré et al. 1987). It is not suggested that ferricyanide is a natural electron acceptor, but it may be a valuable model substrate with which to study this system in vivo.

3.4
Unidentified Antioxidants

It is possible to detect biologically functional, apoplastic anti-oxidants in living maize cell cultures. These anti-oxidants inhibit the peroxidase-catalysed oxidative cross-linking of carbohydrate-bound, radiolabelled phenolics (Kerr and Fry 2004; Sect. 1.3). For experimental convenience, the feruloyl-trisaccharide, FAXX, was used as a test substrate (Encina and Fry 2005). FAXX is small enough to permeate the wall matrix and thus to access the enzymes that normally act on polysaccharide-bound phenolics;

furthermore, FAXX is too hydrophilic to cross the plasma membrane, so it is a specific reporter of *apoplastic* phenolic reactions. Artificially added anti-oxidants (ascorbate, competing phenolics) blocked the cross-linking of radiolabelled FAXX, as expected (Encina and Fry 2005).

More interestingly, the cultures gradually accumulated an endogenous, extracellular, low-M_r factor that blocked the cross-linking of radiolabelled FAXX (Encina and Fry 2005). This heat-labile inhibitor was hydrophilic, remaining in the aqueous phase (at pH 3.5) when shaken with ethyl acetate. These properties were compatible with its being ascorbate; however, ascorbate oxidase did not abolish its inhibitory effect, showing that it was not ascorbate. It will be of interest to identify this natural regulator of wall cross-linking.

4 Oxidative Phenolic Coupling

4.1 Potential Substrates for Oxidative Phenolic Coupling

Numerous *negative* correlations between measured wall peroxidase activity and the rate of cell expansion have been discovered (Fry 1979, 1980; Fry and Street 1980; Sánchez et al. 1996; Arnaldos et al. 2002; Bacon et al. 1997; de Souza and MacAdam 2001). These correlations tend to argue against a role for peroxidase in wall-loosening mediated by peroxidase-generated $^{\bullet}OH$ (Liszkay et al. 2003). Rather, the observations support a principal role for wall peroxidases as wall-tightening agents, probably via their ability to cross-link the phenolic groups that are attached to certain wall hemicelluloses and/or pectins.

The major phenolic present in many primary cell walls is ferulic acid (Fig. 2), esterified to wall polymers (Hartley 1973). It is readily detected by its blue fluorescence under 360-nm UV, intensifying and turning blue-green in the presence of dilute NH_4OH (Harris and Hartley 1976; Fry 1979). Feruloyl esters occur in the primary walls of all commelinoid monocotyledons, dicotyledons of the order Caryophyllales, and all gymnosperms (Carnachan and Harris 2000).

The feruloyl residues are bound to specific – OH groups of specific arabinose or galactose residues of hemicelluloses or pectins. In dicots the major sites of feruloylation are certain galactose and arabinose side-chains of pectic polysaccharides (Fry 1982a), digestion of which yielded specific oligosaccharides such as (6-*O*-feruloyl-β-D-galactopyranosyl)-(1$\rightarrow$4)-D-galactose. In gramineous monocots ferulate groups form an ester-linkage at O-5 of arabinosyl side-chains of arabinoxylan. In maize shoot cell walls, the main ferulate-containing fragment released on enzymic hydrolysis of

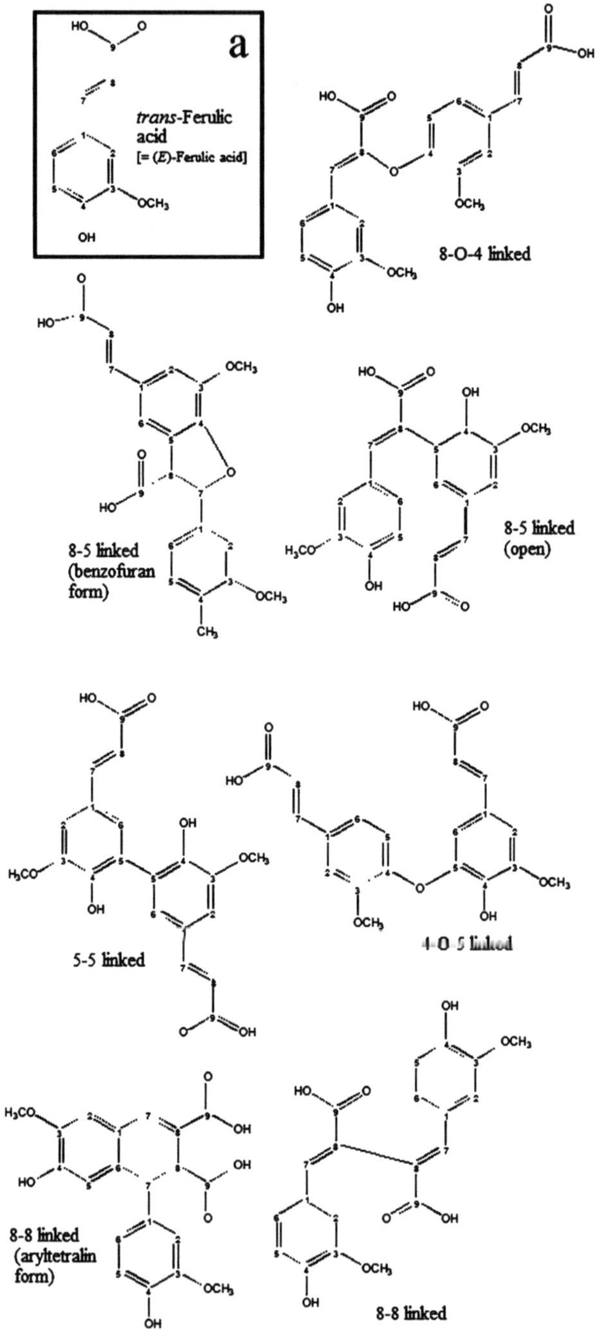

Fig. 2 Ferulic acid and its dehydro-dimers. Ferulic acid (*boxed*), showing numbering of carbon atoms. The remainder of the diagram shows seven isomeric dehydrodimers of ferulic acid (after Ralph et al. 1994)

the polysaccharides was (5-*O*-feruloyl-α-L-arabinofuranosyl)-(1→3)-β-D-xylopyranosyl-(1→4)-D-xylose (FAXX) (Kato and Nevins 1985). In addition, a feruloylated xyloglucan disaccharide, (4-*O*-feruloyl-α-D-xylopyranosyl)-(1→6)-D-glucose, has been isolated from bamboo shoot cell walls (Ishii et al. 1990). Ishii (1997) recorded 17 different feruloylated or *p*-coumaroylated wall fragments. Since then more have been identified, some of which contain two ferulate groups on a single oligosaccharide (Levigne et al. 2004).

4.2 Oxidative Coupling of Feruloyl-Polysaccharides

Ferulic acid residues are able to form covalent bonds with each other, for example, generating dehydrodiferulates. In vitro, and probably also in vivo, this occurs in the presence of peroxidase and H_2O_2 and the resulting dehydrodiferulates have been speculated to form cross-links between hemicelluloses in monocots and between pectic polysaccharides in dicots.

The dimerisation of feruloyl residues (ΦH) proceeds via tautomeric free-radical intermediates ($\Phi^{\bullet}$)

$$2\Phi H + H_2O_2 - [\text{peroxidase}] \rightarrow \Phi^{\bullet} + 2H_2O \qquad (9)$$

which then non-enzymically pair off in a variety of ways, according to the simplified scheme

$$2\Phi^{\bullet} - [\text{non-enzymic}] \rightarrow \Phi\text{–}\Phi \qquad (10)$$

to form a range of dimers (Φ–Φ).

It has long been suggested that dehydrodiferulate "cross-links" play a role in inhibiting cell wall extensibility, for example in spinach cultures (Fry 1979), maize leaves (de Souza and MacAdam 2001), wheat leaves (MacAdam and Grabber 2002) and rice coleoptiles (Tan et al. 1991; Kamisaka et al. 1990).

Geissmann and Neukom (1971) were the first to suggest the involvement of esterified feruloyl residues in the formation of covalent cross-links between cell wall polysaccharides. They suggested that the gelation occurring when water-soluble wheat arabinoxylans were incubated with H_2O_2 and peroxidase was caused by the oxidative coupling of feruloyl side-chains. They were also able to demonstrate that some of the feruloyl residues had coupled to form dehydrodiferulates.

Dehydrodiferulates, released by the alkaline hydrolysis of cell walls, were later found in the cell walls of many Poaceae (Hartley and Jones 1976; Markwalder and Neukom 1976) and Caryophyllales (Brett et al. 1999). For many years 5–5′-dehydrodiferulate was simply called "diferulate" as no other dimers had been identified. However, Ralph et al. (1994) chemically synthesised seven isomeric dimers of ferulic acid (Fig. 2) all of which were found in grass cell walls.

The detection of these seven dimers has meant that the techniques used to measure the "total" dehydrodiferulate content of the cell wall have had to be rethought. 5–5′-Dehydrodiferulate accounts for only about 5–17% of the total diferulate content and it may therefore be a poor indicator of the degree of ferulate dimerisation in vivo (Grabber et al. 1995; MacAdam and Grabber 2002; Ralph et al. 1994).

Recently, polysaccharide-esterified ferulate derivatives larger than dimers have been detected in the cell walls of maize cell-suspension cultures (Fry et al. 2000). The amounts of these larger ferulate-derivatives exceeded those of dehydrodiferulates. This led to the proposal that polysaccharide-esterified material larger than dimers may play an important role in the architecture of primary cell walls. In vitro peroxidase-catalysed coupling of free ferulic acid produced a tetramer of ferulic acid (Ward et al. 2001), and similar oligomerisation seems plausible in vivo. Three specific trimers (dehydrotriferulic acids) have now been isolated from gramineous cell walls and identified as (5–5′, 8′–O–4″)-dehydrotriferulate, (8–O–4′, 8′–8″)-dehydrotriferulate and (8–O–4′, 8′–O–4″)-dehydrotriferulate (Fig. 3) (Rouau et al. 2003; Bunzel et al. 2003; Funk et al. 2005). It seems likely that further trimers as well as still larger coupling products will in the future be found in vivo—probably small quantities of each individual molecular species but adding up to large total quantities because of the enormous diversity of conceivable chemical structures larger than dimers. The (5–5′, 8′–O–4″)-trimer, found in saponified maize bran, represented 12% of the total recorded dimers plus trimers (Rouau et al. 2003). In comparison, 5-5′-dehydrodiferulate accounts for only 5–15% of the total dimers in the cell wall.

Although dehydrodiferulate groups have long been known to exist in vivo (Markwalder and Neukom 1976; Hartley and Jones 1976; Ishii 1991), there is still no unequivocal proof as to whether this type of group form a cross-link between two independent polysaccharide chains (intermolecular linkage) or a bridge between different parts of the same polysaccharide (intramolecular linkage).

When primary cell walls of bamboo were hydrolysed with "Driselase" and the products fractionated chromatographically, small amounts of an extremely interesting di-(feruloyl-trisaccharide) were found (Ishii 1991) (Fig. 4). This fragment consists of two identical arabinoxylan-trisaccharides (Ara-Xyl-Xyl), with their Ara residues linked by a 5-5′-dehydrodiferulate group. Is this intermolecular, as arbitrarily shown in Fig. 5? There is no definitive answer to this question. The di-(feruloyl-trisaccharide) could represent a cross-link between two polysaccharide chains. Equally, however, it could arise from an intra-molecular bridge. As Ishii (1991) notes, "it is impossible to distinguish these diferuloyl arabinoxylan oligosaccharides derived from intermolecular cross-linkages from those associated with a single polysaccharide chain". There should therefore be a careful distinction in the language used to describe the role of diferulate groups within the cell wall. The term

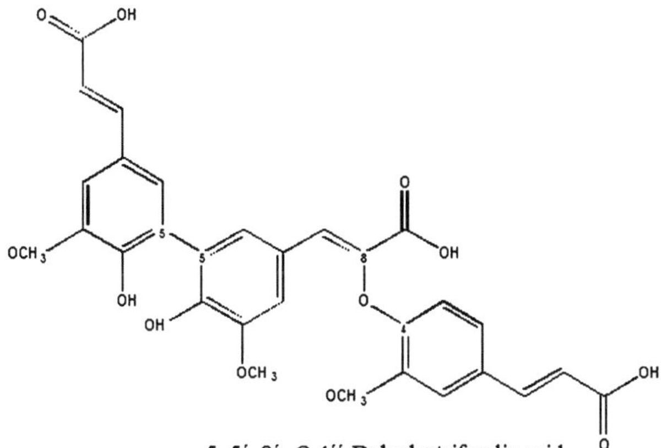

5–5′, 8′–*O*-4′′-Dehydrotriferulic acid

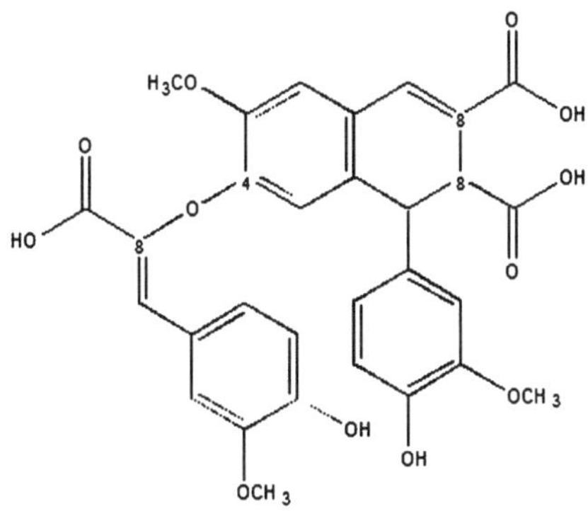

8–*O*-4′, 8′–8′′ -Dehydrotriferulic acid

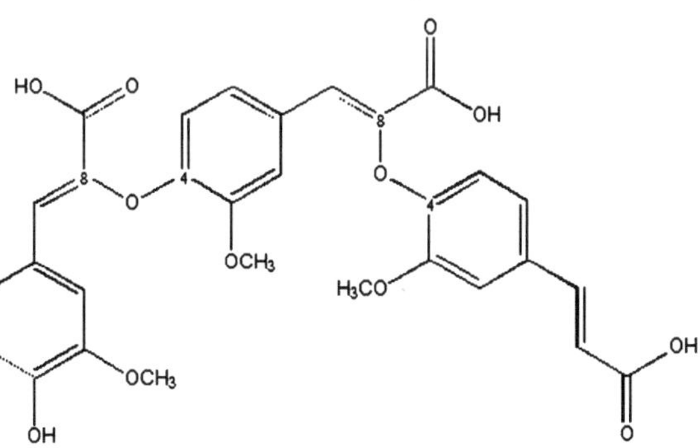

8–*O*-4′, 8′–*O*-4′′ -Dehydrotriferulic acid

Fig. 3 Three dehydrotrimers of ferulic acid. After Rouau et al. (2003), Bunzel et al. (2003) and Funk et al. (2005)

"*cross-link*" should be used only for an inter-polysaccharide linkage (Fig. 6); if the arrangement of the dehydrodiferulate groups is unknown, then a general term such as "linkage" or "bridge" is more appropriate.

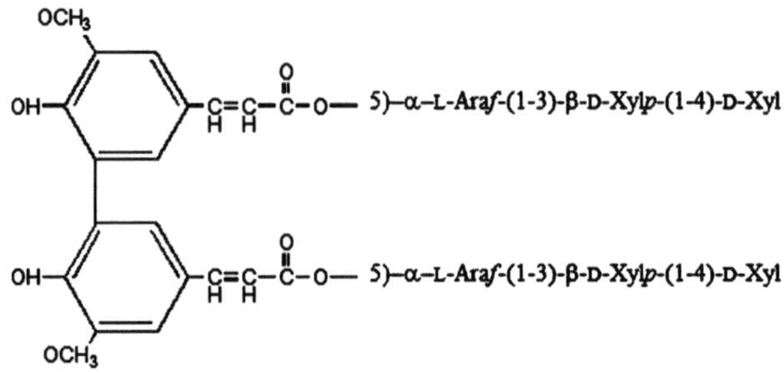

Fig. 4 A di-(feruloyl-trisaccharide) isolated by enzymic digestion of bamboo shoot cell walls (Ishii 1991)

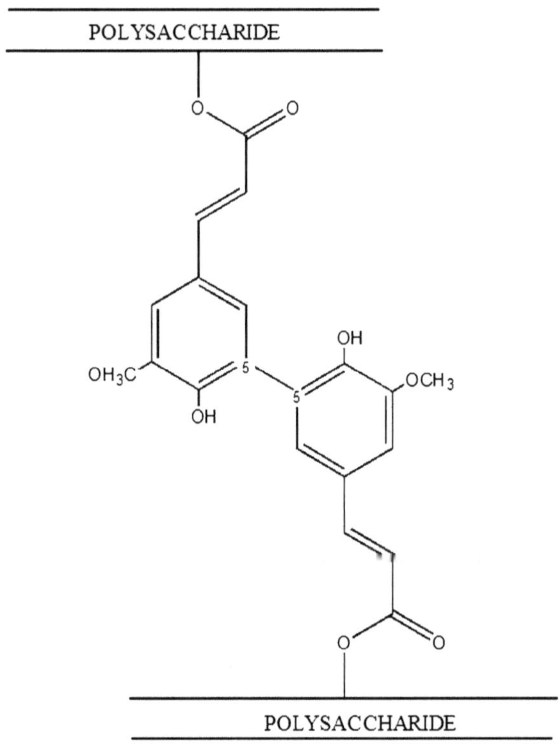

Fig. 5 5–5′-Dehydrodiferulate speculatively shown as a potential inter-polysaccharide cross-link

Many researchers (Bunzel et al. 2003; Ralph et al. 1994; Renger and Steinhart 2000; Saulnier et al. 1999) do not make this distinction and some quote the discovery of the di-(feruloyl-trisaccharide) by Ishii (1991) as proof of the existence of an inter-molecular diferuloyl cross-link.

Since 1991, two other instances of a diferulate group where both carboxy groups were esterified to sugars have been discovered (Saulnier et al. 1999).

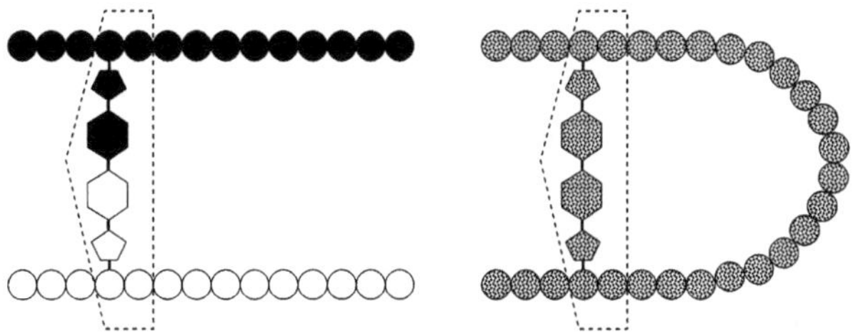

a b

Fig. 6 Diagram to show how a di-(feruloyl-trisaccharide) (Fig. 4) could theoretically be generated from either an inter-polysaccharide cross-link or an intra-polysaccharide bridge. In **a**, two feruloyl-arabinoxylan molecules (*one shown in black, one in white*) have been oxidatively cross-linked; in **b** a single feruloyl-arabinoxylan molecule (*stippled*) has been oxidatively coupled forming an intra-polysaccharide loop [the loop may be wider than shown in this schematic diagram]. In both cases, enzymic digestion yields the same di-(feruloyl-trisaccharide) (*dashed box*). *Key*: *circle* = xylose residue; *pentagon* = arabinose residue; *hexagon* = ferulic acid residue

The two molecules, Ara-Fer-Fer-Ara and Ara-Fer-Fer-Ara-Xyl, like Ishii's di-(feruloyl-trisaccharide), fail to provide information as to whether the dehydrodiferuloyl bridge formed is inter- or intra-polysaccharide.

In conclusion, there are many pieces of research which suggest the existence of a dehydrodiferulate intermolecular linkage, but there has been no unequivocal proof as yet. Dual isotopic labelling work (cf. Thompson and Fry 2001) is currently under way in our laboratory to try to resolve this question.

4.3 Lignins and Lignans

Lignification usually begins in the primary walls of cells that also have a secondary wall. Lignin is a complex three-dimensional mass of sub-units formed by the oxidative coupling of three main monomers—*p*-coumaryl, coniferyl and syringyl alcohols. Small amounts of lignin occur in the primary walls of certain growing cells. Quantities increase during the cessation of growth, for example, of the epidermis and cortex of maize mesocotyl in response to blue light (Schopfer et al. 2001). This suggests a role of non-vascular lignification in the negative control of cell expansion.

4.4 Oxidative Coupling of Tyrosine Residues in Extensins and Other Structural Glycoproteins

Structural proteins in the cell wall are grouped into five main classes (Showalter 1993). Among these, extensins are high-pI, hydroxyproline-rich glycoproteins that are also rich in tyrosine. Tyrosine is the only genetically encoded phenolic amino acid (Fig. 7). Like ferulic acid, tyrosine can oxidatively couple to form dimers—namely dityrosine, the principal dimer formed in vitro and in animals in vivo, and isodityrosine, the only dimer detected to date in plants in vivo (Fry 1982b). Isodityrosine has been shown to occur in part as a short intramolecular loop in extensin, $-\frac{1}{2}$Idt–Lys–$\frac{1}{2}$Idt– (where Idt = isodityrosine; Epstein and Lamport 1984). It probably also occurs as an intermolecular cross-link (Biggs and Fry 1990), though as with ferulic acid, unequivocal proof of the existence of an intermolecular cross-link has yet to be found.

Proteins are secreted into the cell wall, where some of them become inextractable over time (Showalter 1993). The mechanism of this loss of extractability has been studied in relation to the plant's response to pathogenic attack. An increase in extensin intermolecular cross-linking would increase the wall's impenetrability, preventing further intrusion by a potential pathogen. Such cross-linking may also limit cell expansion.

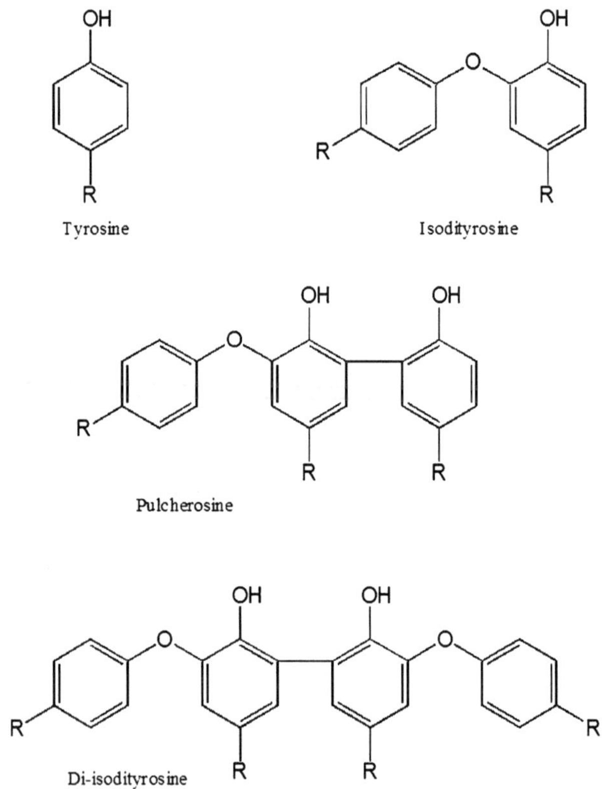

Fig. 7 Tyrosine and those of its oxidative coupling products that have been found in plant cell wall glycoproteins. The **-R** group represents $- CH_2 - CH(NH_2) - COOH$. After Brady et al. (1996, 1998)

In cell walls which have been compromised, either by wounding or by pathogenic attack, some wall proteins rapidly become inextractable. This cross-linking was suggested to be caused by H_2O_2-induced oxidative coupling as treatment of cell walls with H_2O_2 caused the "disappearance" of specific proteins (Otte and Barz 1996). This disappearance was inhibited by the addition of catalase or ascorbate, which would remove H_2O_2. This would suggest that the proteins may have been oxidatively coupled via their tyrosine residues to form a covalent network. However, this evidence does not provide direct evidence of an intermolecular tyrosine–tyrosine linkage.

Trimers and tetramers of tyrosine have also been identified (Fig. 7) (Brady et al. 1996, 1998). Pulcherosine, a trimer, was isolated from the cell walls of tomato cell-suspension cultures. It is unlikely that pulcherosine can form a tight intramolecular loop with all three tyrosine residues present as near-neighbours on a polypeptide chain. Neither is it likely that a looser, intramolecular loop could form, with the three tyrosine residues separated by longer blocks of amino acids, as the intervening blocks are usually rigid. The structural conformation required to form pulcherosine would be prevented by these intervening blocks (Brady et al. 1998).

Di-isodityrosine is a tetramer of tyrosine which is formed through either the coupling of two isodityrosine molecules or the addition of two individual tyrosine molecules to isodityrosine (Brady et al. 1996). The discovery of pulcherosine may point to the stepwise addition of single tyrosine residues. Owing to structural constraints, as discussed for pulcherosine, it is unlikely that di-isodityrosine forms only tight intramolecular loops (with all four Tyr residues occurring in a single polypeptide chain) in the way that Epstein and Lamport have shown that isodityrosine sometimes can. Thus, the discovery of di-isodityrosine also provides strong evidence for Tyr coupling as a means of forming inter-polymeric cross-links, a concept strongly supported by studies on synthetic polypeptides in vivo (Held et al. 2004).

5 Non-Enzymic Scission of Polysaccharides by Hydroxyl Radicals

5.1 Introduction

As the most reactive known molecule, $^{\bullet}OH$ can attack most organic substances—including cell wall polysaccharides, which thus undergo non-enzymic scission. Such scission is not by hydrolysis, transglycosylation or β-elimination, but by a variety of different processes depending on exactly where in the polysaccharide chain the $^{\bullet}OH$ attacks. There are many remaining questions about the mechanism of $^{\bullet}OH$-mediated scission, but it is clear that the process involves the abstraction by OH of a carbon-bonded H atom from the polysaccharide, leaving a modified sugar residue with a carbon-centred radical (Fig. 8, reactions a1, b1, c1, d1). In an aerobic environment (which the plant cell wall usually is), this radical reacts readily with O_2 to form a peroxyl radical (reactions a2, b2, c2, d2), which itself then rapidly eliminates $HO_2^{\cdot}$ (reactions a3, b3, c3, d3), leaving the sugar residue with an unusual $>C=O$ group (Schuchmann and von Sonntag 1978; von Sonntag 1987).

If the $^{\cdot}OH$ initially abstracts an H atom from one of the carbons involved in the glycosidic bond [e.g. positions 1 or 4 in a (1→4)-linked polysaccharide], the polysaccharide backbone will be cleaved and the newly formed terminus is likely to be a "reducing" terminal aldonic acid (or lactone) moiety or a non-reducing terminal glycosulose residue (Fig. 8, reactions c4, d4). If the H is initially abstracted from C-5 of a pyranosyl sugar (or C-4 of a furanosyl sugar), then the glycosidic bond may be converted to a moderately stable aldonic acid ester (reaction a4), which may subsequently undergo esterase-catalysed or non-enzymic hydrolysis, thus again ultimately leading to scission of the polysaccharide's backbone (Miller and Fry, 2001). If the H atom is abstracted from one of the remaining carbons [positions 2, 3 or 6 in a (1→4)-linked polysaccharide built up of pyranose rings], then the modified sugar formed will be a relatively stable glycosulose residue (Fig. 8—N.B., no step beyond reaction b3).

Thus, $^{\cdot}OH$ can cleave polysaccharide chains non-enzymically, while at the same time causing "collateral damage" by introducing a certain proportion of relatively stable oxo groups. Such groups potentially form the basis of a fingerprinting method by which $^{\cdot}OH$-attacked polysaccharides can be recognised in vivo (Fry et al. 2001; see Sect. 5.4).

$^{\cdot}OH$ is widely regarded as biologically hazardous because it can react with lipids, proteins and DNA, causing membrane permeabilisation, denatura-

Fig. 8 Proposed mechanisms by which a hydroxyl radical ($^{\cdot}OH$) attacks polysaccharide chains. The diagram shows some of the proposed mechanisms by reference to a polysaccharide with a backbone of pyranosidically linked pentose residues, such as β-(1→4)-D-xylan (*box in centre*; - - - = continuation of polysaccharide chain). It illustrates the likely consequences occurring when $^{\cdot}OH$ abstracts a hydrogen atom from carbons 1 (*dotted square*), 3 (*dotted circle*), 4 (*grey square*) or 5 (*grey circle*) in the presence of O_2. The dot (·) indicates a free-radical position, i.e. an atom with an unpaired electron. Abstraction of H from carbon 1 or carbon 4 quickly causes polysaccharide chain scission; note that abstraction of H from either C-1 or C-4 may produce the same range of products because of the tautomerisation reaction (↔) shown near the bottom of the diagram. Abstraction of H from position 5 does not immediately cause scission, but can convert a glycosidic bond into a [more labile] ester bond. Abstraction of H from positions 2 or 3 (or 6 in a hexose residue) introduces a relatively stable oxo group, creating a glycosulose residue without cleaving the polysaccharide chain. A by-product formed during each of these pathways is $HO_2^{\cdot}$, which immediately equilibrates with superoxide ($O_2^{\cdot -}$). Reactions ***a1*** to ***d4*** are discussed in the text ▶

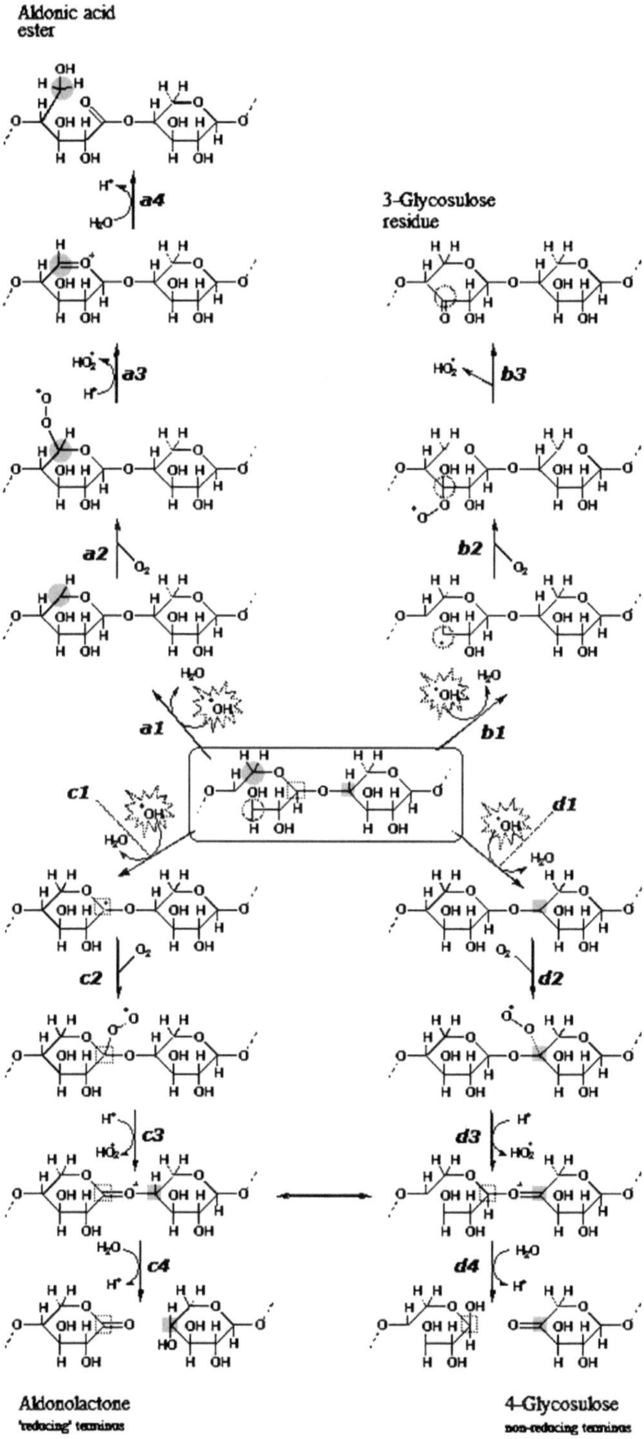
Aldonic acid ester
a4
3-Glycosulose residue
a3
b3
a2
b2
a1
b1
c1
d1
c2
d2
c3
d3
c4
d4
Aldonolactone
'reducing' terminus
4-Glycosulose
non-reducing terminus

tion and mutation, respectively. However, $^{\cdot}OH$ is exceedingly short-lived in the presence of organic matter: it seems unlikely to last more than ~ 1 ns nor to travel more than ~ 1 nm before reacting (for review, see Vreeburg and Fry 2005). Therefore, if produced in the interior of a plant cell wall (typically 100 nm thick), an $^{\cdot}OH$ molecule would react with a wall component (or an apoplastic solute) before reaching the plasma membrane or any genes.

We propose that such reactions of $^{\cdot}OH$ with wall matrix polysaccharides is one, hitherto underestimated, mechanism of wall loosening—with possible significance in cell expansion (Fry 1998; Schopfer 2001; Schopfer et al. 2002; Liszkay et al. 2003; Rodríguez et al. 2004) or fruit softening (Fry 1998; Fry et al. 2001; Dumville and Fry 2003). Several compounds (cyanide, azide, sulphide, hydroxylamine, 1,10-phenanthroline, 2,2′-bipyridyl, adenine, dimethylsulphoxide), which chelate Fenton-active metal ions and/or are $^{\cdot}OH$ scavengers, have been found to block the wall-loosening action of auxin in maize coleoptiles (Schopfer et al. 2002) and to decrease the steady-state concentrations of $^{\cdot}OH$ in similar tissues (Liszkay et al. 2003). This observation is compatible with the hypothesis that $^{\cdot}OH$ plays a role in auxin-induced wall loosening.

5.2 Model Experiments In Vitro

It is very straightforward to demonstrate that polysaccharides are attacked by $^{\cdot}OH$ radicals in vitro. Such radicals are readily formed in a freshly prepared aqueous solution containing ascorbate (e.g. 1 mM), O_2 (i.e., the test-tube is not capped), and traces of Cu^{2+} (e.g. 1 μM $CuSO_4$, or even smaller traces that are unavoidably present as a contaminant) at pH 3–6 [50 mM acetate (Na^+) is a suitable buffer because it has an unusually low reactivity with $OH^{\cdot}$]. For very rapid $^{\cdot}OH$ production, H_2O_2 can also be added (e.g. 1 mM), though this is not necessary. A suitable reaction mixture (ascorbate–$O_2 - Cu^{2+}$, pH 4.5) will cause noticeable scission of polysaccharides within seconds. In the case of polysaccharides that form viscous aqueous solutions (e.g. xyloglucan or pectin), scission can be detected by viscometry: the time taken for the solution to flow out of a 1-ml pipette decreases (Fry 1998). Further evidence of scission can be obtained by gel-permeation chromatography on a column of Bio-Gel or Sephadex: a decrease in M_r is observed (Fry et al. 2001).

This type of experiment is convenient for demonstrating the effects of radical scavengers, metal ion chelators etc., on non-enzymic scission of polysaccharides. Evidence that the observed scission is due to the action of $^{\cdot}OH$ can be acquired from correlations between different scavengers' effects on scission and their published rate-constants for reaction with $^{\cdot}OH$ (for a review see Vreeburg and Fry 2005).

5.3 Detection of ˙OH In Vivo

The considerations discussed earlier suggest that at least two mechanisms could contribute to the production of ˙OH in the cell wall—potentially also in vivo. However, it is very difficult to demonstrate the presence of ˙OH in the apoplast in vivo. The low steady-state concentrations likely to occur in vivo would probably be undetectable by the most characteristic feature of the ˙OH radical, its electron paramagnetic resonance (EPR) spectrum. An alternative strategy is by spin-trapping, in which ethanol plus α-(4-pyridyl 1-oxide)-*N*-*tert*-butylnitrone (4-POBN) are added to the medium in which plant cells are bathed. The ˙OH reacts with the ethanol to form a radical ($^{\cdot}CH_2 - CH_2OH$) which then reacts with the 4-POBN, forming a stable radical detectable by EPR. This approach has given evidence strongly suggesting the presence of ethanol-reactive radicals, probably ˙OH, in living plant cells—most likely in the apoplast. Examples are elicited rice cell cultures (Kuchitsu et al. 1995) and potentially growing maize coleoptiles (Schopfer et al. 2002), although the effect of the applied 3.9% (w/v) ethanol on the continuation of growth requires checking.

Another strategy for detection of ˙OH in vivo involves the application of probes which react with ˙OH to produce characteristic products. One such probe is benzoate (Schopfer et al. 2001), which reacts with ˙OH to form fluorescent products, although benzoate's ability to penetrate the plasma membrane and thus to be acted on by intraprotoplasmic oxygenases needs to be considered. If addition of a membrane-impermeant scavenger of ˙OH, such as mannitol, blocks the production of the fluorescent products, this gives stronger evidence for apoplastic presence of ˙OH (Schopfer et al. 2001). A membrane-impermeant though less sensitive probe for ˙OH is deoxyribose, which by oxidation to thiobarbiturate-reactive products also provided evidence for apoplastic ˙OH (Schopfer et al. 2001).

Other membrane-impermeant probes tested include ^{3}H-labelled benzoyl groups that had been artificially amide-bonded to oligopeptides (Miller and Fry 2004) or to polyallylamine (a cationic polymer) (Fry et al. 2002). These radioactive probes react with ˙OH to release tritiated water ($^{3}H_2O$), which can be assayed with great sensitivity by scintillation-counting. Experiments with these probes demonstrated apoplastic ˙OH production in growing segments of maize coleoptiles and mesocotyls or the stems, petioles or styles of various dicots, but ˙OH production was not enhanced when growth was promoted by auxin (Fry et al. 2002).

5.4 Evidence for ˙OH-Attacked Polysaccharides In Vivo

In view of the difficulty of detecting apoplastic ˙OH radicals, an alternative strategy for testing their occurrence and action in the cell wall is to look for the "collateral damage" which they do to polysaccharides (Sect. 5.1). This damage, which is highly diagnostic of $^{\bullet}OH$ attack and accompanies scission, manifests itself by the introduction of oxo groups, especially the formation of glycosulose residues (Fig. 8; e.g. reaction b3).

Such oxo groups can be detected by radio-labelling with tritiated borohydride (NaB^3H_4), which reduces them to tritiated alcohol groups. If the oxo group is a ketone rather than an aldehyde, the NaB^3H_4 generates two epimeric ^3H-labelled sugar residues. For example, in the case of the xylose-derived 3-glycosulose residue shown in Fig. 8, NaB^3H_4 is expected to yield a mixture of [^3H]xylose and [^3H]ribose residues. In the case of the corresponding 2-glycosulose, the products would be [^3H]xylose and [^3H]lyxose residues. Subsequent hydrolysis of the radiolabelled polysaccharide would then yield free [^3H]xylose, [^3H]ribose and [^3H]lyxose, the last two of which are not normal cell wall components and are therefore highly diagnostic of $^{\bullet}OH$ attack. These techniques have been applied in model in vitro studies, and also in in vivo experiments with pear fruits, in which preliminary evidence for $^{\bullet}OH$ attack was obtained (Fry et al. 2001). It should be noted that the products are sugars, not sugar alcohols (xylitol, ribitol etc.); the latter would be formed by reaction of NaB^3H_4 with ordinary reducing termini of polysaccharides.

Alternative approaches to the detection of $^{\bullet}OH$-induced damage involve fluorescent labelling at the oxo groups by application of reactive fluorescent probes. Development of this approach is in progress in this laboratory (R.A.M. Vreeburg and S.C. Fry, in preparation).

It is hoped that by application of techniques of this type, it will be possible to obtain direct evidence whether the polysaccharides of the plant cell wall are, or are not, being attacked by $^{\bullet}OH$ radicals in vivo, and whether the intensity of any such attack is related to the occurrence of wall-loosening during cell expansion.

6 Concluding Remarks

It seems clear that numerous diverse mechanisms are available to the plant for re-structuring its primary cell walls. For wall loosening, these mechanisms include the enzyme-catalysed hydrolysis and potentially also lyase-mediated cleavage of polysaccharide chains; enzyme-mediated re-structuring via transglycosylation; expansin-mediated un-tethering of adjacent microfibrils; and non-protein-mediated scission of polysaccharides by hydroxyl radicals. For wall tightening, they include the enzymic de-esterification of pectin such that Ca^{2+}-bridges can form; the peroxidase-mediated cross-linking of phenol–polysaccharide conjugates; and the peroxidase-mediated cross-linking of tyrosine-containing glycoproteins. Several of these mechanisms involve redox reactions, and low-M_r oxidants and anti-oxidants in the apoplast are thus in a position to regulate loosening and tightening reactions. Apoplastic ascorbate is a particularly intriguing topic for further investigation because of its "split personality"—its ability to act as both an anti-oxidant and a pro-oxidant. Many of the wall loosening and tightening mechanisms are mainly known from model experiments in vitro, and a major emphasis of future work will be to explore the relative contributions of each proposed mechanism in the walls of living plant cells. To this end, a clear distinction is required between enzyme activity and enzyme action; and further methods need to be developed by which to monitor the *action* of enzymes, as well as the action of non-enzymic processes, in vivo.

Acknowledgements We are very grateful to Drs R.A.M. Vreeburg, A. Encina and A. Kärkönen for valuable discussions and preliminary data. We thank the BBSRC for financial support of our work.

References

Alonso AP, Vigeolas H, Raymond P, Rolin D, Dieuaide-Noubhani M (2005) A new substrate cycle in plants. evidence for a high glucose-phosphate-to-glucose turnover from in vivo steady-state and pulse-labeling experiments with [^{13}C]glucose and [^{14}C]glucose. Plant Physiol 138:2220–2232

Arnaldos TL, Ferrer MA, García AAC, Muñoz R (2002) Changes in peroxidase activity and isoperoxidase pattern during strawberry (*Fragaria* × *ananassa*) callus development. J Plant Physiol 159:429–435

Arwell BJ, Waters I, Greenway H (1982) The effects of oxygen and turbulence on elongation of coleoptiles of submergence-tolerant and -intolerant rice cultivars. J Exp Bot 33:1030–1044

Asard H, Horemans N, Caubergs RJ (1992) Transmembrane electron-transport in ascorbate-loaded plasma-membrane vesicles from higher plants involves a *b*-type cytochrome. FEBS Letts 306:143–146

Asard H, May JM, Smirnoff N (eds) (2004) Vitamin C: its functions and biochemistry in animals and plants. Taylor & Francis, Abingdon, 304 pp

Auh CK, Murphy TM (1995) Plasma membrane redox enzyme is involved in the synthesis of O_2^- and H_2O_2 by *Phytophthora* elicitor-stimulated rose cells. Plant Physiol 107:1241–1247

Bacon MA, Thompson DS, Davies WJ (1997) Can cell wall peroxidase activity explain the leaf growth response of *Lolium temulentum* L. during drought? J Exp Bot 48:2075–2085

Bergman L, Rennenberg H (1978) Efflux und Produktion von Glutathion in Suspensionskulturen von *Nicotiana tabacum*. Z Pflanzenphysiol 88:175–185

Biggs KJ, Fry SC (1990) Solubilization of covalently-bound extensin from *Capsicum* cell walls. Plant Physiol 92:197–204

Bindschedler LV, Minibayeva F, Gardner SL, Gerrish C, Davies DR, Bolwell GP (2001) Early signalling events in the apoplastic oxidative burst in suspension-cultured French bean cells involve cAMP and Ca^{2+}. New Phytol 151:185–194

Brady JD, Sadler IH, Fry SC (1996) Di-isodityrosine, a novel tetrameric derivative of tyrosine in plant cell wall proteins: a new potential cross-link. Biochem J 315:323–327

Brady JD, Sadler IH, Fry SC (1998) Pulcherosine, an oxidatevely coupled trimer of tyrosine in plant cell walls: its role in cross-link formation. Phytochemistry 47:349–353

Brett CT, Wende G, Smith AC, Waldron KW (1999) Biosynthesis of cell wall ferulate and diferulates. J Sci Food Agric 79:421–424

Bunzel M, Ralph J, Funk C, Steinhart H (2003) Isolation and identification of a ferulic acid dehydrotrimer from saponified maize bran insoluble fiber. Eur Food Res Technol 217:128–133

Carnachan SM, Harris PJ (2000) Ferulic acid is bound to the primary cell walls of all gymnosperm families. Biochem Syst Ecol 28:865–879

Carpita NC (1984) Cell wall development in maize coleoptiles. Plant Physiol 76:205–212

Chen SX, Schopfer P (1999) Hydroxyl radical production in physiological reactions—a novel function of peroxidase. Eur J Biochem 260:726–735

Crawford RM (1989) Studies in Plant Survival: Ecological Case Histories of Plant Adaptation to Adversity. Blackwell Scientific, Oxford

Darvill AG, Smith CJ, Hall MA (1978) Cell wall structure and elongation growth in *Zea mays* coleoptile tissue. New Phytol 80:503–516

de Souza IRP, MacAdam JW (2001) Gibberellic acid and dwarfism effects on the growth dynamics of B73 maize (*Zea mays* L.) leaf blades: a transient increase in apoplastic peroxidase activity precedes cessation of cell elongation. J Exp Bot 52:1673–1682

Doke N, Miura Y (1995) *In-vitro* activation of NADPH-dependent O_2^- generating system in a plasma membrane-rich fraction of potato tuber tissues by treatment with an elicitor from *Phytophthora infestans* or with digitonin. Physiol Molecular Plant Pathol 46:17–28

Dumville JC, Fry SC (2003) Solubilisation of tomato fruit pectins by ascorbate: a possible non-enzymic mechanism of fruit softening. Planta 217:951–961

Edelmann HG, Fry SC (1992) Kinetics of integration of xyloglucan into the walls of suspension-cultured rose cells. J Exp Bot 43:463–470

Encina A, Fry SC (2005) Oxidative coupling of a feruloyl-arabinoxylan trisaccharide (FAXX) in the walls of living maize cells requires endogenous hydrogen peroxide and is controlled by a low-M_r apoplastic inhibitor. Planta 223:77–89

Epstein L, Lamport DTA (1984) An intramolecular linkage involving isodityrosine in extensin. Phytochemistry 23:1241–1246

Federico R, Angolini R (1986) Occurrence of diamine oxidase in the apoplast of pea epicotyls. Planta 167:300–302

Franz G (1972) Polysaccharidmetabolismus in den Zellwänden wachsender Keimlinge von *Phaseolus aureus*. Planta 102:334–347

Fry SC (1979) Phenolic components of the primary cell wall and their possible rôle in the hormonal regulation of growth. Planta 146:343–351

Fry SC (1980) Gibberellin-controlled pectinic acid and protein secretion in growing cells. Phytochemistry 19:735–740

Fry SC (1982a) Phenolic components of the primary cell wall: feruloylated disaccharides of D-galactose and L-arabinose from spinach polysaccharide. Biochem J 203:493–504

Fry SC (1982b) Isodityrosine, a new cross-linking amino acid from plant cell wall glycoprotein. Biochem J 204:449–455

Fry SC (1998) Oxidative scission of plant cell wall polysaccharides by ascorbate-induced hydroxyl radicals. Biochem J 332:507–515

Fry SC (2000) The Growing Plant Cell Wall: Chemical and Metabolic Analysis. Reprint Edition, The Blackburn Press, Caldwell, New Jersey

Fry SC (2004) Tansley Review: Primary cell wall metabolism: tracking the careers of wall polymers in living plant cells. New Phytol 161:641–675

Fry SC, Street HE (1980) Gibberellin-sensitive suspension cultures. Plant Physiol 65:472–477

Fry SC, Willis SC, Paterson EJ (2000) Intraprotoplasmic and wall-localised formation of arabinoxylan-bound diferulates and larger ferulate coupling-products and maize cell-suspension cultures. Planta 211:679–692

Fry SC, Dumville JC, Miller JG (2001) Fingerprinting of polysaccharides attacked by hydroxyl radicals in vitro and in the cell walls of ripening pear fruit. Biochem J 357:729–737

Fry SC, Miller JG, Dumville JC (2002) A proposed role for copper ions in cell wall loosening. Plant Soil 247:57–67

Fujii T (1978) Effects of IAA on oxygen-sensitive growth and on hydroxyproline protein level in cell wall. Plant Cell Physiol 19:927–933

Funk C, Ralph J, Steinhart H, Bunzel M (2005) Isolation and structural characterisation of 8–O–4/8–O–4- and 8–8/8–O–4-coupled dehydrotriferulic acids from maize bran. Phytochemistry 66:363–371

Geissmann T, Neukom H (1971) Vernetzung von Phenolcarbonsaureestern von Polysacchariden durch oxydative phenolische Kupplung. Helv Chim Acta 54:1108–1112

Gibeaut DM, Pauly M, Bacic A, Fincher GB (2005) Changes in cell wall polysaccharides in developing barley (*Hordeum vulgare*) coleoptiles. Planta 221:729–738

González-Reyes JA, Hidalgo A, Caler JA, Palos R, Navas P (1994) Nutrient uptake changes in ascorbate free radical-stimulated onion roots. Plant Physiol 104:271–276

Grabber JH, Hatfield RD, Ralph J, Zon J, Amrhein N (1995) Ferulate cross-linking in cell walls isolated from maize cell suspensions. Phytochemistry 40:1077–1082

Green MA, Fry SC (2005a) Vitamin C degradation in plant cells via enzymatic hydrolysis of 4-*O*-oxalyl-L-threonate. Nature 433:83–88

Green MA, Fry SC (2005b) Apoplastic degradation of ascorbate: novel enzymes and metabolites permeating the plant cell wall. Plant Biosystems 139:2–7

Harris PJ, Hartley RD (1976) Detection of bound ferulic acid in cell walls of Gramineae by ultraviolet fluorescence microscopy. Nature 259:508–510

Hartley RD (1973) Carbohydrate esters of ferulic acid as components of cell walls of *Lolium multiflorum*. Phytochemistry 12:661–665

Hartley RD, Jones EC (1976) Diferulic acid as a component of cell walls of *Lolium multiflorum*. Phytochemistry 15:1157–1160

Held MA, Tan L, Kamyab A, Hare M, Shpak E, Kieliszewski MJ (2004) Di-isodityrosine is the intermolecular cross-link of isodityrosine-rich extensin analogs cross-linked in vitro. J Biol Chem 279:55474–55482

Hidalgo A, García-Herdugo G, González-Reyes JA, Morré DJ, Navas P (1991) Ascorbate free-radical stimulates onion root growth by increasing cell elongation. Bot Gaz 152:282–288

Ishii T (1991) Isolation and characterization of a diferuloyl arabinoxylan hexasaccharide from bamboo shoot cell walls. Carbohydr Res 219:15–22

Ishii T (1997) Structure and function of feruloylated polysaccharides. Plant Sci 127:111–127

Ishii T, Hiroi T, Thomas JR (1990) Feruloylated xyloglucan and *p*-coumaroyl arabinoxylan oligosaccharides from bamboo shoot cell walls. Phytochemistry 29:1999–2003

Ito H, Nishitani K (1999) Visualization of EXGT-mediated molecular grafting activity by means of a fluorescent-labeled xyloglucan oligomer. Plant Cell Physiol 40:1172–1176

Kamisaka S, Takeda S, Takahashi K, Shibata K (1990) Diferulic acid in the cell wall of *Avena* coleoptiles—Their relationship to mechanical properties of the cell wall. Physiol Plant 78:1–7

Kärkönen A, Fry SC (2006) Effect of ascorbate and its oxidation-products on H_2O_2 production in cell-suspension cultures of *Picea abies* and in the absence of cells. J Exp Bot 57:1633–1644

Kato Y, Nevins DJ (1985) Isolation and identification of *O*-(5-*O*-feruloyl-α-L-arabinofuranosyl)-(1→3)-*O*-β-D-xylopyranosyl-(1→4)-D-xylopyranose as a component of *Zea* shoot cell walls. Carbohydr Res 137:139–150

Keppler LD, Baker CJ, Atkinson MM (1989) Active oxygen production during a bacteria-induced hypersensitive reaction in tobacco suspension cells. Phytopathology 79:974–978

Kerr EM, Fry SC (2003) Pre-formed xyloglucans and xylans increase in molecular weight in three distinct compartments of a maize cell-suspension culture. Planta 217:327–339

Kerr EM, Fry SC (2004) Extracellular cross-linking of xylan and xyloglucan in maize cell-suspension cultures: the role of oxidative phenolic coupling. Planta 219:73–83

Kuchitsu K, Kosaka H, Shiga T, Shibuya N (1995) EPR evidence for generation of hydroxyl radical triggered by *N*-acetylchitooligosaccharide elicitor and a protien phosphatase inhibitor in suspension-cultured rice cells. Protoplasma 188:138–142

Labavitch JM, Ray PM (1974) Turnover of cell wall polysaccharides in elongating pea stem segments. Plant Physiol 53:669–673

Lane BG, Dunwell JM, Ray JA, Schmitt MR, Cuming AC (1993) Germin, a protein marker of early plant development, is an oxalate oxidase. J Biol Chem 268:12239–12242

Levigne SV, Ralet M-C.J, Quéméner BC, Pollet BN-L, Lapierre C, Thibault J-FJ (2004) Isolation from sugar beet cell walls of arabinan oligosaccharides esterified by two ferulic acid monomers. Plant Physiol 134:1173–1180

Liszkay A, Kenk B, Schopfer P (2003) Evidence for the involvement of cell wall peroxidase in the generation of hydroxyl radicals mediating extension growth. Planta 217:658–667

MacAdam JW, Grabber JH (2002) Relationship of growth cessation with the formation of diferulate cross-links and *p*-coumaroylated lignins in tall fescue leaf blades. Planta 215:785–793

Markwalder HU, Neukom H (1976) Diferulic acid as a possible crosslink in hemicelluloses from wheat germ. Phytochemistry 15:836–37

McDougall GJ, Fry SC (1991) Xyloglucan nonasaccharide, a naturally-occurring oligosaccharin, arises in vivo by polysaccharide breakdown. J Plant Physiol 137:332–336

Miller JG, Fry SC (2001). Characteristics of xyloglucan after attack by hydroxyl radicals. Carbohydr Res 332:389–403

Miller JG, Fry SC (2004) *N*-[^{3}H]Benzoylglycylglycylglycine as a probe for hydroxyl radicals. Anal Biochem 335:126–134

Morré DJ, Auderset G, Penel C, Canut H (1987) Cytochemical localization of NADH-ferricyanide oxidoreductase in hypocotyl segments and isolated membrane vesicles of soybean. Protoplasma 140:133–140

Morrison JC, Greve LC, Richmond PA (1993) Cell wall synthesis during growth and maturation of *Nitella* internodal cells. Planta 189:321–328

Nishitani K, Masuda Y (1982) Acid pH-induced structural changes in cell wall xyloglucans in *Vigna angularis* epicotyl segments. Plant Sci Lett 28:87–94

Otte O, Barz W (1996) The elicitor-induced oxidative burst in cultured chickpea cells drives the rapid insolubilization of two cell wall structural proteins. Planta 200:238–246

Polle A, Chakrabarti K, Schürmann W, Rennenberg H (1990) Composition and properties of hydrogen peroxide decomposing systems in extracellular and total extracts from needles of Norway spruce (*Picea abies* L., Karst.). Plant Physiol 94:312–319

Ralph J, Quideau S, Grabber JH, Hatfield RD (1994) Identification and synthesis of new ferulic acid dehydrodimers present in grass cell walls. J Chem Soc Perkin Trans 1:3485–3498

Rayle DL, Cleland RE (1992) The acid growth theory of cell elongation is alive and well. Plant Physiol 99:1271–1274

Renger A, Steinhart H (2000) Ferulic acid dehydrodimers as structural elements in cereal dietary fibre. Eur Food Res Technol 211:422–428

Rodríguez AA, Cordoba AR, Ortega L, Taleisnik E (2004) Decreased reactive oxygen species concentration in the elongation zone contributes to the reduction in maize leaf growth under salinity. J Exp Bot 55:1383–1390

Rouau X, Cheynier V, Surget A, Gloux D, Barron C, Meudec E, Louis-Montero J, Criton M (2003) A dehydrotetrimer of ferulic acid from maize bran. Phytochemistry. Planta 63:899–903

Roulin S, Buchala AJ, Fincher GB (2002) Induction of (1→3,1→4)-β-D-glucan hydrolases in leaves of dark-incubated barley seedlings. Planta 215:51–59

Sánchez M, Peña MJ, Revilla G, Zarra I (1996) Changes in dehydrodiferulic acids and peroxidase activity against ferulic acid associated with cell walls during growth of *Pinus pinaster* hypocotyl. Plant Physiol 111:941–946

Saulnier L, Crépeau M, Lahaye M, Thibault J, Garcia-Conesa MT, Kroon PA, Williamson G (1999) Isolation and structural derternination of two 5,5′-diferuloyl oligosaccharides indicates that maize heteroxylans are covalently cross-linked by oxidatively coupled ferulates. Carbohydr Res 320:82–92

Schinkel H, Streller S, Wingsle G (1998) Multiple forms of extracellular superoxide dismutase in needles, stem tissues and seedlings of Scots pine. J Exp Bot 49:931–936

Schopfer P (2001) Hudroxyl radical-induced cell wall loosening in vitro and in vivo: implications for the control of elongation growth. Plant J 28:679–688

Schopfer P, Lapierre C, Nolte T (2001) Light-controlled growth of the maize seedling mesocotyl: Mechanical cell wall changes in the elongation zone and related changes in lignification. Physiol Plant 111:83–92

Schopfer P, Liszkay A, Bechtold M, Frahry G, Wagner A (2002) Evidence that hydroxyl radicals mediate auxin-induced extension growth. Planta 214:821–828

Schuchmann MM, von Sonntag C (1978) Effect of oxygen on OH radical-induced scission of glycosidic linkage of cellobiose. Int J Radiat Biol 34:397–400

Shetty NP, Kristensen BK, Newman MA, Møller K, Gregersen PL, Jørgensen HJL (2003) Association of hydrogen peroxide with restriction of *Septoria tritici* in resistance wheat. Physiol Mol Plant Path 62:333–346

Showalter AM (1993) Structure and function of plant cell wall proteins. Plant Cell 5:9–23

Takahama U (1993) Redox state of ascorbic acid in the apoplast of stems of *Kalanchoë daigremontiana*. Physiol Plant 89:791–798

Takahama U, Oniki T (1994) Effects of ascorbate on the oxidation of derivatives of hydroxycinnamic acid and the mechanism of oxidation of sinapic acid by cell wall-bound peroxidases. Plant Cell Physiol 35:593–600

Talbott LD, Ray PM (1992) Changes in size of previously deposited and newly synthesized pea cell wall matrix polysaccharides. Effects of auxin and turgor. Plant Physiol 98:369–379

Tan KS, Hoson T, Masuda Y, Kamisaka S (1991) Correlation between cell wall extensibility and the content of diferulic and ferulic acids in cell walls of *Oryza sativa* coleoptiles grown under water and in air. Physiol Plant 83:397–403

Thompson JE, Fry SC (2001) Restructuring of wall-bound xyloglucan by transglycosylation in living plant cells. Plant J 26:23–34

Thompson JE, Smith RC, Fry SC (1997) Xyloglucan undergoes inter-polymeric transglycosylation during binding to the plant cell wall in vivo: evidence from $^{13}C/^{3}H$ dual labelling and isopycnic centrifugation in caesium trifluoroacetate. Biochem J 327:699–708

Vanacker H, Foyer CH, Carver TLW (1999) Changes in apoplastic antioxidants induced by powdery mildew attack in oat genotypes with race non-specific resistance. Planta 208:444–452

Vissenberg K, Martinez-Vilchez IM, Verbelen JP, Miller JG, Fry SC (2000) In vivo colocalization of xyloglucan endotransglycosylase activity and its donor substrate in the elongation zone of *Arabidopsis* roots. Plant Cell 12:1229–1237

Vissenberg K, Fry SC, Pauly M, Höfte H, Verbelen JP (2005) XTH acts at the microfibril-matrix interface during cell elongation. J Exp Bot 56:673–683

von Sonntag C (1987) The Chemical Basis of Radiation Biology. Taylor & Francis, Abingdon

Vreeburg RAM, Fry SC (2005) Reactive oxygen species in cell walls. In: Smirnoff N (ed) Antioxidants and Reactive Oxygen Species in Plants. Blackwell, Oxford, p 215–249

Wakabayashi K, Hoson T, Kamisaka S (1997) Abscisic acid suppresses the increases in cell wall-bound ferulic and diferulic acid levels in dark-grown wheat (*Triticum aestivum* L) coleoptiles. Plant Cell Physiol 38:811–817

Ward G, Hadar Y, Bilkis I, Konstantinovsky L, Dosoretz CG (2001) Initial steps of ferulic acid polymerization by lignin peroxidase. J Biol Chem 276:18734–18741

Wojtaszek P (1997) Oxidative burst: an early plant response to pathogen infection. Biochem J 322:681–692

Plant Cell Monogr (5)
J.-P. Verbelen · K. Vissenberg: The Expanding Cell
DOI 10.1007/7089_2006_076/Published online: 25 October 2006
© Springer-Verlag Berlin Heidelberg 2006

Mechanics of the Expanding Cell Wall

Ingo Burgert (✉) · Peter Fratzl

Department of Biomaterials, Max-Planck-Institute of Colloids and Interfaces, Research Campus Golm, 14424 Potsdam, Germany
ingo.burgert@mpikg.mpg.de

Abstract Cell enlargement is one of the essential facets in plant growth and development, but the underlying mechanisms of primary cell wall expansion still remain partly elusive. The primary cell wall, only ∼ 100 nm thick, has to be both strong enough to withstand high stresses due to hydrostatic pressure and external loads as well as flexible enough to allow a tremendous expansion of the cell. Evidently, to shed light on this mechanical paradox, an interdisciplinary combination of biophysical (biomechanical), biochemical and physiological approaches is required. Here we deal with the mechanical constraints of cell wall expansion from a plant biomechanics perspective. Possible mechanisms of increasing cell wall volume are introduced. Current models on cell wall structure and its expansion are reviewed with regard to the capability of the polymer network to undergo sufficient elongation. Methods to determine the mechanical properties of living tissues and isolated cells are briefly reviewed focussing on how the deformation behaviour of the cell wall is influenced by various chemical and biochemical treatments. The crucial role of cellulose microfibril orientation and fibre and matrix interaction in the course of cell wall expansion is discussed.

1 Introduction

Primary cell wall expansion is one of the basic requirements in plant growth and development. Cell differentiation in primary as well as secondary tissues depends on the capacity of cells to increase their volume and modulate their shape. Thereby, the balanced counterforce of primary wall stress to turgor pressure transforms cells into "hydraulic machines" (Peters et al. 2000). For example, Carpita and Gibeaut (1993) estimated a 250 MPa tensile stress in the 100 nm thick cell wall, assuming a turgor pressure of 1 MPa in a spherical cell of 50 μm radius. Cell volume typically undergoes a 10 to 1000-fold increase (Veytsmann and Cosgrove 1998) but a more than 30 000-fold enlargement was reported for cells functioning as xylem vessel elements (Cosgrove 2005). This implies a drastic increase in cell wall surface area, which requires a highly optimized interplay of cell wall expansion and new material synthesis in order to avoid a loss of wall integrity (Veytsmann and Cosgrove 1998). So far, the principal mechanisms of cell wall enlargement remain partly elusive. In fact, primary cell walls have to cope with two conflicting goals, because they have to be rigid to withstand the internal and external stresses and have to be compliant to allow cell wall expansion during growth (Cleland 1971a; Taiz 1984). To fulfil both requirements, the cell wall assembly has to be reasonably stiff and tough but at the same time readily modifiable for a drastic increase in surface area. The major challenge from a biomechanical point of view is to accommodate these conflicting mechanical imperatives with the existing models of cell wall structure and cell wall loosening, and thereby identify the crucial parameters in polymer assembly relevant for cell wall expansion.

The plant cell wall is a complex assembly of polysaccharides and structural proteins (McCann and Roberts 1991; McNeil et al. 1984; Schindler 1998) which can be characterized on the lines of a fibre-reinforced composite (Fratzl et al. 2004a; Kerstens et al. 2001). Stiff semi-crystalline cellulose microfibrils are embedded in a soft amorphous matrix consisting of hemicelluloses, pectins and structural proteins. The deposition of the cellulose microfibrils is controlled by geometrical constraints of cell shape and lateral distance between fibrils (Emons and Mulder 1999, 2000). It is mainly the orientation of the microfibrils that dictates the degree of anisotropy of the system and thereby determines the shape of the cell during growth (Baskin 2005). In terms of the mechanical behaviour of the cell wall, the interaction between cellulose microfibrils and matrix, particularly hemicelluloses, is believed to be the major parameter. Interestingly, cell walls of grasses and other flowering plants show the same behaviour in cell wall expansion, even though their matrix composition is completely different (Bacic et al. 1998; Carpita 1987). The type I cell walls of most flowering plants contain xyloglucan in high amounts, whereas in type II cell walls of grasses, glucoronoarabinoxylans are the major component (Carpita and Gibeaut 1993). These findings indicate that the capability of the primary cell wall to expand is dictated by the structural design of the fibre-reinforced composite rather than the specific polymer composition. From a biomechanical perspective this allows for a significant simplification of the structural organization of the cell wall and therefore a general concept of fibre and matrix interaction may help to shed light on the essential structure–function relationships that determine primary wall properties and growth capability.

The main objective of this chapter is to briefly review the current knowledge on the mechanics of cell wall expansion and to examine more closely the possible role of fibre and matrix interaction at the nanoscale of primary cell walls, starting from the deformation imperatives that have been recently found for secondary cell walls (Fratzl et al. 2004b; Keckes et al. 2003). The macromolecular organization and the mechanical interaction of cell wall polymers as supposed in various models are related to potential principles of cell wall deformation with regard to cell growth.

2 Structural Models of the Primary Cell Wall

A better understanding of the mechanisms of cell wall expansion demands a detailed knowledge of the structure and composition of the cell wall. Even though the individual cell wall polymers and their architecture are well characterized (Bacic et al. 1998; Carpita and Gibeaut 1993; McCann and Roberts 1991; McNeil et al. 1984; Vincken et al. 1997), little is known about their intrinsic spatial orientation and interplay during mechanical deformation. Various models of the cell wall assembly have been proposed in the last decades, which mainly differ in the structural composition of the matrix polymers as well as in the bonding characteristics between the molecules. Detailed reviews of their structural and chemical specifications are given by Cosgrove (2000a; 2001) who discussed the model of Keegstra et al. (1973), the "tethered network" model by Fry (1989) and Hayashi (1989), the "multicoat" model by Talbott and Ray (1992) and the "stratified" cell wall model by Ha et al. (1997). Here, we mention just a few aspects with regard to the cellulose–xyloglucan network, which we consider of crucial relevance to the mechanical performance and expansion of the walls, thereby taking over the instructive classification formulated by Cosgrove (2000a; 2001).

All models coincide regarding the non-covalent cross-linking of the cellulose and hemicellulose network. Xyloglucans are attached to the cellulose microfibril surface via a multitude of hydrogen bonds (Fry 1986). However, parts of the xylogucan chains may not be attached to the microfibril surface but segments may be entrapped within the microfibril during its formation (Hayashi 1989). The length of the xyloglucan chains is up to 400 nm, which largely exceeds the distance between individual cellulose microfibrils (approx. 30 nm) (McCann et al. 1992). In the tethered model, cellulose microfibrils are interconnected by long and continuous xyloglucan chains, which are oriented perpendicular to the microfibrils and may interconnect several of them. In terms of the "multicoat" model, polysaccharides coat the cellulose microfibril surfaces in a more parallel fashion with a gradient in the non-covalent bonding pattern, resulting in dead ends in the matrix. Pauly et al. (1999) distinguished three domains of xyloglucan: (i) "free" xyloglucan chains that form loops and dead ends or which function as cross-linking segments between cellulose; (ii) xyloglucan that is bonded to the cellulose microfibril surface and (iii) xyloglucan chains that are entrapped within or between cellulose microfibrils. Their calculations based on the ratio of the domains suggest a structural organization in a tethered manner. However, recently, Thompson (2005) discussed the implications of the tethered network on cell wall expansion and formulated evidence against it, arguing that xyloglucans act like spacers or struts between the cellulose microfibrils rather than as tethers.

3 Enlarging Cell Wall Area

Growth implies an increase in cell volume and thereby an enlargement of its wall area. Therefore, cell expansion depends on the capability of the cell wall to increase its surface area. Theoretically, three possible processes of cell volume increase can be defined: (i) mechanical deformation, (ii) wall swelling and (iii) incrustation of new material. Cell wall loosening per se is not a mechanism of cell wall enlargement, but is considered in this context since it facilitates extension for a given extrinsic force. When taking the cell wall as the frame of reference, both internal hydrostatic pressure (turgor pressure) and external stresses are regarded as extrinsic forces. However, an applied stress cannot be directly converted into turgor pressure equivalent because the boundary conditions are different, and longitudinal extension of the cell in multiaxial stress is much less than under equivalent uniaxial longitudinal stress (Sellen 1983).

3.1 Mechanical Deformation

Tensile deformation typically results in a volume increase. Under uniaxial loading conditions the applied force results in an elongation along the axis and a transverse contraction. The ratio of the transverse contraction to the longitudinal extension is called the Poisson ratio (Fig. 1).

For Poisson ratios below 0.5, the volume of the material does not stay constant but increases due to uniaxial stretching. Because of the anisotropic

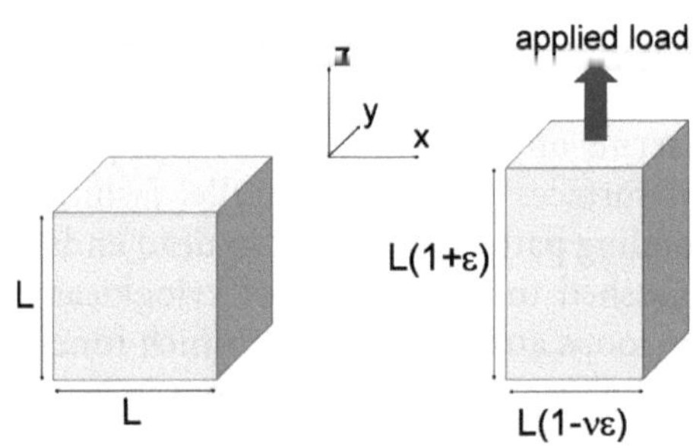

Fig. 1 When a cubic piece of material (cell wall) is subjected to load along the vertical (z-direction) only, its length L is increased by Lε. The relative elongation ε is called strain. In most cases, the dimensions of the cube perpendicular to the load direction will contract at the same time. The ratio ν of the contraction in the x-direction (or in the y-direction) relative to the elongation in z-direction is called the Poisson ratio. In a material where the volume stays constant during deformation, the Poisson ratio is equal to $\nu = 1/2$. The relative increase of the volume during uniaxial stretching is $1-2\nu$. In many solids, the Poisson ratio has a value close to $\nu = 1/3$

nature of the cell wall, longitudinal elongation results in different Poisson ratios in the two directions orthogonal to elongation. The almost transverse orientation of cellulose microfibrils (Baskin 2005) facilitates a longitudinal extension, stabilizes the circumferential direction and conjecturally results in a contraction in the direction of cell wall thickness.

A material may respond to external load by fully reversible (elastic) or by irreversible (plastic) deformation (Fig. 2). In some cases, a material is elastic but needs time to recover its original shape (viscoelastic behaviour). For cell growth, it is essential that the wall extends in an irreversible manner. Such a plastic deformation is usually mediated by a reorganization of the molecular bonds inside the material. A process of this kind will also have to operate in cell walls.

In stress–strain diagrams, the material response to straining can be analysed in more detail. Upon loading and unloading cycles, viscoelastic materials show a hysteresis loop with a characteristic profile. However, the shape of this curve depends strongly on the rate of deformation. When the strain rate is increased, the material appears "stiffer" and the hysteresis loop changes its shape. Since most of the biological materials deform viscoelastically and/or plastically, the strain rate must be considered when analysing the mechanical behaviour (Niklas 1992; Vincent 1990).

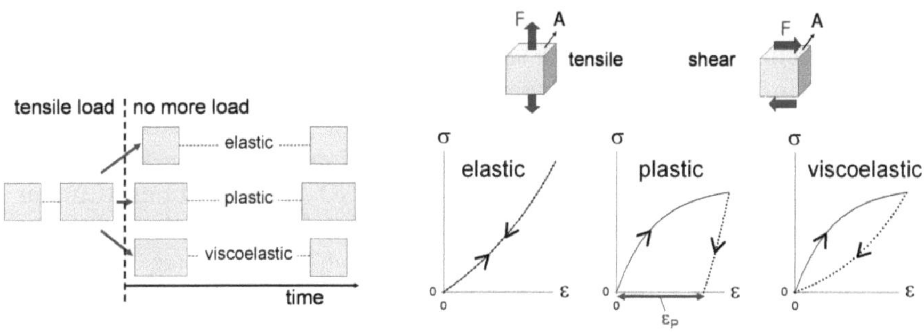

Fig. 2 Materials behave differently when the load is removed after deformation (*left*). An elastic material returns immediately to its original shape, a plastic material keeps the deformed shape forever, and a viscoelastic material returns slowly to its original shape. Materials have often a combination of these properties: An elastoplastic material relaxes partially and retains only part of the deformation. Similarly, a viscoplastic material gradually loses a part of the deformation but a fraction of it stays forever. Tensile stress is defined as the force perpendicular to the surface of a cube of material divided by its area, $\sigma = F/A$. Shear stress is a force parallel to the surface per unit area, $\tau = F/A$. In a stress–strain experiment (*right*), the stress σ is measured as a function of strain ε. *Solid lines* in the $\sigma - \varepsilon$ curves show increasing stress and *dotted lines* decreasing stress. For elastic materials, these two lines coincide. For plastic materials, a permanent strain ε_P remains. For viscoelastic materials, there is a hysteresis. The elastic modulus E is defined as the slope of the *solid line* close to the origin

3.2
Water Swelling

In theory, swelling of the cell wall by water is a further possible way of causing volume increase. Water can be pumped into the cell wall by osmotic pressure or can simply enter by passive flow. Since cellulose chains swell only marginally in their longitudinal direction, a swellable matrix is a prerequisite, which is fulfilled by the polysaccharide components. Pectins behave like a polyelectrolyte hydrogel, which means that the swelling ability can be modulated by various cations (Jarvis 1992; Zwieniecki et al. 2001). Due to the fibre-reinforced structure of the primary cell wall (Kerstens et al. 2001), the level of swelling anisotropy can be adjusted by the orientation of the cellulose microfibrils. A random distribution of fibrils would result in an almost isotropic swelling behaviour, whereas a preferential orientation of microfibrils causes anisotropic swelling. Hence, a parallel alignment of cellulose fibrils in the transverse direction should cause a longitudinal expansion due to swelling.

3.3
Addition of Material

A simple addition of material can result in volume increase and therefore promote cell wall expansion. As shown by Brummell and Hall (1985), a sustained auxin-induced growth depends upon the incorporation of new matrix cell wall components into the wall. A study on growth-promotion by incubating pea shoots with xyloglucan oligosaccharide solution showed that growth rate could be increased by higher XGO concentrations (Cutillas-Iturralde and Lorences 1997). Takeda et al. (2002) found that the integration of whole xyloglucan suppressed cell elongation, whereas the integration of xyloglucan oligosaccharides promoted it. The authors proposed that xyloglucan fragments may loosen the cellulose–xyloglucan network by cleaving the xyloglucan tethers. However, even though cell wall expansion might depend on xyloglucan metabolism, in contrast to mechanical deformation or swelling, the secretion of biopolymers can not act as a driving force to cell expansion. Therefore, one may consider synthesis of new material mainly as a compensation of the ongoing wall thinning during expansion, which is essential to maintain structural integrity of the wall. This process can take place as a decoration of the inner wall surface, but a direct material incorporation into the highly stressed composite as proposed by Takeda et al. (2002) is also possible. Likewise Proseus and Boyer (2005) showed for growing cell walls of *Chara* that even large polymers can be moved into small voids of the primary cell wall by turgor pressure.

3.4 Cell Wall Loosening

Cell wall loosening can ease the extension by a specific alteration of cell wall properties. In terms of biomechanics, this biochemical/biophysical process can be regarded as a relaxation of wall stresses since it takes place in a high stress environment. When growing cells loosen their walls they simultaneously reduce turgor pressure and water potential. The weakening of the cell wall is a basic requirement of cell growth because only the reduction of turgor can cause water uptake and by that an extension of the cell wall (Cosgrove 1987).

Cell wall loosening is accompanied by an occurrence of hydrogen ions in the cell wall (Métraux et al. 1980; Rayle and Cleland 1972, 1992). Growth hormones such as auxin can cause the living cell to excrete protons in the apoplast. This wall acidification results in the cleavage of acid-labile linkages (Rayle and Cleland 1992) or leads to the activation of further possible wall loosening agents, which have been reviewed by Cosgrove (2000b, 2005). Every one of them can increase cell wall extensibility but, in terms of their capability to function as cell wall loosening agents, it is an open question whether they can directly catalyse cell wall loosening, or whether they are available in appropriate amounts to induce sufficient extension in a highly specific manner.

The wall loosening capacities of oxygen radicals controlled by auxin were recently discussed by Schopfer et al. (2002) (see Lindsay and Fry, in this volume). The hydroxyl radical generation is catalysed in the cell wall by peroxidase activity (Liszkay et al. 2003), enabling them to unspecifically cleave cell wall polysaccharides leading to an irreversible wall extension (Schopfer 2001).

Various enzymes are located in the growing cell wall that may play a crucial role in cell wall expansion (Fry 1995). For instance, xyloglucan endotransglycosylase (XET; see Nishitani and Vissenberg, in this volume) is an enzyme with a pH optimum under acid conditions (pH ~ 5.5). Cell wall loosening is correlated with a peak in XET activity (Tomos and Pritchard 1994) and its ability to cut and rejoin xyloglucan chains may promote cell wall loosening (Fry et al. 1992; Pritchard et al. 1993). Expansins are proteins which are secreted by the cell to unlock the network of wall polysaccharides by mediating acid growth through pH-dependent wall extension and stress relaxation (McQueen-Mason and Cosgrove 1995, see also McQueen-Mason, in this volume). Two protein families are described as inducing cell wall extension without hydrolytic breakdown and, hence, do not alter the molecular mass distribution or the viscosity of the solutions of matrix polysaccharides (McQueen-Mason et al. 1992). Expansins are believed to bind at the interface between cellulose microfibrils and matrix polysaccharides and induce extension by reversibly cleaving non-covalent bonds between hemicelluloses and the cellulose microfibril. According to Cosgrove (1999), enzymes such as xyloglucan endotransglycosylase and endo-(1,4)-β-D-glucanase may act as secondary agents that indirectly promote wall extension, whereas expansins are the primary cell wall loosening agents since they can directly cause wall stress relaxation.

4
Exploring the Mechanics of Cell Wall Expansion

Expansion starts with a selective loosening of load-bearing linkages between cellulose microfibrils. First, wall stress relaxation results in a reduced turgor pressure. The enlargement occurs secondarily as a consequence of cellular water uptake. The interrelation of cell enlargement as a function of wall expansion and water uptake was treated in a biophysical model by Lockhart (1965), which proved to be useful for characterising growth responses in various studies. In the most simplified form of the equation, the rate of wall expansion (r) is given by the formula:

$$r = \phi(P - Y) \tag{1}$$

where ϕ is the yield rate coefficient (extensibility of the wall), P is the turgor pressure and Y is the yield threshold of the wall.

By modulating turgor pressure and observing growth rate, one can gather insight into the yield rate coefficient or extensibility of the cell wall. According to this equation, growing tissues should show a linear correlation between turgor pressure and growth rate, with the slope of the curve as a measure of the extensibility of the wall. Indeed, a linear interrelation was shown in some studies (Bunce 1977; Pritchard et al. 1990) but non-linear dependence was also observed (Cosgrove 1993), which might be better expressed by a power function. By this, the wall expansion becomes almost equal to the equation generally used in materials science to describe plastic flow (Haasen 1986). Therefore, the yield rate coefficient (or extensibility) of the cell wall is a measure of its (visco)plastic deformation behaviour. Consistently, Cosgrove (1993) defined cell wall extensibility of growing cells as an irreversible deformation in a time-dependent manner. In literature, cell wall extensibility has been often used in a rather imprecise way, describing elastic, viscoelastic, plastic and viscoplastic deformation properties.

The time-dependent behaviour of materials is typically studied in two types of transient experiments. In relaxation tests, stress relaxation is measured at a given elongation, whereas in creep tests sample extension under a given force is recorded (Fig. 3).

In the transient experiments, in vivo and in vitro tests are distinguished. Creep tests on isolated cell walls can provide valuable information on the intrinsic mechanical properties of the cell wall. Prior to measurements, tis-

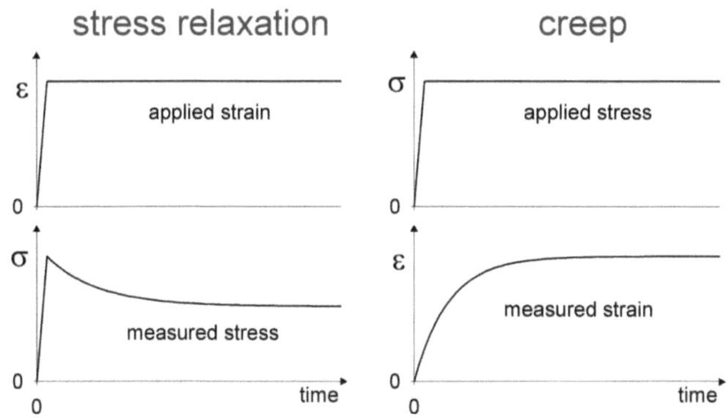

Fig. 3 In viscoelastic or viscoplastic materials, the time constants of viscous flow are determined by creep or stress relaxation experiments, where the strain or the stress is kept constant, respectively

sues are treated in a freeze–thaw cycle (Kutschera and Schopfer 1986a) or are killed in methanol (O'Looney and Fry 2005) and are finally rehydrated. When the protoplast is disrupted by freeze–thaw cycles, cell wall biosynthesis is inhibited and enzymatic activity minimized. However, hydrolytic enzymes may stay active in the cell wall and the non-covalent bonding pattern may be altered by the treatments (Cosgrove 1993). For in vivo creep tests, additional compensatory responses of the living tissue as well as hydraulic conductance have to be considered.

4.1 In Vivo and In Vitro Relaxation

In vivo relaxation tests are based on the functional interrelations of cell growth as depicted in the "Lockhart equation". Obviously, wall stress cannot be measured directly in the growing tissue because it is dependent on several parameters such as turgor pressure, wall thickness and cell geometry. Hence, one simplifying approach is to take turgor as a measure of wall stress and to derive wall extensibility from its functional relation to wall stress (Cosgrove 1985, 1986, 1987). Practically, in vivo stress relaxation of the cell wall is achieved by preventing water uptake and thereby inhibiting cell expansion. Tissues are excised or cut off from the water supply and turgor pressure is continuously measured, or a "pressure block instrument" is utilized where an adjusted gas pressure is applied to a plant so that water uptake is prevented (Cosgrove 1987). Using the pressure block method, Boyer (2001) could show that external pressure applied to the entire plant had only little effect on cell expansion, whereas pressure applied to the shoot alone decreased elongation. Boyer also showed that stress relaxation does not occur in the growing region when mature tissue remains, presumably because that tissue acts as a source of water. This is interpreted as meaning that the deficit in turgor pressure, compared to osmotic potential typically observed in growing tissues (usually several tenths of a MPa; references in Boyer 2001), arises from a limitation on water influx rather than from stress relaxation.

If water uptake is strictly prevented, then the resulting wall relaxation progressively reduces cell turgor, which should asymptotically decay to the yield threshold (Y) of the wall. Measurements on excised tissues of pea stems showed the influence of growth hormones on cell wall relaxation. Those stems which had grown with auxin relaxed much faster than slowly growing tissues, but the yield threshold was not affected (Cosgrove et al. 1984; Cosgrove 1985). Nakahori et al. (1991) found also an increased extensibility due to auxin treatment, however, the effective turgor was changed due to a reduction in the yield threshold.

In vitro stress relaxation tests are performed as uniaxial tests on isolated strips of tissue that have gone through a freeze–thaw cycle to eliminate the cell content and consequently cell turgor. The samples are rapidly stretched while the strain is kept constant. By monitoring stress decay, different relaxation times of the cell wall composite were identified (Yamamoto et al. 1970) and a different stress decay was found upon growth hormone treatments (Yamamoto et al. 1970; Yamamoto and Masuda 1971) and variation of pH (Cosgrove 1989). In particular, the fastest relaxation time of the cell wall component is significantly decreased by the auxin treatment, leading to the assumption that auxin is likely to partly degrade wall matrix polysaccharides (Yamamoto and Masuda 1971).

Even though the time courses of in vitro and in vivo stress relaxation tests generally correspond, the stress relaxation of isolated and growing tissues differs substantially. Cosgrove (1993) identified three possible reasons: (i) the different loading conditions of the cell wall (multilateral/unilateral), (ii) the stress history of the wall (long-term prestress for in vivo stress relaxation and rapid straining for in vitro measurements) and (iii) differences in cell wall metabolism.

4.2
In Vivo and In Vitro Creep

In contrast to the relaxation assays, in vivo creep tests on growing organs and in vitro creep tests on isolated tissue strips are based on the same methodical procedure. However, it is more difficult to get the stress during in vivo tests. Tissue samples are loaded at a constant stress level and the rate of extension is monitored either for living tissues or for isolated tissue (Fig. 3).

Creep tests have been widely used to study the effect of specific cell wall treatments on the time-dependent mechanical behaviour. Thereby it was found that in vivo and in vitro studies detect a rather similar material response, particularly with regards to the creep extension under acid conditions and growth hormone treatments (Kutschera and Schopfer 1986a/b; Tepfer and Cleland 1979). At a neutral pH the rate of extension falls quickly, whereas in acidic buffer the wall extends rapidly and continuously (Cleland et al. 1987; Cosgrove 1989). This so-called acid-induced creep has its maximum at a pH between 2.5 and 4.0 (Cosgrove 1989). The drastic changes in deformation behaviour upon pH, for both isolated and living cells, indicate a modification of the intrinsic property of the primary cell wall. Likewise, auxin treatments can increase the creep rate (Cleland 1971b; Hohl and Schopfer 1992; Kutschera and Briggs 1987; Kutschera and Schopfer 1986a, 1986b) since auxin prompts cells to excrete protons to the cell wall (Rayle and Cleland 1992). For both in vivo and in vitro tests, the plant material creep could be inhibited by the removal of protons. However, acid-induced loosening without tensile pre-stress could only occur in living but not in isolated tissue (Tepfer and Cleland 1979).

Cosgrove (1989) investigated whether acid-induced creep alone is sufficient to maintain cell extension or whether additional cell wall loosening agents are required. His findings support a crucial role of enzymatic activities since he found a temperature optimum for acid creep and an inhibition at high temperatures, or upon denaturation by boiling in water or by specific metal cations. In creep studies aimed at shedding light on specific wall loosening agents, McQueen-Mason et al. (1992) investigated the possible role of expansins and found that they promote cell wall expansion without hydrolytic degradation. Whitney et al. (2000) added expansins to a network of cell wall analogues consisting of bacterial cellulose and xyloglucans and could show that extension rate was substantially increased. However, no effect was observed when networks of bacterial cellulose and glucomannan or galactomannan were examined. Schopfer (2001) could show in in vivo and in vitro creep tests that hydroxyl radials are capable of inducing cell wall extension.

In a test protocol added to standard creep assays, unloading and reloading of the samples were performed during the ongoing creep test aiming at sensing the elastic and plastic components of the deformation. In vivo and in vitro creep assays with a loading/unloading cycle showed that auxin increased the "plastic" deformation whereas the "elastic" deformation remained almost unaltered (Kutschera and Schopfer 1986a,b). Likewise, low water potentials due to growth under water stress largely influenced the plastic properties of the cell walls, whereas the elastic properties were only slightly affected (Nonami and Boyer 1990).

A drawback of these testing protocols is that the reversible (elastic) and the irreversible (plastic) portions in the deformation behaviour cannot be unambiguously determined without knowing the time constants of viscous flow in loading and unloading. Therefore, when assigning deformation in creep tests to plastic and elastic deformation of the plant material, the time after stress release has to be taken into consideration.

Hohl and Schopfer (1995) could show in complete loading and unloading cycles during the creep assay that the hysteresis loops are closed. Therefore, no irreversible deformation remained but the material showed a viscoelastic behaviour with a long time constant of viscous flow (Nolte and Schopfer 1997). However, recently, Suslov and Verbelen (2006) found elastic and plastic deformation when performing in vitro creep assays on onion epidermal cell walls. The contradiction can be explained by the different definitions of the basic stress state of the cell. Suslov and Verbelen (2006) argue that a complete unloading would correspond to a cell without turgor pressure and this in fact implies the complete absence of stresses, which is physiologically irrelevant. Hence, relative reversibility of deformation might be defined by comparing cell wall extensibility at different stress states. Consistently, the authors could show that acid conditions increased the creep rate but did not affect the initial deformation when the load was applied (Suslov and Verbelen 2006). However, to calculate absolute values of (visco)elastic and plastic deformations the basic stress state of the cell has to be defined, which is hardly obtainable in in vitro uniaxial creep tests considering that under in vivo conditions the cells face multiaxial and long-term prestress.

5
Cell Wall Deformation Models and Effects of Cellulose Orientation

Clearly, the composite nature of the cell wall comprising stiff cellulose fibrils embedded in a matrix of hemicelluloses and pectin is essential for cell expansion. Cellulose fibrils have early been found to be essentially perpendicular to the cell growth direction in primary walls of growing cells (Fry-Wyssling 1953). Given the stiffness and strength of cellulose, it is much easier to expand the cell wall in the direction perpendicular to the cellulose orientation. This is schematically explained in Fig. 4.

From this simple analysis, it also follows that when the cellulose fibrils are not exactly perpendicular to the growth direction, they will eventually tilt towards the cell axis. This was recognized very early on and it is an experimental observation that cellulose fibrils in the outer layers of the primary wall are more random, or even longitudinal (Gertel and Green 1977; Green 1960; Roelofson 1965). Such a passive rotation has also been directly observed during stretching of cellulose tissues (though not directly in growing cells) (Astley et al. 2003; Wilson et al. 2000). Preston (1982) supported this multi-net growth theory by calculating the passive reorientation of microfibrils based on strain. It was recently shown, however, that the reorientation expected from the multi-net growth hypothesis cannot be evidenced in in vitro creep experiments on cucumber hypocotyls for longitudinal strains up to 30% (Marga et al. 2005).

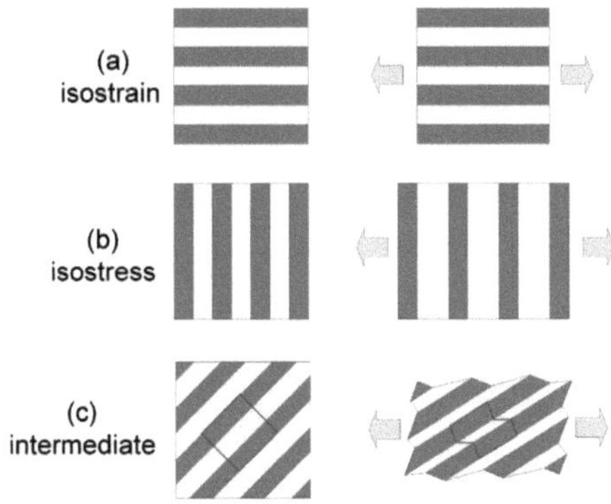

Fig. 4 Tensile deformation of a hypothetic composite of extremely stiff (essentially undeformable) fibres (*grey*) in a soft matrix (*white*). **a** In the isostrain case, both fibres and matrix must have the same strain, which means that no extension is possible. **b** In the isostress case, fibres and matrix are subjected to the same stress and an extension is possible by stretching the soft matrix. **c** In the intermediate case, deformation occurs by matrix shear and, depending on the angle of the fibres, even compression. The *red box* (*left*), which deforms into a more complex *red shape* (*right*), was drawn to better visualize the deformation. Clearly, the *grey* fibres inside the *red box* stay totally undeformed, while the *white matrix* is sheared and compressed. Note that the fibres have tilted towards the stretching direction. The composite in the intermediate case **c** is deformable, but much stiffer that in the isostress case **b**. The stiffness depends on the angle of the fibres with respect to the stretching direction

5.1 Geometrical Constraints Introduced by Long Cellulose Fibrils

To get a quantitative picture of the reorientation of (essentially inextensible) cellulose fibrils, we have plotted the change in angle as a function of cell elongation in Fig. 5. The basic assumption is that cellulose fibrils are long enough to dominate the behaviour of the composite in the fibril direction (Fig. 4a). Under this assumption, there will be no permanent deformation of the cell wall in the direction of the fibrils. Indeed, cellulose fibrils will stretch elastically when load is applied, but the deformation will not be permanent as long as the cellulose is not broken. The matrix between the fibrils, on the other hand, may undergo permanent (plastic) deformation by mechanisms which are discussed in the following Sect. 5.2.

When the cell elongates from L_0 to L with essentially inextensible cellulose fibrils, which are tilted by an angle α_0 with respect to the perpendicular to the cell axis before elongation, a tilting of the cellulose angle will occur. In this case, the cell elongation can be calculated from the simple relation $L/L_0 = \sin(\alpha)/\sin(\alpha_0)$, where α is the tilt angle after elongation. The sketches on top of Fig. 5 also show that the perimeter of the cell also changes when the cellulose fibrils tilt. This leads to simple relation for the cell diameter D,

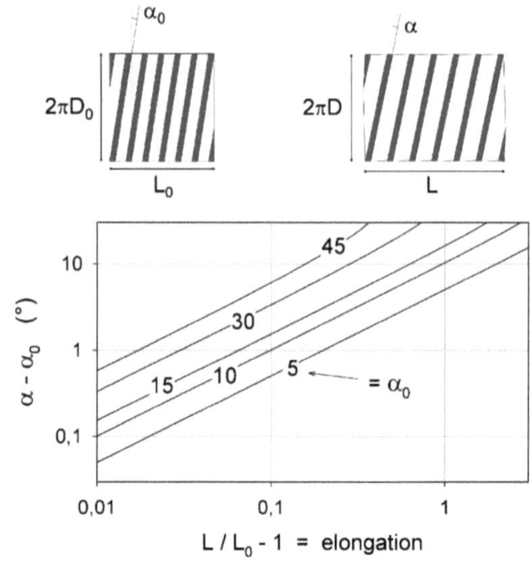

Fig. 5 Change of cellulose orientation as a function of cell elongation in a model with inextensible cellulose fibrils. The angle of cellulose fibrils with the direction perpendicular to the cell axis is called α (with the value α_0 before elongation). The length of the cell changes from L_0 to L, and its diameter from D_0 to D. The numbers in the figure indicate the cellulose angle α_0 before cell elongation. Both axes are drawn with logarithmic scales

given as $D/D_0 = \cos\alpha / \cos\alpha_0$. It should be noted here that the angle α is defined here for convenience with respect to the normal to the cell axis, while it is customary in secondary cell walls to define the microfibril angle μ with respect to the cell axis (Keckes et al. 2003; Preston 1974; Reiterer et al. 1998). With these definitions, the angle α is related to the microfibril angle μ by the simple relation $\alpha = 90° - \mu$.

Figure 5 shows the change in the cellulose angle with the longitudinal extension of the cell. One of the central observations that can be made from the figure is that the cellulose reorientation is very small when the initial tilt angle is small. For example, when the initial angle is less than 10°, a 10% elongation leads to an increase of the cellulose angle by less than 1°. Such a small reorientation would be very difficult to detect experimentally. In this sense, the fact that no reorientation was detected in in vitro creep experiments for strains up to 30% may not fully contradict the classical multi-net growth hypothesis, as has been recently argued (Marga et al. 2005).

Figure 5 also shows that the reorientation of cellulose fibrils becomes significant when the cell extension becomes very large, such as 100% and more. This agrees with the fact that a reorientation of cellulose fibrils to the longitudinal axis was observed in living *Nitella* during cell wall expansion by means of polarized light microscopy (Richmond 1983). In this study, cellulose microfibril synthesis was shown to be decoupled from cell expansion, whose directionality is governed by the inner part of the cell wall only. Figure 5 is also in good agreement with the fact that larger cellulose angles are typically found in the outer layers of the cell wall, which are the oldest and presumably have to undergo the largest extension (Preston 1982).

Further insights into the geometrical constraints imposed on the growing cell by (supposedly inextensible) cellulose fibrils can be gained by plotting the change in cell diameter and volume. Figure 6 shows that a constant length of the cellulose fibrils requires a reduction of cell diameter D when the cell length increases. Most informative is the change in cell volume $V = \pi LD^2/4$. From the variation of L and D (see discussion of Fig. 5) one can conclude that the volume of the cell will change with α as $V/V_0 = \sin\alpha\cos^2\alpha/(\sin\alpha_0\cos^2\alpha_0)$. As a consequence, the cell volume turns out to be maximum when the cellulose angle α is such that $\tan^2\alpha = 1/2$, which means $\alpha \approx 35°$. The reason for the existence of a maximum in the volume is the reduction in cell diameter that accompanies any cell elongation (Fig. 6). As the

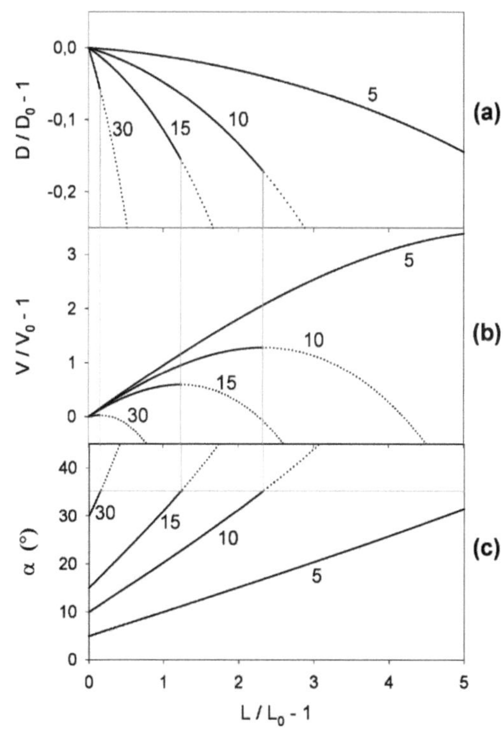

Fig. 6 Changes in cell geometry as a function of elongation. **a** Shows the change in cell diameter, **b** the change in cell volume, and **c** the cellulose angle α with respect to the normal to the cell axis. The volume of the cell increases up to an angle α of about 35° (*horizontal grey line* in **c**). Further elongation would lead to a reduction of the cell volume (*dotted lines* in all three graphs). The numbers next to each line indicate the value of α before cell elongation

turgor pressure should always lead to an increase in cell volume, this geometrical constraint clearly limits the possible cell elongation. This limit is larger when the initial cellulose angle α_0 is smaller. For example, with a cellulose angle of $\alpha_0 = 10°$ before elongation, the cell could elongate by up to 232% before coming to rest.

The second important message of Fig. 6 is the fact that the cell diameter decreases with cell elongation. This means that there is always a lateral compression of the wall matrix (by the cellulose fibrils), which accompanies the longitudinal extension of the cell (at least, as long as $\alpha_0 \neq 0$). This is quite interesting in relation to the alternative cell growth model by Thompson (2005), which proposes that non-cellulosic wall components hold microfibrils apart rather than sticking fibrils together. Thompson argues that the cell wall components can neither be arranged according to the isostrain case nor to the isostress case (Fig. 4) because the axial elastic stiffness would be too high in the first case and too low in the second. However, he did not explicitly discuss the intermediate case (Fig. 4) with tilted fibrils which has, indeed, an intermediate stiffness. Figure 6 shows that the matrix between cellulose fibrils needs to act in two ways in this intermediate case: first, the matrix has to extend longitudinally (keeping the contact between fibrils) and, second, it should compress laterally (keeping the space between fibrils). Hence, the matrix has to act both as spacer (laterally) and as tether (longitudinally) between cellulose fibrils.

No definitive data exist on the mechanisms by which the non-cellulosic matrix mediates the deformations needed to allow cell elongation. As shown in Fig. 5, for example, the elongation is associated primarily with a large permanent deformation of the hemicellulose and pectin matrix, which is stabilized by many hydrogen bonds. For a review on hydrogen bonding in polymers see (Coleman and Painter 1995). A widespread hypothesis is that that the main contribution to the cell wall stress is caused by the interpenetrated hydrogen-bonded network of cellulose fibrils and xyloglucan threads. Cell wall yield threshold and strength are determined mostly by the concentration of cellulose and xyloglucan and not by the strength of the hydrogen bonding. At a given level of glucans, either an increase or a decrease of the concentration of microfibrils leads to a decrease in yield threshold (Veytsmann and Cosgrove 1998). The yield threshold would then represent the stress needed for unzipping the hydrogen bonds (Passioura 1994). A problem is that the yield threshold and the wall extensibility are not constant parameters but, because they change during elongation, they behave more like variables.

5.2 Deformation of the Wall Matrix During Cell Elongation

As already mentioned, there is a need for a (permanent) longitudinal elongation of the matrix and a lateral contraction (at least if the cellulose fibrils are practically inextensible). Several models have been proposed in the literature (see Fricke and Chaumont, in this volume). Xyloglucans join fibrils by sticking them together, forming either a continuous bridge, a discontinuous bridge held together by hydrogen bonds, or held separate by spacers, see Fig. 7 from left to right. The first two ideas are based on the fact that an elongation of the matrix is necessary in the longitudinal direction and the second that a compression is needed in lateral direction. At present, there is no direct evidence that would allow exclusion of any of these possibilities. Creep experiments clearly show that the matrix between fibrils is able to elongate considerably (with no measurable change in cellulose angle, as discussed earlier) when non-covalent bonds are weakened between either xyloglucan and cellulose or between matrix polymers (Marga et al. 2005). This supports models of the type shown in Fig. 7a and b. On the other hand, Thompson (2005) has argued that the mechanical work done during cell wall extension is greater than the total hydrogen bond energy of all interactions between hemicellulose and cellulose microfibrils so that these interactions should not limit growth rates. His conclusion was that cell growth should be limited by spacers between fibrils (Fig. 7c) rather than by tethers. Another observation is that reduced relative permittivity inhibits both extension and contraction of cell wall material, whereas the tethered model would predict only inhibition of extension. The discussion of Fig. 6 suggests the conclusion that both processes could play a role successively in the growth process. Indeed, with increasing angle α, a gradual transition between both concepts would be favourable. At very small α (that is, at the beginning of cell extension), the dominant process would be the stretching of tethers (loosened by cell wall alteration) between fibrils (Marga et al. 2005). At larger α (that is, at later stages in cell growth), the matrix needs to be compressed laterally and softening of the spacers would be crucial (Thompson 2005).

A very interesting perspective is also the possible existence of a geometrically determined maximum cell extension (Fig. 6). The general view is that

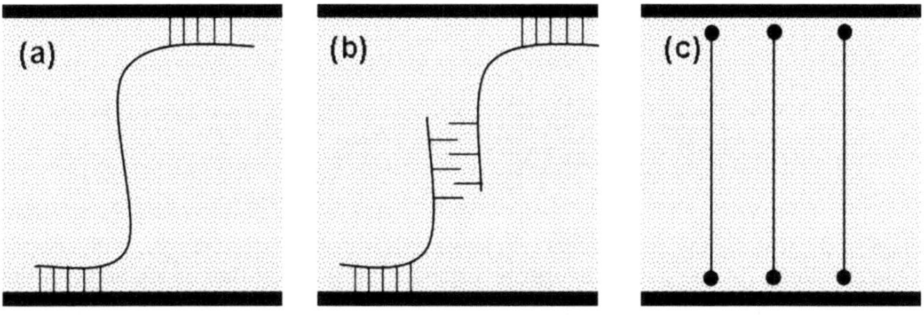

Fig. 7 Xyloglucan: Tethers or struts? Xyloglucans join fibrils by sticking them together, forming either a continuous bridge (**a**), a discontinuous bridge held together by hydrogen bonds (**b**), or held separate by spacers (**c**)

creep of the cell wall matrix starts as soon as the turgor pressure exceeds the yield stress of the matrix. This is not done by increasing the turgor pressure but by lowering the yield stress by biochemical modification of the cell wall. The question, which is usually not addressed, is how the creep process is finally brought to an end (before the failure of the cell wall). One possibility is again a reduction of turgor pressure or the addition of newly synthesized material on the inside of the cell. All this requires a great deal of control over these complex processes. Here the cellulose fibrils could play an interesting role. No matter how low the yield stress of the matrix might be, the cellulose fibrils would eventually constrain the longitudinal growth by tilting into the growth direction and by compressing the cell laterally. Hence, the final length of the cell would be determined solely by the initial cellulose angle α_0 (Fig. 6).

Finally, the ideas presented above depend to a large extent on the (plastic) inextensibility of the cell wall in the direction of the fibrils. Given the fact that cellulose fibrils are rather short compared to the perimeter of the cell (a length of $\sim 7\,\mu m$ has been estimated by Somerville et al. 2004), there is the question of how strong the geometrical boundary condition imposed by the cellulose fibrils really is. Usually, one would not expect a possible plastic elongation of cellulose fibrils. However, when the fibrils are significantly shorter than the cell perimeter, a (plastic) elongation of the cell wall material in the direction of the cellulose fibrils becomes at least conceivable by the mechanism outlined in Fig. 8.

The conclusion would be that the geometric constraints (as discussed in Figs. 6 and 7) would have to be somewhat relaxed because the length of the

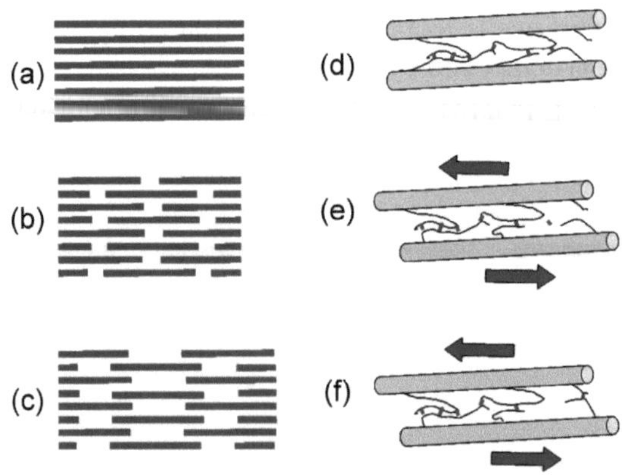

Fig. 8 A set of very long and parallel cellulose fibrils is hardly extensible (**a**). With shorter cellulose fibrils (**b**) the composite can be extended by shearing the matrix between the fibrils. Such a (permanent) shearing (**c**) might be possible by opening and closing of hydrogen bonds (*red bars*) between the fibrils (**d** to **f**). Note that most parts of the matrix have deformed by bending of the tethers, while one bond on the right side has opened and reformed in a different position

fibrils could then increase. This applies to both the lateral compression of the cell and to the maximum elongation imposed by the cellulose. However, there is no reason to assume that the yield stress for the elongation of the cell wall material in the fibril direction (via the shear flow of the matrix between fibrils) should be the same as for the longitudinal creep of the matrix. Currently there is no clear experimental evidence which supports or disproves such a hypothesis.

6
Conclusion and Outlook

The various mechanical test assays reviewed can provide valuable information on the intrinsic properties of the primary cell wall but the crucial question on the mechanism of cell wall expansion can not be fully answered. The plant cell wall shows viscoelastic rather than plastic material behaviour in all mechanical test protocols, but cell wall expansion must be based on viscoplastic deformation behaviour. If changes in mechanical properties due to auxin treatment or acidification were related to the viscoelastic behaviour of the wall, they would not necessarily account for the irreversible deformation during wall extension (Nolte and Schopfer 1997). However, Suslov and Verbelen (2006) recently argued that the relative reversibility of deformation can be determined when different stress states of the cell wall are compared. Regardless of this, a current dilemma of mechanical tests on plant cell walls is that "extensibility" of the cell wall, which according to the Lockhart equation is a measure of pure plastic deformation behaviour, is still difficult to be filtered out of the mechanical test assays. Likewise, blue light irradiation does not influence the stiffness of isolated tissue but decreased "wall extensibility" as measured by the pressure block technique (Cosgrove 1988). Hence, the specific cell wall loosening processes by means of biochemical alteration of the wall (Cleland 1984; Cosgrove 1993) must turn long-term viscoelastic behaviour into plastic deformability.

An approach that should find further consideration is to alter turgor pressure at different temperature levels while simultaneously recording growth (Proseus et al. 1999, 2000). In so doing the authors were able to distinguish between growth and elastic deformation of the cell wall of internode cells of *Chara* in the course of cell enlargement.

Moreover, mechanical test protocols gain only little insight into the structural functionality and mechanical interaction of the cell wall polymers. Here loading/unloading cycles (Fig. 2) at different strain rates may help to shed light on the interrelation between mechanical properties and cell wall composition at the molecular level. Further knowledge on the structure–function relationships of the cell wall composite may be expected from mechanical investigations on genetically modified plant material. Micromechanical tests on *Arabidopsis* hypocotyls with an altered xyloglucan side chain structure showed that defects in the xyloglucan-specific fucosyltransferase (*mur2* mutant) (Vanzin et al. 2002) caused only minor changes in stiffness and strength, whereas a defect in galactosyltransferase (*mur3* mutant) (Madson et al. 2003) showed a drastic decrease in the mechanical performance of the cell wall (Pena et al. 2004; Ryden et al. 2003).

In terms of the cell wall deformation models it seems evident that the geometrical constraints imposed by (plastically) inextensible cellulose fibrils have a profound effect on the cell growth behaviour. This would not be evident at small extensions (up to $\sim 30\%$) where the fibrils are not yet expected to tilt appreciably into the cell direction. Most probably, the plastic flow of non-cellulosic matrix (where the yield point was reduced by a biochemical alteration of the cell wall) dominates the deformation at this stage (Marga et al. 2005). For much larger extensions (beyond 100%), the tilting of the cellulose becomes important, leading to a stiffening of the cell in the longitudinal direction and a lateral compression of the cell. It is predicted here that these constraints should even lead to a maximum extension of the cell, depending only on the initial orientation of the cellulose. This prediction has not yet been checked in experiments, however. Finally, there might be a mechanism for plastic extension of the cell wall in the fibril direction, if the length of the cellulose fibrils is small enough. New types of experiments would be needed to clarify whether such a mechanism plays an important role in the physiological cell extension process.

Acknowledgements We thank Tobias I. Baskin for helpful comments and suggestions.

References

Astley OM, Chanliaud E, Donald AM, Gidley MJ (2003) Tensile deformation of bacterial cellulose composites. Int J Biol Macromol 32:28–35

Bacic A, Harris PJ, Stone BA (1998) Structure and function of plant cell walls. In: Priess J (ed) The biochemistry of plants. Academic, New York, pp 297–371

Baskin TI (2005) Anisotropic expansion of the plant cell wall. Annu Rev Cell Dev Biol 21:203–222

Boyer JS (2001) Growth-induced water potentials originate from wall yielding during growth. J Exp Bot 52:1483–1488

Brummell DA, Hall JL (1985) The role of cell wall synthesis in sustained auxin-induced growth. Physiol Plant 63:406–412

Bunce JA (1977) Leaf elongation in relation to leaf water potential in soybean. J Exp Bot 28:156–161

Carpita NC (1987) The biochemistry of "growing" cell walls. In: Cosgrove DJ, Knievel DP (eds) Physiology of cell expansion during plant growth. Am Soc Plant Physiol, Rockville MD, pp 28–45

Carpita NC, Gibeaut DM (1993) Structural models of primary cell walls in flowering plants: consistency of molecular structure with the physical properties of the walls during growth. Plant J 3:1–30

Cleland RE (1971a) Cell wall extension. Ann Rev Plant Physiol 22:179–222
Cleland RE (1971b) The mechanical behavior of isolated *Avena* coleoptile walls subjected to constant stress. Plant Physiol 47:805–811
Cleland R (1984) The Instron technique as a measure of immediate-past wall extensibility. Planta 160:514–520
Cleland R, Cosgrove DJ, Tepfer M (1987) Long-term acid-induced wall extension in an in-vitro system. Planta 170:379–385
Coleman MM, Painter PC (1995) Hydrogen bonded polymer blends. Prog Polym Sci 20:1–59
Cosgrove DJ (1985) Cell wall yield properties of growing tissue. Plant Physiol 78:347–356
Cosgrove DJ (1986) Biophysical control of plant cell growth. Ann Rev Plant Physiol 37:377–405
Cosgrove DJ (1987) Wall relaxation and the driving force for cell expansive growth. Plant Physiol 84:561–564
Cosgrove DJ (1988) Mechanism of rapid suppression of cell expansion in cucumber hypocotyls after blue-light radiation. Planta 176:109–116
Cosgrove DJ (1989) Characterization of long-term extension of isolated cell walls from growing cucumber hypocotyls. Planta 177:121–130
Cosgrove DJ (1993) Wall extensibility: Its nature, measurement and relationship to plant cell growth. New Phytol 124:1–23
Cosgrove DJ (1999) Enzymes and other agents that enhance cell wall extensibility. Annu Rev Plant Physiol Plant Mol Biol 50:391–417
Cosgrove DJ (2000a) Expansive growth of plant cell walls. Plant Physiol Biochem 38:109–124
Cosgrove DJ (2000b) Loosening of plant cell walls by expansins. Nature 407:321–326
Cosgrove DJ (2001) Wall structure and wall loosening. A look backwards and forwards. Plant Physiol 125:131–134
Cosgrove DJ (2005) Growth of the plant cell wall. Nat Rev 6:850–861
Cosgrove DJ, Van Volkenburgh E, Cleland RE (1984) Stress relaxation of cell walls and the yield threshold for growth: demonstration and measurement by micro-pressure probe and psychrometer techniques. Planta 162:46–52
Cutillas-Iturralde A, Lorences EP (1997) Effect of xyloglucan oligosaccharides on growth, viscoelastic properties, and long-term extension of pea shoots. Plant Physiol 113:103–109
Emons AM, Mulder BM (1999) The making of the architecture of the plant cell wall: How cells exploit geometry. Plant Biol 95:7215–7219
Emons AM, Mulder BM (2000) How the deposition of cellulose microfibrils builds cell wall architecture. Trends Plant Sci 5:35–40
Fratzl P, Burgert I, Gupta HS (2004a) On the role of interface polymers for the mechanics of natural polymeric composites. Phys Chem Chem Phys 6:5575–5579
Fratzl P, Burgert I, Keckes J (2004b) Mechanical model for the deformation of the wood cell wall. Z Metallkd 95:579–584
Fricke W, Chaumont F (2007) Solute and water relations of growing plant cells. In: Verbelen JP, Vissenberg K (eds) The expanding cell. Plant Cell Monographs, vol 5. Springer, Berlin Heidelberg New York (in press)
Fry-Wyssling A (1953) Submicroscopic morphology of protoplasm. Elsevier, Amsterdam, pp 279–292
Fry SC (1986) Cross-linking of matrix polymers in the growing cell walls of angiosperms. Ann Rev Plant Physiol 37:165–186
Fry SC (1989) Cellulases, hemicelluloses and auxin-stimulated growth – a possible relationship. Physiol Plant 75:532–536

Fry SC (1995) Polysaccharide-modifiying enzymes in the plant cell wall. Annu Rev Plant Physiol Plant Mol Biol 46:497–520

Fry SC, Smith RC, Renwick KF, Martin DJ, Hodge SK, Matthews KJ (1992) Xyloglucan endotransglycosylase, a new wall-loosening enzyme activity from plants. Biochem J 282:821–828

Gertel ET, Green PB (1977) Cell growth pattern and wall fibrillar arrangement: experiments with *Nitella*. Plant Physiol 60:247–254

Green PB (1960) Multinet growth in the cell wall of *Nitella*. J Biophys Biochem Cytol 7:289–296

Ha MA, Apperley DC, Jarvis MC (1997) Molecular rigidity in dry and hydrated onion cell walls. Plant Physiol 115:593–598

Haasen P (1986) Physical metallurgy, 2nd edn. Cambridge University Press, Cambridge

Hayashi T (1989) Xyloglucans in the primary cell wall. Annu Rev Plant Physiol Plant Mol Biol 40:139–168

Hohl M, Schopfer P (1992) Physical extensibility of maize coleoptile cell walls: apparent plastic extensibility is due to elastic hysteresis. Planta 187:498–504

Hohl M, Schopfer P (1995) Rheological analysis of viscoelastic cell-wall changes in maize coleoptiles as affected by auxin and osmotic-stress. Physiol Plant 94:499–505

Jarvis MC (1992) Control of thickness of collenchyma cell walls by pectins. Planta 187:218–220

Keckes J, Burgert I, Frühmann K, Müller M, Kölln K, Hamilton M, Burghammer M, Roth SV, Stanzl-Tschegg SE, Fratzl P (2003) Cell-wall recovery after irreversible deformation of wood. Nat Mater 2:810–814

Keegstra K, Talmadge KW, Bauer WD, Albersheim P (1973) The structure of plant cell walls III: a model of the walls of suspension-cultured sycamore cells based on the interconnections of the macromolecular components. Plant Physiol 51:188–196

Kerstens S, Decraemer WF, Verbelen JP (2001) Cell walls at the plant surface behave mechanically like fiber-reinforced composite materials. Plant Physiol 127:381–385

Kutschera U, Briggs WR (1987) Differential effect of auxin on in vivo extensibility of cortical cylinder and epidermis in Pea internodes. Plant Physiol 84:1361–1366

Kutschera U, Schopfer P (1986a) Effect of auxin and abscisic acid on cell wall extensibility in maize coleoptiles. Planta 167:527–535

Kutschera U, Schopfer P (1986b) In-vivo measurement of cell-wall extensibility in maize coloeptiles: effects of auxin and abscisic acid. Planta 169:437–442

Lindsay SE, Fry SC (2007) Redox and wall-restructuring. In: Verbelen JP, Vissenberg K (eds) The expanding cell. Plant Cell Monographs, vol 5. Springer, Berlin Heidelberg New York (in press)

Liszkay A, Kenk B, Schopfer P (2003) Evidence for the involvement of cell wall peroxidase in the generation of hydroxyl radicals mediating extension growth. Planta 217:658–667

Lockhart JA (1965) An analysis of irreversible plant cell elongation. J Theor Biol 8:264–275

Madson M, Dunand C, Li XM, Verma R, Vanzin GF, Calplan J, Shoue DA, Carpita NC, Reiter WD (2003) The *mur3* gene of *Arabidopsis* encodes a xyloglucan galactosyltransferase that is evolutionarily related to animal exostosins. Plant Cell 15:1662–1670

Marga F, Grandbois M, Cosgrove DJ, Baskin TI (2005) Cell wall extension results in the coordinate seperation of parallel microfibrils: evidence from scanning electron microscopy and atomic force microscopy. Plant J 43:181–190

McCann MC, Roberts K (1991) Architecture of the primary cell wall. In: Loyd CW (ed) The cytosceletal basis of plant growth and form. Academic, New York, pp 109–129

McCann MC, Wells B, Roberts K (1992) Complexitiy in the spatial localization and length distribution of plant cell-wall matrix polysaccharides. J Microsc 166:123–136
McNeil M, Darvill AG, Fry SC, Albersheim P (1984) Structure and function of the primary cell walls of plants. Ann Rev Biochem 53:625–663
McQueen-Mason SJ, Cosgrove DJ (1995) Expansin mode of action on cell walls. Plant Physiol 107:87–100
McQueen-Mason SJ, Durachko DM, Cosgrove DJ (1992) Two endogenous proteins that induce cell wall extension in plants. Plant Cell 4:1425–1433
McQueen-Mason SJ, Le NT, Brocklehurst D (2007) Expansins. In: Verbelen JP, Vissenberg K (eds) The expanding cell. Plant Cell Monographs, vol 5. Springer, Berlin Heidelberg New York (in press)
Métraux JP, Richmond PA, Taiz L (1980) Control of cell elongation in *Nitella* by endogenous cell wall pH gradients. Plant Physiol 65:204–210
Nakahori K, Katou K, Okamoto H (1991) Auxin changes both the extensibility and the yield threshod of the cell wall of *Vigna* hypocotyls. Plant Cell Physiol 32:121–129
Niklas KJ (1992) Plant biomechanics. University of Chicago Press, Chicago
Nishitani K, Vissenberg K (2007) Roles of the XTH protein family in the expanding cell. In: Verbelen JP, Vissenberg K (eds) The expanding cell. Plant Cell Monographs, vol 5. Springer, Berlin Heidelberg New York (in press)
Nolte T, Schopfer P (1997) Viscoelastic versus plastic cell wall extensibility in growing seedling organs: a contribution to avoid some misconceptions. J Exp Bot 48:2103–2107
Nonami H, Boyer JS (1990) Wall extensibility and cell hydraulic conductivity decrease in enlarging stem tissues at low water potentials. Plant Physiol 93:1610–1619
O'Looney N, Fry SC (2005) A simple apparatus for measuring long-term extension of plant cell walls subjected to tensile stress. Plant Biosyst 139:102–106
Passioura JB (1994) The physical chemistry of the primary cell wall: implications for the control of expansion rate. J Exp Bot 45:1675–1682
Pauly M, Albersheim P, Darvill AG, York WS (1999) Molecular domains of the cellulose/xyloglucan network in the cell walls of higher plants. Plant J 20:629–639
Pena MJ, Ryden P, Madson M, Smith AC, Carpita NC (2004) The galactose residues of xyloglucan are essential to maintain mechanical strength of the primary cell walls in Arabidopsis during growth. Plant Physiol 134:443–451
Peters WS, Hagemann W, Tomos D (2000) What makes plants different? Principles of extracellular matrix function in soft plant tissues. Comp Biochem Physiol A 125:151–167
Preston R (1974) The physical biology of plant cell walls. Chapmann and Hall, London, UK
Preston RD (1982) The case of multinet growth in growing walls of plant cells. Planta 155:356–363
Pritchard J, Hetherington PR, Fry SC, Tomos AD (1993) Xyloglucan endotransglycosylase activity, microfibril orienation and the profiles of cell wall properties along growing regions of maize roots. J Exp Bot 44:1281–1289
Pritchard J, Jones RGW, Tomos AD (1990) Measurement of yield threshold and cell wall extensibility of intact wheat roots under different ionic, osmotic and temperature treatments. J Exp Bot 41:669–675
Proseus TE, Boyer JS (2005) Turgor pressure moves polysaccharides into growing cell walls of *Chara corallina*. Ann Bot 95:967–979
Proseus TE, Zhu G-L, Boyer JS (2000) Turgor, temperature and the growth of plant cells: using *Chara corallina*. as a model system. J Exp Bot 51:1481–1494
Proseus TE, Ortega JKE, Boyer JS (1999) Separating growth from elastic deformation during cell enlargement. Plant Physiol 119:775–784

Rayle DL, Cleland RE (1972) The *in-vitro* acid-growth response: relation to *in-vivo* growth responses and auxin action. Planta 104:282–296

Rayle DL, Cleland RE (1992) The acid growth theory of auxin-induced elongation is alive and well. Plant Physiol 99:1271–1274

Reiterer A, Jakob HF, Stranzl-Tschegg SE, Fratzl P (1998) Spiral angle of elementary cellulose fibrils in cell walls of *Picea abies* determined by small-angle X-ray scattering. Wood Sci Technol 32:335–345

Richmond PA (1983) Patterns of cellulose microfibril deposition and rearrangement in *Nitella*: in vivo analysis by a birefringance index. J Appl Polym Sci: Applied Polymer Symposium 37:107–122

Roelofson PA (1965) Ultrastructure of the wall in growing cells and its relation to the direction of the growth. Adv Bot Res 2:69–149

Ryden P, Sugimoto-Shirasu K, Smith AC, Findlay K, Reiter W-D, McCann MC (2003) Tensile properties of *Arabidopsis* cell walls depend on both a xyloglucan cross-linked microfibrillar network and rhamnogalacturonan II-borate complexes. Plant Physiol 123:1033–1040

Schindler TM (1998) The new view of the primary cell wall. Z Pflanzenernähr Bodenk 161:499–508

Schopfer P (2001) Hydroxyl radical-induced cell wall loosening in vitro and in vivo: implications for the control of elongation growth. Plant J 28:679–688

Schopfer P, Liszkay A, Bechtold M, Frahry G, Wagner A (2002) Evidence that hydroxyl radicals mediate auxin-induced extension growth. Planta 214:821–828

Sellen DB (1983) The response of mechanically anisotropic cylindrical cells to multiaxial stress. J Exp Bot 34:681–687

Somerville C, Bauer S, Brininstool G, Facette M, Hamann T, Milne J, Osborne E, Paredez A, Persson S, Raab T, Vorwerk S, Youngs H (2004) Toward a systems approach to understanding plant cell walls. Science 306:2206–2211

Suslov D, Verbelen JP (2006) Cellulose orientation determines mechanical anisotropy in onion epidermis cell walls. J Exp Bot 57:2183–2192

Taiz L (1984) Plant cell expansion: regulation of cell wall mechanical properties. Ann Rev Plant Physiol 35:585–657

Takeda T, Furata Y, Awano T, Mizuno K, Mitsuishi Y, Hayashi T (2002) Suppression and acceleration of cell elongation by integration of xyloglucans in pea stem segments. Proc Natl Acad Sci USA 99:9055–9060

Talbott LD, Ray PM (1992) Molecular size and separability features of Pea cell wall polysaccharides. Plant Physiol 98:357–368

Tepfer M, Cleland RE (1979) A comparison of acid-induced cell wall loosening in *Valonia ventricosa* and in oat coleoptiles. Plant Physiol 63:898–902

Thompson DS (2005) How do cell walls regulate plant growth. J Exp Bot 56:2275–2285

Tomos D, Pritchard J (1994) Biophysical and biochemical control of cell expansion in roots and leaves. J Exp Bot 45:1721–1731

Vanzin GF, Madson M, Carpita NC, Raikhel NV, Keegstra K, Reiter WD (2002) The *mur2* mutant of *Arabidopsis thaliana* lacks fucosylated xyloglucan because of a lesion in fucosyltransferase AtFUT1. Proc Natl Acad Sci USA 99:3340–3345

Veytsmann BA, Cosgrove DJ (1998) A model of cell wall expansion based on thermodynamics of polymer networks. Biophysical J 75:2240–2250

Vincent JFV (1990) Structural biomaterials. Princeton University Press, Princeton, New Jersey

Vincken JP, York WS, Beldman G, Voragen AGJ (1997) Two general branching patterns of xyloglucan, XXXG and XXGG. Plant Physiol 114:9–13

Whitney SEC, Gidley MJ, McQueen-Mason SJ (2000) Probing expansion action using cellulose/hemicellulose composites. Plant J 22:327–334
Wilson RH, Smith AC, Kacurakova M, Saunders PK, Wellner N, Waldron KW (2000) The mechanical properties and molecular dynamics of plant cell wall polysaccharides studied by Fourier-transform infrared spectroscopy. Plant Physiol 124:397–405
Yamamoto R, Masuda Y (1971) Stress relaxation properties of the *Avena* coleoptile cell wall. Physiol Plant 25:330–335
Yamamoto R, Shinozaki K, Masuda Y (1970) Stress-relaxation properties of plant cell walls with special reference to auxin action. Plant Cell Physiol 11:947–956
Zwieniecki MA, Melcher PJ, Holbrook NM (2001) Hydrogel control of xylem hydraulic resistance in plants. Science 291:1059–1062

Plant Cell Monogr (5)
J.-P. Verbelen · K. Vissenberg: The Expanding Cell
DOI 10.1007/7089_2006_077/Published online: 31 October 2006
© Springer-Verlag Berlin Heidelberg 2006

The Cytoskeleton and Co-Ordination of Directional Expansion in a Multicellular Context

Geoffrey O. Wasteneys[1] (✉) · David A. Collings[2]

[1]Department of Botany, University of British Columbia, 3529-6270 University Blvd., Vancouver, BC V6T 1Z4, Canada
geoffwas@interchange.ubc.ca

[2]Plant Cell Biology Group, Research School of Biological Sciences, The Australian National University, GPO Box 475, ACT 2601 Canberra, Australia

Abstract The cytoskeleton governs many critical processes in expanding plant cells, including the delivery of wall components and the establishment and maintenance of growth direction. This work describes how cytoskeletal arrays assemble, and how their spatial organization and dynamics regulate the anisotropic properties of plant cell walls. We describe the mechanisms that construct and organize transverse microtubule arrays, and explore how these arrays, and the direction of elongation, are influenced by hormones. We then consider how cortical microtubules regulate the mechanical properties of the load-bearing cellulose microfibrils, through interacting with cellulose synthase complexes, and by coordinating the secretion of wall proteins. Actin microfilaments form part of the machinery that controls polar auxin transport, and have critical functions in vesicle transport. In recent years, it has become increasingly clear that microtubules and actin microfilaments work in concert to coordinate cell expansion. This microfilament-microtubule coordination is mediated through the activity of Rop GTPase signalling switches. We highlight this process in the growth of pavement cells found in the epidermal layers of leaves.

1 Introduction

The cytoskeleton is essential for all processes governing plant cell expansion. These processes include the transport of wall material and complexes that secrete or synthesize wall material at the plasma membrane, to the cell periphery, a process that relies on robust networks of actin microfilaments along which myosin-bound Golgi stacks and Golgi-derived vesicles track. Extensive arrays of microtubules also line the plasma membrane, and play a key role in coordinating the direction of cell expansion in diffusely expanding cells. This role is intimately linked to regulating the mechanical properties of load-bearing cellulose microfibrils, which, like microtubules, are almost always oriented perpendicular to the elongation axis. It would be too simplistic, however, to conclude that this is the extent of cytoskeletal involvement in cell expansion or that microfilaments and microtubules work separately. Indeed, as microscopy improves and molecular-genetic studies uncover numerous accessory proteins and regulatory pathways, we are coming to realize that many aspects of cell expansion are integrated by the cytoskeleton and we start to appreciate how the activities of the actin filament and microtubule networks are finely balanced and integrated.

The hallmark of plant development is axis formation and, although this is most apparent in stems and roots, it is common to most organs and to many tissues such as leaf veins. Axis formation involves the collective elongation of cells in the tissue layers that make up organs. The bulk of growth takes place in the lateral walls, which may extend to many times their original length, while end walls expand relatively little or in some cases not at all. To achieve growth anisotropy at the cellular level, end walls and lateral walls need distinct mechanical properties, and thus the physiology of cell ends must be fundamentally different to the lateral regions. These differences are likely to include variations in membrane properties (see Fricke and Chaumont 2006, in this volume), exo- and endocytic activities, proton pump activity, and in the cell wall polysaccharide (see Hématy and Höfte 2006, in this volume; Obel et al. 2006, in this volume; Verhertbruggen and Knox 2006, in this volume) and protein composition (see Nishitani and Vissenberg 2006, in this volume; McQueen-Mason et al. 2006, in this volume). In considering plant cell expansion in a multicellular context, our primary emphasis in this work will concern how microtubule- and actin microfilament-dependent mechanisms coordinate these differences to regulate axial growth. In addition to the narrow cylindrical cells that comprise elongating organs like roots and stems, we also look at interdigitating pavement cells at the surface of leaves and leaf-like organs.

2
Cortical Microtubule Organization in the Expansion Domain

Before the onset of rapid cell expansion, a cortical microtubule array is established. When cytokinesis is complete, microtubules form a radial pattern extending from the nucleus to the cell cortex (Schmit 2002; Wasteneys 2002). Isolated nuclei also retain the capacity to nucleate microtubules (Stoppin et al. 1994), a property that convinced many researchers that microtubules are exclusively nucleated at the nuclear envelope. However, stringent, semi-in vitro assembly assays (Wasteneys et al. 1989) and the imaging of microtubule assembly in living cells (Shaw et al. 2003; Wasteneys et al. 1993; Yuan et al. 1994) demonstrate that microtubules are also assembled at the plasma membrane. As in all eukaryotic cells, microtubule initiation in plant cells is highly regulated both spatially and temporally by nucleating complexes that contain γ-tubulin and its associated proteins (Schmit 2002). But unlike stereotypical eukaryotic cells that have well-defined microtubule organizing centres (centrosomes in animal cells, spindle pole bodies in yeast cells), the microtubule initiating material in higher plant cells is highly dispersed (Wasteneys 2002). Centrosomes are not found in plant cells, with the exception of the motile sperm cells of lower plants and some gymnosperms.

2.1 Microtubule-Dependent Microtubule Assembly May Be the Key to Dispersal of Cortical Microtubule Arrays

Early investigations of microtubule organization patterns during cell cycle transitions (Bajer and Mole-Bajer 1986) and during recovery from drug-induced disassembly (Falconer et al. 1988; Wasteneys and Williamson 1989a), described the appearance of microtubules in small clusters or foci. These have been variously described as fir trees (Bajer and Mole-Bajer 1986), V-shaped configurations (Falconer et al. 1988), branching clusters (Wasteneys and Williamson 1989a), converging centres (Bajer and Mole-Bajer 1986), spinifex texture (Wasteneys 1992) and fractal trees (Wasteneys 2002). Identification of equivalent microtubule patterns in non-perturbed cells (Wasteneys and Williamson 1989a) led to the hypothesis that cortical microtubules normally initiate along pre-existing microtubules where "microtubule initiating factors" are positioned (Wasteneys 1992; Wasteneys and Williamson 1989a). This branch-form assembly is illustrated in Fig. 1A,B. Immunolabelling indicates that γ-tubulin is located in a punctate manner along cortical microtubules rather than being concentrated at one end (Drykova et al. 2003; Liu et al. 1994; McDonald et al. 1993; Stoppin-Mellet et al. 2000), and γ-tubulin appears to colocalize with the Spc98p homologue in the cell cortex (Erhardt et al. 2002). These observations led to the prediction that the initiation factors found along existing microtubules would contain γ-tubulin and associated proteins (Wasteneys 2002). This prediction was supported by the suppression of branch-form assembly on exposed plasma membrane ghosts by γ-tubulin-specific antibodies (Murata et al. 2005). It remains to be determined how γ-tubulin complexes are recruited to the sides of microtubules, though one model predicts the recycling and transport of minus-end complexes as part of an efficient and stochastic dispersal mechanism (Wasteneys 2002).

Mutational and transgenic analyses confirm that γ-tubulin is essential for the establishment and maintenance of cortical microtubule arrays. In *Arabidopsis thaliana*, two genes, TUBG1 and TUBG2, encode γ-tubulins. Their functions appear to be redundant (Pastuglia et al. 2006). All single homozygous mutant lines develop normally but knocking out both TUBG genes is lethal. The combination of the *tubg1-1* null allele and the low expressing *tubg2-2* allele produces viable embryos but seedlings die within three weeks of germination after severe disruption of microtubule arrays, meristem loss and radial swelling (Pastuglia et al. 2006). Similarly, dexamethasone-inducible RNAi reveals that when knockdown is nearly complete, micro-

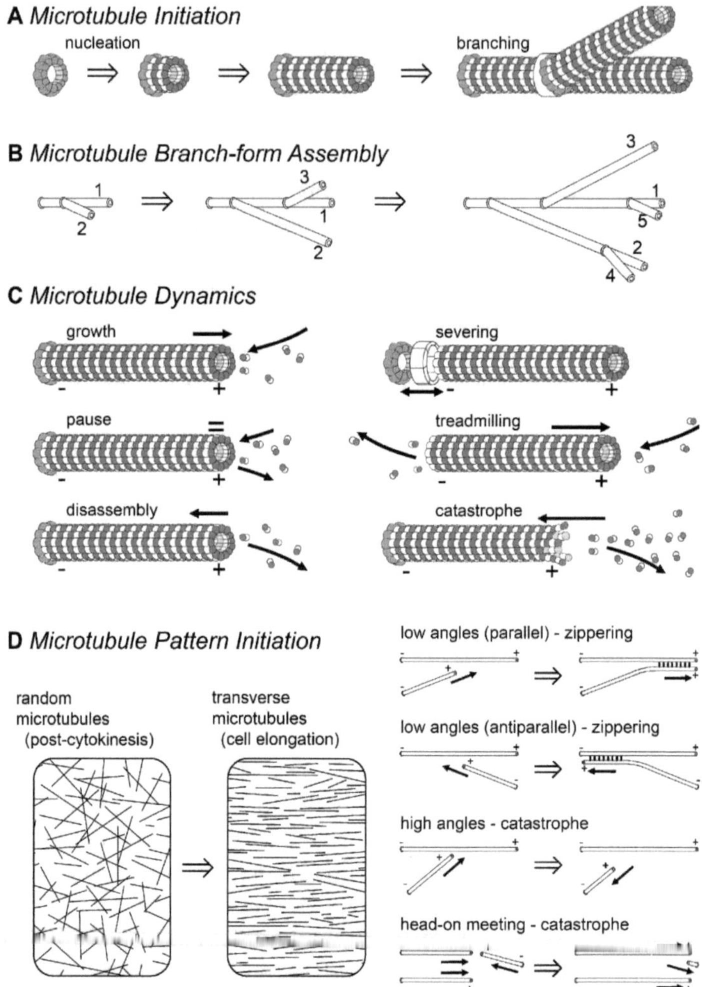

Fig. 1 Establishing a transverse, parallel array of cortical microtubules is regulated by microtubule assembly dynamics and microtubule interactions. **A** Microtubule nucleation in the cortex requires γ-tubulin-containing complexes. In addition to microtubule assembly at isolated sites, nucleating complexes can also be found along existing microtubules, leading to branch-form assembly. **B** Branch-form assembly produces new microtubules in a variety of orientations, distal to the original nucleation event that produced the microtubule labelled 1. **C** Microtubule dynamics is dependent on the exchange of tubulin dimers, which is most active at the plus end. Plus-end dynamics are regulated by microtubule associated proteins and the hydrolysis of β-tubulin bound GTP. Net subunit exchange determines the rate of growth, disassembly, and the incidence of pausing and catastrophe. Katanin ATPase-dependent microtubule severing near the minus end may allow tubulin dimer loss from the minus end, resulting in treadmilling of microtubules. **D** From a disordered post-cytokinetic microtubule array, transverse orientation is established and maintained by microtubule encounters. Low-angle encounters may result in zippering by cross-linking proteins, whereas high-angle encounters may result in catastrophe

tubules are lost, whereas partial depletion of γ-tubulin causes loss of parallel order in the cortical microtubule array (Binarova et al. 2006). Thus, γ-tubulin is not only essential for microtubule formation but may also play other roles in the organization of microtubules.

2.2
Consolidation of Microtubules Into a Transverse Array—Selective Stabilization

After nucleation and assembly, the next level of cortical microtubule organization involves formation and maintenance of parallel arrays that lie perpendicular to the cell's long axis during the most rapid phase of cell expansion (Sugimoto et al. 2000). Behavioural analysis of the cortical microtubule array, in fixed and living cells, has led to a series of models that explain how a transverse order is established and maintained (Fig. 1D).

An important consideration is whether microtubule orientation depends on the direction in which they nucleate and assemble. An early immunofluorescence-based study using Nitella giant internodal cells established that microtubule assembly and orientation can be distinct processes (Wasteneys and Williamson 1989b). During recovery from drug-induced microtubule disruption, microtubules initially assembled in predominantly transverse directions. Branching clusters soon developed from the first formed microtubules, causing microtubule orientation to become increasingly dispersed. Restoration of parallel order only began once maximum microtubule polymer levels were reached. After this point, microtubule branching clusters were less apparent, and increasing transverse order was observed until cortical microtubule organization was the same as in untreated control cells. This analysis led to two alternative models for achieving transverse order from microtubules assembled at various angles (Wasteneys and Williamson 1989b). The biased turnover model considered that microtubules assembled in unfavourable orientations would disassemble, while those assembling in transverse orientations would be selectively stabilized. The alternative reorientation model suggested microtubules could be reoriented by motor proteins after assembly.

The ability to observe fluorescently labelled cortical microtubules in living cells, first by microinjection of fluorescent tubulin and later by transgenic expression of fluorescent fusion-tagged tubulins or microtubule-binding proteins, has provided a clearer picture of the dynamic processes of cortical microtubules. Early glimpses of cortical microtubules in living plant cells revealed just how dynamic microtubules were, with the incorporation of brain tubulin throughout the full length of microtubules within minutes of injection (Wasteneys et al. 1993), and fluorescence recovery after photobleaching measuring microtubule half-life times similar if not shorter than those recorded in animal cells (Hush et al. 1994).

There is no evidence in plant cells for any large scale or rapid microtubule translocation that would suggest motor proteins assist microtubule movement. Photobleaching regions of GFP-tubulin-containing cortical microtubules reveals that tubulin subunits remain in place relative to the plasma membrane, suggesting that the slow migration of cortical microtubules is a form of treadmilling in which tubulin subunits are lost from the minus end of microtubules, while the plus ends can either extend or undergo rapid disassembly (Fig. 1C). These observations in intact plant tissues (Shaw et al. 2003), along with similar ones in suspension cultured cells (Vos et al. 2004), included many instances of encounters between microtubules that resulted in microtubule bundling. Dixit and Cyr (2004a) predicted that encounters between cortical microtubules play a major role in the organizational state of the cortical array (Fig. 1D). Their model incorporates several testable hypotheses including predictions that bundling will be favoured by microtubule encounters at shallow angles whereas a microtubule encountering the side of the nearby microtubule at a steeper angle may undergo catastrophe. Encounter frequency is predicted to be reduced when microtubules no longer lie in the same plane (i.e., should they become detached from the plasma membrane) or when microtubules are less dynamic. Drug treatments and mutations affecting the function of tubulin or microtubule-associated proteins that yield short microtubules will also reduce the frequency of encounters and therefore inhibit the maintenance of parallel order. Computer simulations demonstrate that self-organization into parallel cortical arrays requires the right combination of polymerization to increase the frequency of encounters, and stabilization events to consolidate the array (Dixit and Cyr 2004b).

Observations in living cells do appear to demonstrate that the angle of cortical microtubule encounters influences the response of individual microtubules (Dixit and Cyr 2004b), yet the inherent congestion of cortical microtubule arrays makes this a non-trivial exercise. More recently, Dixit et al. (2006) used moderate EB1-GFP decoration to track the plus ends of cortical microtubules. Previous observation of microtubule arrays indicated that cortical arrays comprised microtubules of mixed polarity. In this study, however, microtubules showed predominant localized polarities, with 70% or more EB1 comets moving in one direction. Those polymerizing in the opposite, or minority direction, grew at the same rate as those in the majority direction but frequently disappeared when they encountered microtubules polymerizing in the majority direction. The authors suggest that this is evidence for existence of a mechanism that selectively stabilizes co-polar microtubules leading to a self-reinforcement of a majority polarity (Fig. 1D). The same study also followed the establishment of a cortical array following cytokinesis in tobacco BY-2 cells. In the newly formed isodiametric cells, microtubules initially had a highly scattered orientation and less than 40% plus-end co-polarity. As time passed, however, microtubules became increasingly oriented in one direction, coincident with an increase in co-polarity. The tendency for the co-polarity of parallel cortical arrays provides an important new insight into the behaviour of cortical microtubules, with broader implications for understanding how microtubule organization regulates the mechanical properties of the expanding cell wall (see Burgert and Fratzl 2006, this volume).

Dynamic turnover of microtubules is regulated by GTP hydrolysis. Both α- and β-tubulins, which dimerize to form the building blocks of microtubules, are monomeric GTPases and depend on GTP binding for polymer formation. Both have GTP-binding sites but the α-tubulin-bound GTP is protected from hydrolysis by the α–β dimerization process. GTP on the β-tubulin subunit, however, is hydrolyzed soon after it binds the α-tubulin subunit of an incoming free dimer. Energy from this hydrolysis event alters the conformation of tubulin dimers, causing a curved protofilament and weaker polymer bonds, and promotes depolymerization of the microtubule. As long as GTP tubulin dimer addition keeps pace with or exceeds the rate of hydrolysis, microtubules can remain intact or continue growing. Losing the GTP cap causes rapid disassembly, yet this seemingly disruptive process accords microtubules the opportunity to be reconfigured.

With nine α- and six β-tubulin genes in the modest *Arabidopsis thaliana* genome, mutations affecting tubulins are unlikely to cause severe phenotypes unless they are dominant. Single amino acid substitutions at residue 180 in TUA4 and 6 shift the orientation of cortical microtubule arrays to a right-handed helical arrangement, and cause left-handed twisting of organs (Thitamadee et al. 2002). These *lefty* mutants are dominant-negative and can only be rescued by further mutagenesis to delete the defective genes; the null alleles identified in a screen for suppressors of the *lefty* mutants have no obvious phenotypes. Hence, the changes in the conformation of the mutant tubulins mimic the effect of destabilizing herbicides. Low concentrations of the herbicides oryzalin and propyzamide, which also generate a right-handed helically arranged cortical array, increase the time microtubules spend in the paused state and generally decrease the amount of microtubule turnover (Nakamura et al. 2004). Similar analysis of *lefty* and other mutants generating left-handed organ twisting is likely to identify similar effects on polymer dynamics.

Right-handed organ twisting is a side-effect in some transgenic lines developed for visualizing microtubules in living cells (Hashimoto 2002; Wasteneys and Collings 2004). Abe and Hashimoto (2005) determined that adding even very short peptides to the N-terminus of α-tubulin suppresses microtubule dynamics, increases polymer formation and generates shallow-pitched left-handed helical microtubule arrays that in turn promote right-handed organ twisting. Their work suggests that interfering with the putative α-tubulin GTPase-activating domain inhibits the rate of GTP hydrolysis on the β-tubulin of the adjacent dimer, a model that is supported by the increased size of EB1-GFP comets on the dynamically suppressed microtubules (Abe and Hashimoto 2005).

2.3
Microtubule-Associated Proteins Regulate Cortical Microtubule Dynamics and Orientation Patterns

Consolidation of cortical microtubules into specific orientation patterns (Fig. 1D) is mediated by a suite of proteins and their regulatory factors that carry out a combination of processes including microtubule nucleation, assembly, severing, crosslinking, stabilization and disassembly (Hashimoto 2003; Hussey et al. 2002; Wasteneys 2002; Wasteneys and Collings 2004). Many of these proteins are common to eukaryotic cells, while others may be unique to the plant kingdom.

2.3.1
Microtubule Severing is Required for Establishing Transverse Microtubule Organization

The need for microtubule turnover and remodelling is underscored by the role that the microtubule severing protein katanin plays in cortical microtubule organization. Katanins use ATP hydrolysis to sever microtubules and are best known for their activity in severing the minus ends of microtubules at centrosomes, believed to be important for releasing them (Quarmby 2000). Plants have putative homologues of both the 60 kDa (p60) catalytic subunit (McClinton et al. 2001) and the 80 kDa regulatory subunit, p80 (Bouquin et al. 2003). Little is known about the function of katanin p80 in plant cells, though it was shown to interact with the p60 subunit by a yeast 2-hybrid assay (Bouquin et al. 2003). The *Arabidopsis* katanin p60 homologue, AtKSS, has ATP-dependent severing activity in vitro (Stoppin-Mellet et al. 2002), and more recently, evidence for microtubule severing *in planta* has been demonstrated by inducing overexpression of AtKSS with an ethanol-responsive promoter (Stoppin-Mellet et al. 2006).

Katanin mutants have been retrieved from a variety of forward genetics screens in *Arabidopsis* (Bichet et al. 2001; Bouquin et al. 2003; Burk et al. 2001; Webb et al. 2002) and rice (Komorisono et al. 2005). At the onset of elongation growth, katanin mutants fail to establish transverse cortical microtubule arrays, leading to impaired elongation (Bichet et al. 2001), ectopic root hair production (Webb et al. 2002) and weakened cell walls (Burk et al. 2001; Burk and Ye 2002; Ryden et al. 2003). One of the most intriguing phenotypes of katanin mutants is their altered expression of gibberellin biosynthesis genes. The AtKSS *lue1* allele was identified in a screen for high expression of the gene encoding GA-20-oxidase, a critical enzyme in the production of biologically active GA (Bouquin et al. 2003), while the rice *dwarf and gladiolus leaf 1* mutant, *dgl1-1*, also a putative katanin p60 homologue, shows constitutive upregulation of various GA-biosynthesis genes (Komorisono et al. 2005). These studies are discussed in further detail in Sect. 2.4 on hormones, cytoskeleton and wall extensibility.

Structural microtubule-associated proteins (MAPs) play further key roles in organizing the cortical array. These proteins selectively stabilize microtubules by promoting polymerization and by cross-linking microtubules into bundles, and may also link microtubules to the plasma membrane or destabilize them. Probably the most ubiquitous and essential plant MAP is the 217 kDa MOR1, a member of the highly conserved DIS1-TOGp-XMAP215 family of HEAT repeat-containing MAPs. Its function in cortical microtubule organization was discovered through the identification of two alleles, *mor1-1* and *mor1-2*, in which single amino acid substitutions generate temperature-dependent disorganization of the cortical arrays, left-handed twisting of organs and loss of growth anisotropy (Whittington et al. 2001). The putatively null *gemini* alleles generate gametophytic defects and cannot be recovered as homozygotes (Twell et al. 2002), suggesting that *MOR1* is an essential gene. Detailed phenotype and immunolocalization analyses indicate that the MOR1 protein associates along the entire length of cortical microtubules and promotes their elongation (Kawamura et al. 2006). Mechanisms by which MOR1 regulates cortical microtubule assembly have been proposed (Hamada et al. 2004; Hashimoto 2003; Hussey and Hawkins 2001; Wasteneys 2002) though there is no evidence for MOR1 bundling microtubules or interacting with other proteins to regulate their access to microtubules. EB1 is another ubiquitous MAP whose known function in animal and fungal cells as a microtubule plus-end tracker is apparently conserved in plant cells (Bisgrove et al. 2004) and EB1b-GFP labels the growing ends of microtubules, consistent with a role in microtubule plus end dynamics (Dixit et al. 2006).

Other MAPs appear to be plant-specific, and may have functions related to specialized attributes of plant microtubule arrays, such as cortical bundle formation or tropisms. SPR2/TOR1 is a 94 kDa HEAT repeat-containing protein that colocalizes with cortical microtubules and may help regulate their dynamics; loss of function mutations in *Arabidopsis* promote left-handed microtubule helical patterns and right-handed organ twisting (Buschmann et al. 2004; Shoji et al. 2004). The 12 kDa SPR1/SKU6 protein is also important for directional control during rapid expansion (Nakajima et al. 2004; Sedbrook et al. 2004), and *Arabidopsis* mutants display right handed twisting of roots, hypocotyls and petioles (Furutani et al. 2000). Its high expression in rapidly elongating tissue and promotion of elongation and microtubule drug resistance when over-expressed (Nakajima et al. 2004) suggest that SPR1/SKU6 stabilizes cortical microtubule plus ends. Consistent with this, SPR1-GFP labels all microtubule arrays and appears to favour microtubule plus ends, especially those that are polymerizing rapidly (Sedbrook et al. 2004). WAVE-DAMPENED2 (WVD2) and a related protein, WVD2-LIKE1, are also likely to be microtubule-associated. Unlike the tubulin and MAP mutations that generate consistent left or right twisting, overexpression generates right-handed twisting of roots and hypocotyls and left-handed twisting of petioles (Yuen et al. 2003).

The MAP65 proteins comprise a large family; in *Arabidopsis* they are encoded by nine genes. These proteins were first identified from cell cultures by microtubule affinity purification (Chan et al. 1996; Jiang and Sonobe 1993). In vitro, they lack the classic MAP attribute of stimulating microtubule polymerization but instead bundle microtubules through 25 nm cross-links likely formed by protein dimers (Smertenko et al. 2004). At least two MAP65 members have been detected in cortical arrays, where they are likely to be important for bundling and stabilizing a subset of microtubules (Smertenko et al. 2004; Van Damme et al. 2004). When AtMAP65-1-GFP and AtMAP65-5-GFP fusion proteins were expressed at high levels in cultured tobacco BY-2 cells, they preferentially localized with coaligned, apparently bundled microtubules that showed greater resistance to microtubule destabilizing drugs (Van Damme et al. 2004). In the cortical array, MAP65 associations were seen with microtubules co-aligned in both the opposite and same polarity, as judged by tracking elongation events (Fig. 1D).

RIC1 may also play an important role in microtubule bundling, especially in the prominent microtubule band formation that is responsible for the interdigitation of adjacent epidermal pavement cells in many organs (Fu et al. 2005). This protein was identified as a putative effector of the ROP2/4 proteins and is discussed in detail in Sect. 6. Constitutive activity of ROPs or the knock down of RIC expression both lead to the loss of microtubule band formation in pavement cells (Fu et al. 2005). The mechanism for RIC1-dependent microtubule bundling, however, remains to be determined.

2.4
Hormones Influence Cortical Microtubule Organization and Wall Extensibility

Abiotic factors have an enormous influence on cell expansion, and many of their effects are manifested by changes in the patterns of cortical microtubules in the expansion zones. Hormones mediate these changes through diverse and sometimes contrasting effects, depending on the organ or abiotic factor. Ethylene, for example, can alter microtubule and microfibril orientation from transverse to longitudinal to switch the direction of growth from longitudinal to radial (Lang et al. 1982) and in dark-grown seedlings it can promote hypocotyl radial swelling as part of the triple response. Conversely, in the light it can stimulate hypocotyl elongation (Smalle et al. 1997) or, under hypoxic conditions, stimulate stem elongation (Van der Straeten et al. 2001). Gibberellins and brassinosteroids, discussed in further detail below, promote both transverse cortical microtubule orientation and anisotropic expansion by promoting wall loosening in aerial tissues. Auxin also plays a major role in promoting transverse cortical microtubule orientation (Nick et al. 1990; Shibaoka 1994; Wasteneys 2000; Wiesler et al. 2002), although it is better known for its role in wall loosening by stimulating wall synthesis, increasing wall plasticity and activating genes encoding wall enzymes and the proton pumps that acidify the wall (Cosgrove 2001). The involvement of the cytoskeleton in polar auxin transport is considered in Sect. 4.

The relationship between cortical microtubules and gibberellin (GA) signalling is central to research on directional cell expansion (Shibaoka 1994) yet recent studies indicate that this relationship is complex. Kinematic analysis of a GA-deficient barley mutant, for example, suggested that gibberellins act independently on wall loosening and microtubule orientation (Wenzel et al. 2000). One tantalizing link between microtubule organization and GA signalling is the surprising discovery that the *lue1* and *dgl* katanin p60 mutants have altered expression of GA biosynthesis genes (Bouquin et al. 2003; Komorisono et al. 2005). This discovery is also counter-intuitive since GA is thought to control cortical microtubule organization and anisotropic expansion, not vice versa (Shibaoka 1994). Why should the loss of katanin's severing activity increase the expression of GA-biosynthesis genes that ultimately increase the concentration of active GAs? One possibility is that a negative feedback mechanism ties the regulation of AtKSS expression to GA levels. Indeed, Bouquin et al. (2003) did show that AtKSS expression was low in the GA-deficient mutant *ga1-1*, and that GA application increased AtKSS expression. The specificity of the katanin p60-GA biosynthesis connection is compelling because their screen apparently did not identify other microtubule organizing factors. On the other hand, it remains to be demonstrated that other proteins involved in organizing the cortical microtubule array are not also upregulated by GA.

Another explanation is that the disorganization of cortical microtubules in the katanin mutants may be one trigger that cells use to sense low GA levels and stimulate GA biosynthesis genes. The idea that microtubule polymer status and organization may be integrated with signalling networks has been previously discussed (Wasteneys 2003, 2004; Wasteneys and Galway 2003). It is known that GA levels negatively regulate the expression of key GA biosynthesis genes (Meier et al. 2001; Raventos et al. 2000), so low GA levels that lead to cortical microtubule disorganization will favour upregulation of these biosynthesis genes. Katanin mutants may therefore simply be mimicking a low GA condition. While the external application of GA_3 and GA_4 reduces the *lue1* mutant's flowering time to the same extent as in wild-type plants, GA application does not rescue the microtubule disorganization-dependent phenotype (Bouquin et al. 2003). The obvious explanation here is simply that it is not possible to upregulate a nonsense mutant allele.

In light of the connection between GA and AtKSS expression in GA-deficient mutants, how does manipulating katanin's expression affect microtubule function? When constitutively over-expressed, an AtKSS-GFP-GUS reporter fusion protein impaired growth anisotropy, paradoxically resembling loss of function mutants (Bouquin et al. 2003). Ethanol-induced expression of a fluorescent reporter fusion protein generated heavily fragmented microtubule arrays and radial swelling of cells (Stoppin-Mellet et al. 2006). These results imply that protein activity levels and localization, which may be controlled by the p80 subunit, need to be finely tuned in order to achieve the optimal microtubule organization.

Brassinosteroids play a major role in cell elongation and there is some evidence that at least part of this influence is mediated through cortical microtubule organization. Mayumi and Shibaoka (1996) demonstrated that application of brassinolide (BL) to azuki bean epicotyls was correlated with the restoration of transverse microtubule arrays and cell elongation. Studies on two brassinosteroid-deficient *Arabidopsis* mutants, *dim* and *bul1-1*, also reveal a close link between microtubule organization and brassinosteroid levels. Takashi et al. (1995) showed that β-1 tubulin expression was reduced in the *dim* mutant, suggesting that brassinosteroidal control over microtubule organization is related to tubulin production. A later study, however, revealed that the restoration of microtubule organization and elongation growth was not dependent on activation of tubulin gene expression or any increase in tubulin protein concentration in the tissues affected (Catterou et al. 2001). Total tubulin RNA levels were similar to wild-type in the *bul1-1* or *dim* mutants though, in agreement with previous analysis on *dim* mutants (Klahre et al. 1998; Takahashi et al. 1995), specific β-tubulin mRNA levels were substantially lower than in wild-type. Nevertheless, immunoblotting showed no clear differences in tubulin content in the petiole tissues most severely affected by the BL deficiency and application of homoBL did not alter tubulin concentration and even reduced tubulin mRNA levels in the mutant plants. Brassinosteroid effects on microtubule organization are more likely to be mediated through auxin activity. It has recently been determined, for example, that BL treatment enhances transcription of the *PIN* genes, which encode auxin efflux facilitators (Li et al. 2005).

3 Microtubules and the Mechanical Properties of Cellulose Microfibrils

Organizing a structurally complex transverse microtubule array, as described in the preceding sections of this work, is needed for anisotropic expansion. The influence cortical microtubules have over directional cell expansion is undeniable. Yet, despite more than 40 years since the suggestion that proteins of a spindle fibre nature might be active in the control of wall texture and cell form (Green 1962), the exact nature of this activity is still not clear. The strong correspondence between microtubule and microfibril orientation in elongating cells, first described by Ledbetter and Porter (1963), makes microtubules clear contenders for a role in regulating the orientation of cellulose microfibrils. More recently, it has been suggested that cortical microtubules also play an important role in regulating the mechanical properties of cellulose microfibrils (Wasteneys 2004).

The proposed mechanism for microtubule-dependent microfibril orientation seems to have come full circle. The earliest models, developed before the days of fluorescence microscopy, envisaged cellulose synthase complexes tracking directly along microtubules, being towed by motile microtubules, or being moved passively by microtubule-dependent motor paddles. Once it was demonstrated that cellulose crystallization can provide the motive force for the displacement of the plasma membrane-embedded synthase rosette complex (Herth 1980), the cellulose synthase constraint model emerged (Giddings and Staehelin 1991). This model proposed that co-aligned cortical microtubules form mechanical barriers that constrain the movement of cellulose synthase complexes, to deposit cellulose microfibrils in roughly the same orientation as the microtubules. In recent years, however, several challenges to this model have emerged.

Under many conditions there is no obligatory association between microtubule orientation and microfibril deposition (Baskin 2001; Emons and Mulder 2000; Wasteneys 2000). In tip-growing cells, the mainly longitudinal microtubule orientation is distinct from the varied and complex cellulose microfibril texture (Emons and Kieft 1994). In diffusely expanding cells, microtubules and cellulose microfibrils show tight co-alignment during the most rapid phase of elongation, but at other stages, microtubule orientation may be random, oblique or longitudinal in cells that are depositing transverse microfibrils (Sugimoto et al. 2000). This suggests that microtubules may be required for aligning cellulose microfibrils only during the most rapid phase of expansion.

There are some suggestions that microtubule arrays may rely, at least to some extent, on information from the cell wall. Fisher and Cyr (1998) perturbed microtubule organization in tobacco BY-2 culture cells using drugs that reduced cellulose synthesis. In intact *Arabidopsis* seedlings, the drug DCB randomizes microfibril texture, and under these conditions, microtubule orientation becomes increasingly dispersed about the transverse axis (Himmelspach et al. 2003).

Recent evidence shows that microtubules are not required for the production of parallel transverse microfibril orientation. Microtubule disorganization in the temperature-sensitive *mor1-1* mutant, or complete depolymerization using the drug oryzalin, inhibits growth anisotropy without preventing cellulose microfibrils from being deposited in parallel order (Sugimoto et al. 2003), at least at the cellular level (Baskin et al. 2004). Moreover, recovery of well-ordered microfibril texture after its disruption with the drug DCB can occur in the *mor1-1* mutant at its restrictive temperature, also casting doubt on the idea that pre-existing cellulose microfibrils serve as a template for deposition of new microfibrils (Himmelspach et al. 2003) or that a third element forms the critical template (Baskin 2001; Williamson 1990).

How can well-ordered cellulose microfibrils be deposited in the absence of microtubules? There is evidence that a default self-organization of cellulose deposition is dependent on the level of cellulose synthase activity. Mutations or drug treatments that specifically reduce cellulose synthesis without targeting microtubules disrupt the parallel order of cellulose microfibrils (Pagant et al. 2002; Sugimoto et al. 2001). Thus, the influence of microtubules over microfibril deposition may be stage-dependent and not solely related to orientation. On the other hand, the *Arabidopsis fra2* mutant, which has disordered microtubules as a result of the loss of the katanin p60's severing activity, also has misaligned cellulose microfibril patterns (Burk and Ye 2002). Like the *rsw1-1* mutant, however, *fra2* mutants also have significantly reduced cellulose levels, while in the *mor1-1* mutant, cellulose synthesis is not reduced. Interestingly, the KIF4 kinesin mutant *fra1* has normally oriented microtubules but microfibrils that show variable alignment (Zhong et al. 2002). Although the exact function of this cortically located motor protein is not known it may be an important part of the mechanism linking microtubules with cellulose orientation.

Taken together, these recent findings support the view that transverse cortical microtubules and cellulose microfibrils are both essential, but each is insufficient for growth anisotropy. Cellulose microfibrils can still be deposited in surprisingly well-ordered patterns in the absence of microtubules but these microfibrils are inferior to those formed in the presence of microtubules. Thus, microtubule interaction with cellulose synthesis machinery seems to be essential for the deposition of microfibrils capable of withstanding the forces of turgor to prevent lateral expansion (Wasteneys 2004).

The cellulose synthase constraint model has also come under fire for its contention that cellulose synthase rosettes move between, rather than directly along microtubules (Giddings and Staehelin 1988). Recent ground-breaking imaging of fluorescently tagged CesA6 in living cells provides good evidence that cellulose synthase complexes track directly along cortical microtubules (Paredez et al. 2006) thus bringing proposed mechanisms to explain microtubule-dependent microfibril orientation full circle. How can this new information be integrated with the other studies, described above, which demonstrate that parallel deposition of cellulose microfibrils, and therefore the movement of cellulose synthase complexes, can occur when cortical microtubule arrays are disrupted (Baskin et al. 2004; Himmelspach et al. 2003; Sugimoto et al. 2003)? While the direct tracking of cellulose synthase complexes along microtubules, and the ability of parallel transverse microfibrils to be deposited in the absence of microtubules seem paradoxical, these observations are compatible with the microfibril length regulation model (Wasteneys 2004).

The Microfibril Length Regulation Model proposes that long and strong microfibrils deposited in transverse directions provide no scope for lateral separation and will therefore only allow longitudinal wall expansion. Conversely short microfibrils can move apart from each other both perpendicular to and in the direction that they are oriented (Wasteneys 2004). The model predicts that cellulose microfibrils produced without the involvement of cortical microtubules will be mechanically compromised, and can therefore move apart from each other in both directions to allow radial and longitudinal expansion. The key hypothesis to be tested is that microtubule disruption generates short microfibrils that allow radial expansion through end-to-end separation in the lateral direction. Microfibril length will be dependent on both the longevity of synthase complexes and the tensile properties of the microfibrils, which in turn depends on the degree of β-1,4 glucose polymerization in individual cellulose strands as well as the degree of crystallinity in the microfibril (Wasteneys and Fujita 2006). Thus, when cellulose synthase complexes interact with microtubules, they are expected to produce mechanically sound, and relatively long microfibrils. In the event of derailment or loss of microtubule tracks, synthase complexes may continue to move in parallel, transverse directions, but produce inferior microfibrils that can protect the cell from bursting but cannot control growth direction. Cellulose synthase complexes that track on or close to microtubules may also be less vulnerable to endocytic recycling, and therefore survive longer to produce relatively long microfibrils.

3.1 Do Microtubules Regulate Wall Polysaccharide and Protein Composition?

Most studies on microtubules and cell walls have focussed on the role microtubules play in regulating the cellulosic component of cell walls (Baskin 2001). Do microtubules control other wall components, whose synthesis, unlike cellulose, occurs in the Golgi apparatus? A relationship between xyloglucan metabolism, microtubule orientation and the stimulation of growth has been identified (Takeda et al. 2002), yet we are far from understanding how this regulates directional expansion. The glycosylphosphatidylinositol (GPI)-anchored protein COBRA may be one key link between microtubule orientation and directional cell expansion. COBRA is expressed in the root elongation zone and is deposited in transverse bands at the surface of epidermal cells. Drug-induced microtubule disruption causes this banding pattern to dissipate suggesting that its transverse deposition pattern is microtubule-dependent (Roudier et al. 2005). The conditional *cob-1* mutant (Schindelman et al. 2001) and the null allele *cob-4* (Roudier et al. 2005) are unable to maintain anisotropic expansion. Analysis of wall texture in these mutants showed that cellulose microfibrils were disordered. These observations provide strong evidence that COBRA may control microfibril orientation in a microtubule-dependent manner. Putative cellulose binding domains on COBRA suggest that COBRA may have a structural function in the cell wall, though cellulose binding remains to be proven.

COBRA is confined to the lateral walls of root epidermal cells, with no detectable protein at the end walls (Roudier et al. 2005; Wasteneys and Fujita 2006). This clear differentiation between the end wall and the lateral wall in terms of composition, while novel, is not likely to be unique. The distinct properties of lateral and end walls is a critical feature of axial growth. The involvement of the cytoskeleton in end wall-specific processes is explored in the next section.

4
End Walls, Polar Auxin Transport and its Regulation by the Cytoskeleton

The complex biology of the plant hormone auxin underlies much of the development and growth of plants. Since the 1970s, the chemiosmotic theory of polar auxin transport has explained the apical to basal movement of auxin in plants through the weak acid nature of indoleacetic acid (IAA, the most common biologically occurring auxin) and through the asymmetric distribution of auxin influx and efflux carriers. This theory has recently received experimental validation. Auxin efflux carriers from the *PIN* gene family and influx carriers from the *AUX* gene family have been identified (reviewed in Leyser 2006), and more recently, proteins from the p-glycoprotein family have also been shown to independently contribute to auxin efflux and influx (Geisler and Murphy 2006). The temporal and spatial expression patterns and subcellular locations of these carriers are consistent with the paths that auxin takes through the plant. In the vascular cells of *Arabidopsis* roots, for example, the AUX1 influx carrier is located at the basal (towards the base of the root) plasma membrane (Swarup et al. 2001) while the efflux carrier AtPIN1 is located at the apical membrane (Galweiler et al. 1998), consistent with the central flow of auxin from the shoots down through the root. Recent analysis has shown that the regulated flow of auxin through the plant is a highly complex but robust system with numerous feedbacks occurring at multiple levels (Leyser 2006).

Auxin efflux carriers reside predominantly in the plasma membrane, but their distribution can be perturbed by pharmacological agents, suggesting that these proteins can undergo rapid microfilament-dependent cycling to and from the plasma membrane via an endosomal compartment (Fig. 2A) (Boutte et al. 2006; Geldner et al. 2001). Brefeldin A (BFA), which blocks vesicle trafficking and exocytosis by targeting ARF-GEFs, causes PIN proteins to accumulate in the brefeldin compartment, a peri-nuclear aggregate of endosomes and Golgi stacks (Boutte et al. 2006; Geldner et al. 2001; Steinmann et al. 1999). This process is reversible, and after BFA washout, the normal plasma membrane distribution of PIN proteins recovers (Fig. 2Bi). Microfilament disruption with either cytochalasin D or latrunculin B prevents this recovery (Fig. 2Bii) (Geldner et al. 2001), demonstrating that the microfila-

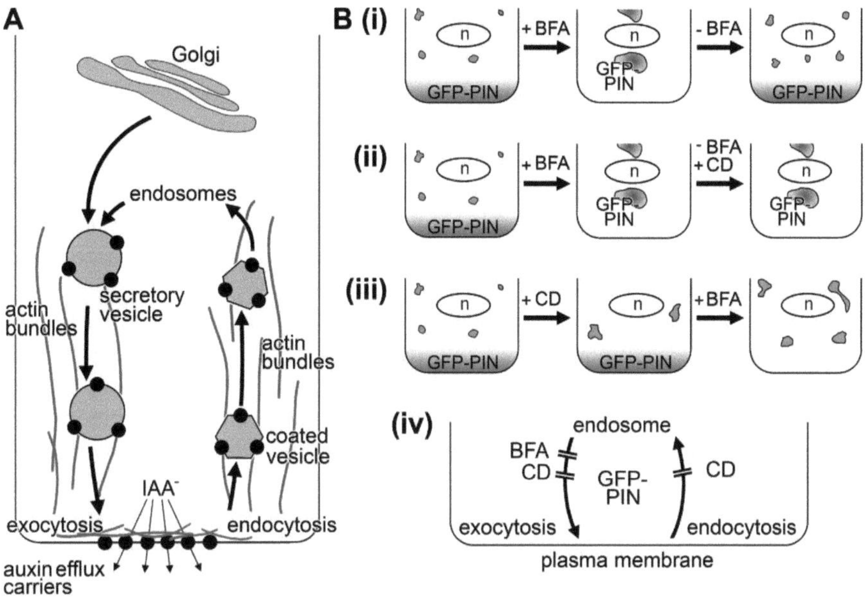

Fig. 2 Microfilament contribution to cell elongation through involvement in polar auxin transport, which is regulated by tissue-specific expression and subcellular distributions of auxin efflux (PIN and PGP proteins) and influx carriers (AUX proteins). **A** The shuttling of vesicles containing PIN proteins (auxin efflux carriers, shown as *black dots*) between the plasma membrane and endosomes is microfilament-dependent. **B** GFP-tagged At-PIN1 is distributed at the basal plasma membrane with comparatively little GFP-PIN in cytoplasmic compartments. Pharmacological experiments in *Arabidopsis* and maize roots demonstrate microfilament involvement (Boutte et al. 2006; Geldner et al. 2001). (**i**) The exocytosis inhibitor brefeldin A (BFA) causes a rapid but reversible accumulation of PIN1 into a perinuclear aggregate of endosomes, referred to as the brefeldin compartment. (**ii**) Microfilament disruption with cytochalasin D or latrunculin B prevents GFP-PIN's return to the plasma membrane after BFA removal. (**iii**) Microfilament disruption alone does not remove GFP-PIN from the plasma membrane, but does cause organelle-associated PIN1 to aggregate. Subsequent BFA treatments generate aberrant brefeldin compartments (Boutte et al. 2006; Geldner et al. 2001). (**iv**) Collectively, these experiments demonstrate that microfilaments serve as tracks for the movement of PIN-carrying vesicles in both the exocytotic and endocytotic pathways. They suggest, however, that microfilaments are not required for exocytosis or endocytosis, and that microfilaments do not stabilize the polar distribution of PIN proteins. Figure modified from Wasteneys and Collings (2004)

ments are necessary for exocytosis and PIN cycling at some stage after the site of BFA blockage (Fig. 2Biv). Microfilament disruption alone, however, has different effects. While the polar location of PIN within the plasma membrane is unaffected (Boutte et al. 2006; Geldner et al. 2001), the organelle-associated PIN forms aggregates (Boutte et al. 2006). Subsequent BFA treatments show varied results. Geldner et al. (2001) reported that prior microfilament disruption "inhibited BFA-induced intracellular PIN1 accumulation" in *Arabidopsis* roots. In contrast, Boutté et al. (2006) showed in maize roots that "BFA induces the disappearance of polar [PIN] staining" although normal brefeldin compartments failed to form. Although this latter result seems to rule out a direct role for microfilaments in the initial stages of endocytosis at the plasma membrane, both experiments demonstrate a role for microfilaments in distributing PIN-containing endocytotic vesicles within the cell. Consistent with these observations, treatments that disrupt microfilaments reduce auxin transport through plant tissues (Butler et al. 1998).

Numerous explanations have been proposed for the purpose of PIN cycling. Certainly, cycling would enable the rapid relocations of AtPIN3 that drive the reorientation of root growth during gravitropic responses (Friml et al. 2002), the polarity flipping of AtPIN7 during embryo development (Friml et al. 2003), and the redistributions of AtPIN1 during primordia development in the inflorescence meristem (Heisler et al. 2005). Rapid cycling might maintain the polar distribution of efflux carriers, either by asymmetric delivery to the plasma membrane or through an asymmetric recovery mechanism. It has even been suggested that vesicle transport along microfilaments may form part of the auxin transport pathway across plant cells or that auxin transport is analogous to neurotransmitter secretion in which microfilaments at the ends of cells play an important regulatory role (Baluska et al. 2003, 2005). As we have noted previously (Wasteneys and Collings 2004), this theory has several problems. First, the majority of auxin carrier proteins are located at the plasma membrane, and not associated with vesicles (Boutte et al. 2006; Galweiler et al. 1998; Geldner et al. 2001; Steinmann et al. 1999). Second, auxin's weak acid nature would cause it to protonate in the acidic environment of small secretory vesicles and to rapidly diffuse back into the cytoplasm. Furthermore, one of auxin's own actions is to inhibit PIN protein cycling by decreasing endocytosis (Paciorek et al. 2005).

How is polar distribution of auxin transport proteins regulated and what is the cytoskeleton's role? These proteins must be distributed *de novo* after cell division, because GFP-tagged PIN proteins demonstrate even insertion of recycled proteins into the cell plate during cytokinesis (Boutte et al. 2006; Geldner et al. 2001). It is tempting to speculate that an autoregulatory system uses positional cues derived from the direction of auxin flow in the tissue around recently divided cells to re-establish transport protein asymmetries. Nevertheless, it remains unclear whether exocytosis or endocytosis or both are targeted to specific domains of the plasma membrane. Microfilaments have a clear role in vesicle trafficking and interact functionally with proteins regulating the auxin efflux carrier (Muday 2000), but specific and direct roles for the cytoskeleton in auxin transport protein asymmetries seem to be ruled out. Neither microfilament nor microtubule disruption disturbs GFP-PIN protein asymmetries in the short term (Boutte et al. 2006; Geldner et al. 2001). However, the cell wall may function in maintaining asymmetries. In tobacco BY-2 cells, GFP-PIN distribution remains polarized when the cell wall is being digested to make protoplasts. But once the cell wall has been fully digested, GFP-PIN rapidly and evenly distributes through the plasma membrane (Boutte et al. 2006). The cell wall's role in PIN protein distribution might also explain why longer term treatments disrupting microtubules generates PIN protein mislocalization. While this was interpreted to mean "that microtubules were indirectly involved in the control of the final cellular location of PIN" (Boutte et al. 2006), it may simply reflect the loss of cell-to-cell contact, and hence the need for auxin transport, in protoplasts.

Some progress has been made in understanding the molecular processes underlying PIN distribution mechanisms. The serine-threonine kinase PINOID (PID) partially controls PIN localizations by acting as a binary switch, with over-expression and knock-out experiments demonstrating that below-threshold levels target PIN proteins to the lower ends of cells (apical in roots) and above-threshold levels target PIN proteins to the tops of cells (basal in roots) (Friml et al. 2004). Several observations, however, suggest there are further layers of regulation. For example, AUX1 and PIN1 distribute to opposite ends of the same vascular cells (Galweiler et al. 1998; Swarup et al. 2001), while PIN2 localizes to the apical plasma membrane of cortical cells but the basal plasma membrane of the adjacent epidermal cells (Muller et al. 1998). Further, when PIN1 expression is driven by the PIN2 promoter, it correctly distributes to the basal membranes of cortical cells, but also moves to the basal membranes of the adjacent epidermal cells. Thus, some targeting information must reside within the PIN protein sequences (Wisniewska et al. 2006).

5
How Does the Actin Cytoskeleton Contribute to Cell Elongation?

In post-mitotic and elongating root cells, microfilaments form into various distinct arrays. GFP fusions with the actin-binding domain of *Arabidopsis* fimbrin (GFP-fABD2), in agreement with immunofluorescence, show large subcortical bundles running lengthwise through cells, providing tracks for cytoplasmic streaming and vesicle transport, and finer microfilaments lying parallel to the transverse cortical microtubules (Collings and Wasteneys 2005; Voigt et al. 2005). A fine microfilament network has also been described adjacent to the apical and basal plasma membranes (eg., Baluska et al. 2001; Voigt et al. 2005) although the detailed organization of this actin array has not been described. Special emphasis has been placed on these polar-distributed microfilaments, in part because of interest in the regulation of polar auxin transporters (Sect. 4) but also because several microfilament-regulating proteins have a similar polar distribution. These include myosin VIII (Reichelt et al. 1999), microfilament nucleating formin-homology proteins (Deeks et al.

2005), and Rop GTPases (Molendijk et al. 2001). Rops, which act as molecular switches, are discussed in more detail in Sect. 6.

How then do microfilaments contribute to cell elongation? Microfilament disruption reduces root elongation, and promotes root tip swelling. Redistribution of auxin transport proteins may in part account for this swelling, but microfilaments may also contribute to cell elongation in ways unrelated to vesicle transport and streaming. Transgenic over- and under-expression of actin-binding proteins such as profilin and actin depolymerization factor alter expansion properties (Dong et al. 2001; Ramachandran et al. 2000). There is also evidence that microfilaments work closely with microtubules. Evidence for direct interactions has been published (Collings et al. 1998; Collings and Wasteneys 2005) but there are also compelling indications for indirect interactions between microtubules and microfilaments. Microtubule disruption by mutation or mild drug treatments has recently been shown to generate hypersensitivity to actin destabilizing drugs and stimulate root swelling (Collings et al. 2006). The apparent cross-talk between the microtubule and microfilament regulatory pathways is even more clearly illustrated in Sect. 6.

6
Microtubule—Actin Network Coordination in Pavement Cell Morphogenesis

The coordination of microtubule and actin-based networks in a multicellular environment is well demonstrated in the formation of epidermal pavement cells. These cells start out polyhedral in shape but as expansion proceeds, they interdigitate with neighbouring cells to form characteristic neck and lobed regions (Panteris and Galatis 2005). In dicots, these cells often resemble the interlocking pieces of a jigsaw puzzle. In monocots, the interdigitation can be very regular, forming a zipper-like connection between adjacent cell files.

Pavement cell morphogenesis and the known molecular mechanisms regulating it are depicted in Fig. 3. The neck regions are characterized by prominent microtubule bands, which mark regions of high vesicle accumulation and intense wall thickenings (Panteris and Galatis 2005). These wall thickenings prevent circumferential expansion. Intervening lobe regions have disordered or absent microtubules, patch-like actin microfilament networks, and are thin-walled. Lobes and necks alternate along the cell periphery, and lobe regions of one cell coincide with the necks of adjacent cells, suggesting that both intra- and intercellular signalling mechanisms regulate the spatial separation of microtubule and actin-based networks in these cells.

As with other expanding cells, perturbing either microtubules or microfilaments in growing pavement cells affects their morphogenesis. Applying drugs that depolymerize microtubules eliminates pavement cell interdigitation (Panteris et al. 1993), and mutants that disorganize microtubules can eliminate (Kotzer and Wasteneys 2006) or impede (Bichet et al. 2001; Burk

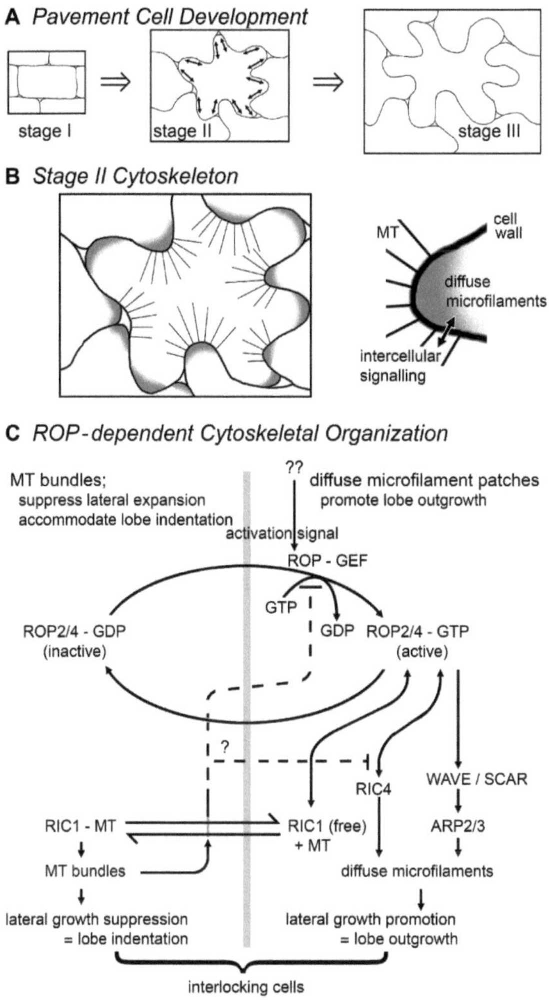

Fig. 3 Pavement cell development requires ROP GTPase-dependent sectoring of the cell periphery into microtubule bands and patches of fine microfilament networks. **A** The regular shape of pavement cells in leaf primordia (stage I) is modified as cells grow by coordinated cell expansion (*arrows*) resulting in the formation of lobes between adjacent cells (stage II) to produce a highly interdigitated morphology (stage III). **B** Microtubule bands (*lines*) mark the neck regions where wall thickenings constrain lateral expansion. Diffuse actin microfilament networks (*shading*) are found in intervening lobe-forming zones, and intercellular signaling between necks and lobes reinforces this pattern. **C** The spatial separation of microtubule-rich necks and actin microfilament-rich lobes is regulated by Rop GTPase interactions with two effector proteins, RIC1 and RIC4. ROP activity in lobe-forming regions promotes microfilament patches through RIC4 and WAVE/SCAR-ARP2/3 pathways. RIC1 promotes microtubule bundling in Rop-free zones and this activity is suppressed by Rop-RIC1 interaction in lobe-forming zones. Microtubule bands, in turn negatively regulate Rop-RIC4 interactions, possibly through intercellular signalling

et al. 2001) the process. Similarly, interdigitation is lost or reduced in mutants that have reduced actin patch formation, as a result of defects in subunits of either the ARP2/3 or its activating WAVE/SCAR complex (Kotzer and Wasteneys 2006).

Rops, members of the plant-specific Rho-like small GTPase family and important regulators of tip growth (Yang 2002), also regulate the spatial separation and activities of microtubule bands and actin patches in expanding pavement cells (Fu et al. 2005). At the onset of rapid leaf expansion, the cell periphery is sectored into zones of high Rop concentration where lobes will form, and zones where Rops are not detected, which become microtubule-rich and form the neck regions. In *Arabidopsis* pavement cells, Rop2 and Rop4 work redundantly. RNAi suppression of Rop2 expression in Rop4 null mutants produces narrow, weakly lobed cells that have extensive microtubule arrays (Fu et al. 2005).

The nature of Rop localization and activation is still not clear but since RhoGTPases are known to integrate extracellular signals, intercellular signalling between adjacent pavement cells is probably very important. Examining the distribution of lobes and necks on the anticlinal surface of pavement cells, it is clear that lobes in one cell lie adjacent to the necks of its neighbour (Fig. 3A). Intercellular symplastic or apoplastic signalling may reinforce the sectoring of Rop activity (Kotzer and Wasteneys 2006). Rop-specific guanine nucleotide exchange factors (GEFs) most likely play a critical role in the distribution of Rop activity. A family of 14 plant-specific RhoGEFs has been recently described (Berken et al. 2005) and the Dock180 homologue SPIKE1 (SPK1) may also be a RopGEF (Qiu et al. 2002). GEFs are critical factors for the activation of GTPases since the rate limiting process is the removal, not the addition, of GTP.

Rops regulate cytoskeletal organization in pavement cells by interacting with at least two effectors, the Rop-interactive Cdc42/Rac-interactive binding (CRIB) motif-containing (RIC) proteins, RIC1 and RIC4 (Wu et al. 2001; Fig. 3C). RIC4 associates with active Rops and promotes actin patch formation. Conversely, RIC1 associates with and promotes microtubule band formation in regions of low Rop activity. This pattern suggests that in the lobe-forming zones, activated Rops may sequester RIC1 to suppress microtubule band formation, and interact with RIC4 to promote actin patch formation. RIC1-dependent microtubule band formation, in turn, seems to work in a positive feedback manner to inhibit Rop interactions with RIC4 (Fu et al. 2005). RIC1 over-expression promotes microtubule assembly throughout the cortex, and reduces Rop2-RIC4 interactions, actin patch, and lobe formation. This cross-talk can also be demonstrated by drug or mutation-induced microtubule disruption, which enhances Rop-RIC4 interaction (Fu et al. 2005).

How microtubule banding can negatively regulate Rop activity is unclear. One possibility is that RIC1-dependent microtubule banding actually promotes Rop activation in nearby areas of neighbouring cells. Receptor Ser/Thr kinases are known to interact with and activate Rops (Trotochaud et al. 1999). Local inhibition of such receptor kinases in microtubule-rich zones presents one simple explanation for the regulation of Rop-targeted kinase activity, and this is consistent with RIC1 over-expression inhibiting lobe formation (Fu et al. 2005). In addition, secretion of a cell wall-diffusible ligand in microtubule-rich neck-forming regions could activate Rop-targeted receptor kinases in adjacent cells. Electron micrographs indicate that vesicle traffic is abundant near the microtubule bands (Panteris and Galatis 2005). Members of the Carboxy-terminal-Binding Protein and Brefeldin A ADP-Ribosylated Substrates (CtBP/BARS) family have dual functions in transcriptional repression and Golgi processing (Chinnadurai 2003). An *Arabidopsis* member of this family, ANGUSTIFOLIA (AN), regulates pavement cell morphogenesis. In *an* mutants (Rédei 1962), pavement cells in the highly elongated leaves have reduced shape complexity (Tsuge et al. 1996) and altered microtubule patterns (Folkers et al. 2002; Kim et al. 2002). AN's connection with microtubules may be via the kinesin-like calmodulin binding protein, KCBP, whose interaction with AN was demonstrated in yeast 2-hybrid assays (Folkers et al. 2002). This interaction has been suggested to be important for stimulating localized wall loosening in the formation of trichome branches (Smith and Oppenheimer 2005). Similarly, KCBP-dependent AN activity in microtubule-band regions might regulate intercellular signalling through the localized secretion of receptor kinase-targeted ligands.

It is still not clear how Rop-GTPase activation of RIC4 promotes actin patch formation, though a formin-dependent actin nucleation mechanism or the suppression of actin depolymerizing factor have been suggested (Smith and Oppenheimer 2005). Rop activity appears to independently stimulate ARP2/3 complex-dependent actin patch formation through activation of the WAVE/SCAR (WASP (Wiscott–Aldrich Syndrome protein) family verprolin-homologous protein/suppressor of cAMP receptor) complex (Mathur 2005; Smith and Oppenheimer 2005; Szymanski 2005). Yeast 2-hybrid assays provide evidence for interaction of Rop2-GTP with the SRA1 subunit of the WAVE/SCAR complex (Basu et al. 2004). WAVE/SCAR-activated ARP2/3 complexes associate with actin filaments to promote branched microfilament networks (May 2001). Many of the WAVE/SCAR and ARP2/3 complex members have been identified in mutant screens for defective trichome (leaf hair) morphology but the identified mutants also have reduced pavement cell interdigitation (summarized in Kotzer and Wasteneys 2006).

7
Summary

In this work, we highlighted the most obvious and best-characterized functions of the cytoskeleton in diffusely expanding cells of multicellular plants.

At the forefront is the highly organized and dynamic cortical microtubule array that controls growth anisotropy. Organizing transverse parallel microtubules requires the coordination of many processes involving a suite of microtubule-associated proteins that regulate nucleation, assembly dynamics and cross-linking. Hormones also influence cortical microtubule organization but recent studies suggest that feedback from the microtubule array can also affect hormone biosynthesis. Cortical microtubules regulate directional expansion through the mechanical properties of the cell wall, especially the load-bearing cellulose microfibrils. Recent work demonstrates that cellulose synthase complexes track closely along microtubules, yet it has also been demonstrated that parallel cellulose microfibril order can be achieved in the absence of microtubules. This paradox may be resolved in future studies to investigate whether microtubule interaction with cellulose synthase complexes generates microfibrils of greater tensile strength and or length.

The actin microfilament cytoskeleton is also critical for cell expansion, providing tracks for the majority of components delivered to the cell wall. The trafficking of auxin transport regulators is currently of great interest, and auxin, in turn, plays a key role in microtubule organization in elongating cells. The activity of Rop GTPases in the polar localization of auxin transport extends the repertoire of these signalling switches from their role in tip growth to axial growth. Indeed, Rops now appear to be the master coordinators of microfilaments and microtubules in most growing cells, and are critical for epidermal pavement cell morphogenesis, where the sectoring of the cell periphery into microtubule- and microfilament-rich zones produces the necks and lobes that characterize the complex shapes of pavement cells. Coordinating microtubule and actin microfilament activities is one of many fascinating processes that future studies on plant cell expansion will need to sort out.

Acknowledgements We thank members of our laboratories and collaborators for stimulating ideas. Research in Geoffrey Wasteneys' laboratory is supported by a Natural Sciences and Engineering Research Council (Canada) Discovery grant. David Collings acknowledges the support of an Australian Research Council Research Fellowship and Discovery Grant DP0208806.

References

Abe T, Hashimoto T (2005) Altered microtubule dynamics by expression of modified α-tubulin protein causes right-handed helical growth in transgenic Arabidopsis plants. Plant J 43:191–204

Bajer AS, Molè-Bajer J (1986) Reorganization of microtubules in endosperm cells and cell fragments of the higher plant *Haemanthus* in vivo. J Cell Biol 102:263–281

Baluška F, Jasik J, Edelmann HG, Salajova T, Volkmann D (2001) Latrunculin B-induced plant dwarfism: Plant cell elongation is F-actin-dependent. Developmental Biology 231:113–124

Baluška F, Šamaj J, Menzel D (2003) Polar transport of auxin: carrier-mediated flux across the plasma membrane or neurotransmitter-like secretion? Trends Cell Biol 13:282–285
Baluška F, Volkmann D, Menzel D (2005) Plant synapses: actin-based domains for cell-to-cell communication. Trends Plant Sci 10:106–111
Baskin TI (2001) On the alignment of cellulose microfibrils by cortical microtubules: a review and a model. Protoplasma 215:150–171
Baskin TI, Beemster GT, Judy-March JE, Marga F (2004) Disorganization of cortical microtubules stimulates tangential expansion and reduces the uniformity of cellulose microfibril alignment among cells in the root of Arabidopsis. Plant Physiol 135:2279–2290
Basu D, El-Assal SED, Le J, Mallery EL, Szymanski DB (2004) Interchangeable functions of *Arabidopsis* PIROGI and the human WAVE complex subunit SRA1 during leaf epidermal development. Development 131:4345–4355
Berken A, Thomas C, Wittinghofer A (2005) A new family of RhoGEFs activates the Rop molecular switch in plants. Nature 436:1176–1180
Bichet A, Desnos T, Turner S, Grandjean O, Höfte H (2001) *BOTERO1* is required for normal orientation of cortical microtubules and anisotropic cell expansion in *Arabidopsis*. Plant J 25:137–148
Binarová P, Cenklová V, Procházková J, Doskocilová A, Volc J, Vrlík M, Bögre L (2006) γ-tubulin is essential for acentrosomal microtubule nucleation and coordination of late mitotic events in *Arabidopsis*. Plant Cell 18:1199–1212
Bisgrove SR, Hable WE, Kropf DL (2004) +TIPs and microtubule regulation. The beginning of the plus end in plants. Plant Physiol 136:3855–3863
Bouquin T, Mattsson O, Næsted H, Foster R, Mundy J (2003) The *Arabidopsis lue1* mutant defines a katanin p60 ortholog involved in hormonal control of microtubule orientation during cell growth. J Cell Sci 116:791–801
Boutté Y, Crosnier MT, Carraro N, Traas J, Satiat-Jeunemaitre B (2006) The plasma membrane recycling pathway and cell polarity in plants: studies on PIN proteins. J Cell Sci 119:1255–1265
Burk DH, Liu B, Zhong R, Morrison WH, Ye ZH (2001) A katanin-like protein regulates normal cell wall biosynthesis and cell elongation. Plant Cell 13:807–827
Burk DH, Ye ZH (2002) Alteration of oriented deposition of cellulose microfibrils by mutation of a katanin-like microtubule-severing protein. Plant Cell 14:2145–2160
Buschmann H, Fabri CO, Hauptmann M, Hutzler P, Laux T, Lloyd CW, Schäffner AR (2004) Helical growth of the *Arabidopsis* mutant *tortifolia1* reveals a plant-specific microtubule-associated protein. Curr Biol 14:1515–1521
Butler JH, Hu S, Brady SR, Dixon MW, Muday GK (1998) In vitro and in vivo evidence for actin association of the naphthylphthalamic acid-binding protein from zucchini hypocotyls. Plant J 13:291–301
Catterou M, Dubois F, Schaller H, Aubanelle L, Vilcot B, Sangwan-Norreel BS, Sangwan RS (2001) Brassinosteroids, microtubules and cell elongation in Arabidopsis thaliana. II. Effects of brassinosteroids on microtubules and cell elongation in the bul1 mutant. Planta 212:673–683
Chan J, Rutten T, Lloyd C (1996) Isolation of microtubule-associated proteins from carrot cytoskeletons: a 120 kDa map decorates all four microtubule arrays and the nucleus. Plant J 10:251–259
Chinnadurai G (2003) CtBP family proteins: more than transcriptional corepressors. Bioessays 25:9–12

Collings DA, Asada T, Allen NS, Shibaoka H (1998) Plasma membrane-associated actin in bright yellow 2 tobacco cells. Evidence for interaction with microtubules. Plant Physiol 118:917–928

Collings DA, Lill AW, Himmelspach R, Wasteneys GO (2006) Hypersensitivity to cytoskeletal antagonists demonstrates microtubule-microfilament cross-talk in the control of root elongation in *Arabidopsis thaliana*. New Phytol 170:275–290

Collings DA, Wasteneys GO (2005) Actin microfilament and microtubule distribution patterns in the expanding root of *Arabidopsis thaliana*. Can J Bot 83:579–590

Cosgrove DJ (2001) Wall structure and wall loosening. A look backwards and forwards. Plant Physiol 125:131–134

Deeks MJ, Cvrcková F, Machesky LM, Mikitová V, Ketelaar T, Zársky V, Davies B, Hussey PJ (2005) Arabidopsis group Ie formins localize to specific cell membrane domains, interact with actin-binding proteins and cause defects in cell expansion upon aberrant expression. New Phytol 168:529–540

Dixit R, Chang E, Cyr R (2006) Establishment of polarity during organization of the acentrosomal plant cortical microtubule array. Mol Biol Cell 17:1298–1305

Dixit R, Cyr R (2004a) The cortical microtubule array: from dynamics to organization. Plant Cell 16:2546–2552

Dixit R, Cyr R (2004b) Encounters between dynamic cortical microtubules promote ordering of the cortical array through angle-dependent modifications of microtubule behavior. Plant Cell 16:3274–3284

Dong CH, Kost B, Xia GX, Chua NH (2001) Molecular identification and characterization of the *Arabidopsis AtADF1*, *AtADF5* and *AtADF6* genes. Plant Mol Biol 45:517–527

Dryková D, Cenklová V, Sulimenko V, Volc J, Dráber P, Binarová P (2003) Plant γ-tubulin interacts with $\alpha\beta$-tubulin dimers and forms membrane-associated complexes. Plant Cell 15:465–480

Emons AMC, Kieft H (1994) Winding threads around plant cells—applications of the geometrical model for microfibril deposition. Protoplasma 180:59–69

Emons AMC, Mulder BM (2000) How the deposition of cellulose microfibrils builds cell wall architecture. Trends Plant Sci 5:35–40

Erhardt M, Stoppin-Mellet V, Campagne S, Canaday J, Mutterer J, Fabian T, Sauter M, Muller T, Peter C, Lambert AM, Schmit AC (2002) The plant Spc98p homologue colocalizes with γ-tubulin at microtubule nucleation sites and is required for microtubule nucleation. J Cell Sci 115:2423–2431

Falconer MM, Donaldson G, Seagull RW (1988) MTOCs in higher plant cells: an immunofluorescent study of microtubule assembly sites following depolymerization by APM. Protoplasma 144:46–55

Folkers U, Kirik V, Schöbinger U, Falk S, Krishnakumar S, Pollock MA, Oppenheimer DG, Day I, Reddy AS, Jürgens G, Hülskamp M, Reddy AR (2002) The cell morphogenesis gene *ANGUSTIFOLIA* encodes a CtBP/BARS-like protein and is involved in the control of the microtubule cytoskeleton. EMBO J 21:1280–1288

Friml J, Vieten A, Sauer M, Weijers D, Schwarz H, Hamann T, Offringa R, Jürgens G (2003) Efflux-dependent auxin gradients establish the apical-basal axis of *Arabidopsis*. Nature 426:147–153

Friml J, Wiśniewska J, Benková E, Mendgen K, Palme K (2002) Lateral relocation of auxin efflux regulator PIN3 mediates tropism in *Arabidopsis*. Nature 415:806–809

Friml J, Yang X, Michniewicz M, Weijers D, Quint A, Tietz O, Benjamins R, Ouwerkerk PB, Ljung K, Sandberg G, Hooykaas PJ, Palme K, Offringa R (2004) A PINOID-dependent binary switch in apical-basal PIN polar targeting directs auxin efflux. Science 306:862–865

Fu Y, Gu Y, Zheng Z, Wasteneys G, Yang Z (2005) *Arabidopsis* interdigitating cell growth requires two antagonistic pathways with opposing action on cell morphogenesis. Cell 120:687–700

Furutani I, Watanabe Y, Prieto R, Masukawa M, Suzuki K, Naoi K, Thitamadee S, Shikanai T, Hashimoto T (2000) The *SPIRAL* genes are required for directional control of cell elongation in *Arabidopsis thaliana*. Development 127:4443–4453

Gälweiler L, Guan C, Müller A, Wisman E, Mendgen K, Yephremov A, Palme K (1998) Regulation of polar auxin transport by AtPIN1 in Arabidopsis vascular tissue. Science 282:2226–2230

Geisler M, Murphy AS (2006) The ABC of auxin transport: the role of p-glycoproteins in plant development. FEBS Lett 580:1094–1102

Geldner N, Friml J, Stierhof YD, Jürgens G, Palme K (2001) Auxin transport inhibitors block PIN1 cycling and vesicle trafficking. Nature 413:425–428

Giddings TH, Staehelin LA (1988) Spatial relationship between microtubules and plasma-membrane rosettes during deposition of primary wall microfibrils in *Closterium* sp. Planta 173:22–30

Giddings TH, Staehelin LA (1991) Microtubule-mediated control of microfibril deposition; a re-examination of the hypothesis. In: Lloyd CW (ed) The cytoskeletal basis of plant growth and form. Academic Press, San Diego, CA, p 85–100

Green PB (1962) Mechanism for plant cellular morphogenesis. Science 138:1404–1405

Hamada T, Igarashi H, Itoh TJ, Shimmen T, Sonobe S (2004) Characterization of a 200 kDa microtubule-associated protein of tobacco BY-2 Cells, a member of the XMAP215/MOR1 family. Plant Cell Physiol 45:1233–1242

Hashimoto T (2002) Molecular genetic analysis of left-right handedness in plants. Philos Trans R Soc Lond B Biol Sci 357:799–808

Hashimoto T (2003) Dynamics and regulation of plant interphase microtubules: a comparative view. Curr Opin Plant Biol 6:568–576

Heisler MG, Ohno C, Das P, Sieber P, Reddy GV, Long JA, Meyerowitz EM (2005) Patterns of auxin transport and gene expression during primordium development revealed by live imaging of the *Arabidopsis* inflorescence meristem. Curr Biol 15:1899–1911

Herth W (1980) Calcofluor white and congo red inhibit chitin microfibril assembly of *Poterioochromonas*: Evidence for a gap between polymerization and microfibril formation. J Cell Biol 87:442–450

Himmelspach R, Williamson RE, Wasteneys GO (2003) Cellulose microfibril alignment recovers from DCB-induced disruption despite microtubule disorganization. Plant J 36:565–575

Hush JM, Wadsworth P, Callaham DA, Hepler PK (1994) Quantification of microtubule dynamics in living plant cells using fluorescence redistribution after photobleaching. J Cell Sci 107:775–784

Hussey PJ, Hawkins TJ (2001) Plant microtubule-associated proteins: the HEAT is off in temperature-sensitive *mor1*. Trends Plant Sci 6:389–392

Hussey PJ, Hawkins TJ, Igarashi H, Kaloriti D, Smertenko A (2002) The plant cytoskeleton: recent advances in the study of the plant microtubule-associated proteins MAP-65, MAP-190 and the *Xenopus* MAP215-like protein, MOR1. Plant Mol Biol 50:915–924

Jiang CJ, Sonobe S (1993) Identification and preliminary characterization of a 65-kDa higher-plant microtubule-associated protein. J Cell Sci 105:891–901

Kawamura E, Himmelspach R, Rashbrooke MC, Whittington AT, Gale KR, Collings DA, Wasteneys GO (2006) MICROTUBULE ORGANIZATION 1 regulates structure and function of microtubule arrays during mitosis and cytokinesis in the Arabidopsis root. Plant Physiol 140:102–114

Kim GT, Shoda K, Tsuge T, Cho KH, Uchimiya H, Yokoyama R, Nishitani K, Tsukaya H (2002) The *ANGUSTIFOLIA* gene of *Arabidopsis*, a plant *CtBP* gene, regulates leaf-cell expansion, the arrangement of cortical microtubules in leaf cells and expression of a gene involved in cell-wall formation. EMBO J 21:1267–1279

Klahre U, Noguchi T, Fujioka S, Takatsuto S, Yokota T, Nomura T, Yoshida S, Chua NH (1998) The Arabidopsis DIMINUTO/DWARF1 gene encodes a protein involved in steroid synthesis. Plant Cell 10:1677–1690

Komorisono M, Ueguchi-Tanaka M, Aichi I, Hasegawa Y, Ashikari M, Kitano H, Matsuoka M, Sazuka T (2005) Analysis of the rice mutant dwarf and gladius leaf 1. Aberrant katanin-mediated microtubule organization causes up-regulation of gibberellin biosynthetic genes independently of gibberellin signaling. Plant Physiol 138:1982–1993

Kotzer AM, Wasteneys GO (2006) Mechanisms behind the puzzle: microtubule-microfilament cross-talk in pavement cell formation. Can J Bot 84:594–603

Lang JM, Eisinger WR, Green PB (1982) Effects of ethylene on the orientation of microtubules and cellulose microfibrils in pea epicotyl cells with polylamellate cell walls. Protoplasma 110:5–14

Ledbetter MC, Porter KR (1963) A "microtubule" in plant cell fine structure. J Cell Biol 19:239–250

Leyser O (2006) Dynamic integration of auxin transport and signalling. Curr Biol 16:R424–433

Li L, Xu J, Xu ZH, Xue HW (2005) Brassinosteroids stimulate plant tropisms through modulation of polar auxin transport in Brassica and Arabidopsis. Plant Cell 17:2738–2753

Liu B, Joshi HC, Wilson TJ, Silflow CD, Palevitz BA, Snustad DP (1994) γ-Tubulin in Arabidopsis: gene sequence, immunoblot, and immunofluorescence studies. Plant Cell 6:303–314

Mathur J (2005) The ARP2/3 complex: giving plant cells a leading edge. Bioessays 27:377–387

May RC (2001) The Arp2/3 complex: a central regulator of the actin cytoskeleton. Cell Mol Life Sci 58:1607–1626

Mayumi K, Shibaoka H (1996) The cyclic reorientation of cortical microtubules on walls with a crossed polylamellate structure—effects of plant hormones and an inhibitor of protein kinases on the progression of the cycle. Protoplasma 195:112–122

McDonald AR, Liu B, Joshi HC, Palevitz BA (1993) γ-tubulin is associated with a cortical-microtubule-organizing zone in the developing guard cells of *Allium cepa* L. Planta 191:357–361

Meier C, Bouquin T, Nielsen ME, Raventos D, Mattsson O, Rocher A, Schomburg F, Amasino RM, Mundy J (2001) Gibberellin response mutants identified by luciferase imaging. Plant J 25:509–519

Molendijk AJ, Bischoff F, Rajendrakumar CS, Friml J, Braun M, Gilroy S, Palme K (2001) *Arabidopsis thaliana* Rop GTPases are localized to tips of root hairs and control polar growth. EMBO J 20:2779–2788

Muday GK (2000) Maintenance of asymmetric cellular localization of an auxin transport protein through interaction with the actin cytoskeleton. J Plant Growth Regul 19:385–396

Müller A, Guan C, Gälweiler L, Tänzler P, Huijser P, Marchant A, Parry G, Bennett M, Wisman E, Palme K (1998) *AtPIN2* defines a locus of *Arabidopsis* for root gravitropism control. EMBO J 17:6903–6911

Murata T, Sonobe S, Baskin TI, Hyodo S, Hasezawa S, Nagata T, Horio T, Hasebe M (2005) Microtubule-dependent microtubule nucleation based on recruitment of γ-tubulin in higher plants. Nat Cell Biol 7:961–968

Nakajima K, Furutani I, Tachimoto H, Matsubara H, Hashimoto T (2004) SPIRAL1 encodes a plant-specific microtubule-localized protein required for directional control of rapidly expanding Arabidopsis cells. Plant Cell 16:1178–1190

Nakamura M, Naoi K, Shoji T, Hashimoto T (2004) Low concentrations of propyzamide and oryzalin alter microtubule dynamics in Arabidopsis epidermal cells. Plant Cell Physiol 45:1330–1334

Nick P, Bergfeld R, Schäfer E, Schopfer P (1990) Unilateral reorientation of microtubules at the outer epidermal wall during photo- and gravitropic curvature of maize coleoptiles and sunflower hypocotyls. Planta 181:162–168

Paciorek T, Zažímalová E, Ruthardt N, Petrášek J, Stierhof YD, Kleine-Vehn J, Morris DA, Emans N, Jurgens G, Geldner N, Friml J (2005) Auxin inhibits endocytosis and promotes its own efflux from cells. Nature 435:1251–1256

Pagant S, Bichet A, Sugimoto K, Lerouxel O, Desprez T, McCann M, Lerouge P, Vernhettes S, Höfte H (2002) *KOBITO1* encodes a novel plasma membrane protein necessary for normal synthesis of cellulose during cell expansion in Arabidopsis. Plant Cell 14:2001–2013

Panteris E, Apostolakos P, Galatis B (1993) Microtubules and morphogenesis in ordinary epidermal cells of *Vigna sinensis* leaves. Protoplasma 174:91–100

Panteris E, Galatis B (2005) The morphogenesis of lobed plant cells in the mesophyll and epidermis: organization and distinct roles of cortical microtubules and actin filaments. New Phytol 167:721–732

Paredez AR, Somerville CR, Ehrhardt DW (2006) Visualization of cellulose synthase demonstrates functional association with microtubules. Science 312:1491–1495

Pastuglia M, Azimzadeh J, Goussot M, Camilleri C, Belcram K, Evrard JL, Schmit AC, Guerche P, Bouchez D (2006) γ -tubulin is essential for microtubule organization and development in *Arabidopsis*. Plant Cell 18:1412–1425

Qiu JL, Jilk R, Marks MD, Szymanski DB (2002) The Arabidopsis SPIKE1 gene is required for normal cell shape control and tissue development. Plant Cell 14:101–118

Quarmby L (2000) Cellular Samurai: katanin and the severing of microtubules. J Cell Sci 113:2821–2827

Ramachandran S, Christensen HEM, Ishimaru Y, Dong CH, Chao-Ming W, Cleary AL, Chua NH (2000) Profilin plays a role in cell elongation, cell shape maintenance, and flowering in arabidopsis. Plant Physiol 124:1637–1647

Raventos D, Meier C, Mattsson O, Jensen AB, Mundy J (2000) Fusion genetic analysis of gibberellin signaling mutants. Plant J 22:427–438

Rédei GP (1962) Single locus heterosis. Z Vererbungs 93:164–170

Reichelt S, Knight AE, Hodge TP, Baluska F, Samaj J, Volkmann D, Kendrick-Jones J (1999) Characterization of the unconventional myosin VIII in plant cells and its localization at the post-cytokinetic cell wall. Plant J 19:555–567

Roudier F, Fernandez AG, Fujita M, Himmelspach R, Borner GH, Schindelman G, Song S, Baskin TI, Dupree P, Wasteneys GO, Benfey PN (2005) COBRA, an Arabidopsis extracellular glycosyl-phosphatidyl inositol-anchored protein, specifically controls highly anisotropic expansion through its involvement in cellulose microfibril orientation. Plant Cell 17:1749–1763

Ryden P, Sugimoto-Shirasu K, Smith AC, Findlay K, Reiter WD, McCann MC (2003) Tensile properties of Arabidopsis cell walls depend on both a xyloglucan cross-linked microfibrillar network and rhamnogalacturonan II-borate complexes. Plant Physiol 132:1033–1040

Schindelman G, Morikami A, Jung J, Baskin TI, Carpita NC, Derbyshire P, McCann MC, Benfey PN (2001) COBRA encodes a putative GPI-anchored protein, which is polarly localized and necessary for oriented cell expansion in Arabidopsis. Genes Dev 15:1115–1127

Schmit AC (2002) Acentrosomal microtubule nucleation in higher plants. Int Rev Cytol 220:257–289

Sedbrook JC, Ehrhardt DW, Fisher SE, Scheible WR, Somerville CR (2004) The Arabidopsis sku6/spiral1 gene encodes a plus end-localized microtubule-interacting protein involved in directional cell expansion. Plant Cell 16:1506–1520

Shaw SL, Kamyar R, Ehrhardt DW (2003) Sustained microtubule treadmilling in Arabidopsis cortical arrays. Science 300:1715–1718

Shibaoka H (1994) Plant hormone-induced changes in the orientation of cortical microtubules—alterations in the cross-linking between microtubules and the plasma membrane. Annu Rev Plant Physiol Plant Mol Biol 45:527–544

Shoji T, Narita NN, Hayashi K, Asada J, Hamada T, Sonobe S, Nakajima K, Hashimoto T (2004) Plant-specific microtubule-associated protein SPIRAL2 is required for anisotropic growth in Arabidopsis. Plant Physiol 136:3933–3944

Smalle J, Haegman M, Kurepa J, Van Montagu M, Van Der Straeten D (1997) Ethylene can stimulate *Arabidopsis* hypocotyl elongation in the light. Proc Natl Acad Sci USA 94:2756–2761

Smertenko AP, Chang HY, Wagner V, Kaloriti D, Fenyk S, Sonobe S, Lloyd C, Hauser MT, Hussey PJ (2004) The Arabidopsis microtubule-associated protein AtMAP65-1: molecular analysis of its microtubule bundling activity. Plant Cell 16:2035–2047

Smith LG, Oppenheimer DG (2005) Spatial control of cell expansion by the plant cytoskeleton. Annu Rev Cell Dev Biol 21:271–295

Steinmann T, Geldner N, Grebe M, Mangold S, Jackson CL, Paris S, Gälweiler L, Palme K, Jürgens G (1999) Coordinated polar localization of auxin efflux carrier PIN1 by GNOM ARF GEF. Science 286:316–318

Stoppin V, Vantard M, Schmit AC, Lambert AM (1994) Isolated plant nuclei nucleate microtubule assembly: the nuclear surface in higher plants has centrosome-like activity. Plant Cell 6:1099–1106

Stoppin-Mellet V, Gaillard J, Vantard M (2002) Functional evidence for in vitro microtubule severing by the plant katanin homologue. Biochem J 365:337–342

Stoppin-Mellet V, Gaillard J, Vantard M (2006) Katanin's severing activity favors bundling of cortical microtubules in plants. Plant J 46:1009–1017

Stoppin-Mellet V, Peter C, Lambert AM (2000) Distribution of γ-tubulin in higher plant cells: Cytosolic γ-tubulin is part of high molecular weight complexes. Plant Biology 2:290–296

Sugimoto K, Himmelspach R, Williamson RE, Wasteneys GO (2003) Mutation or drug-dependent microtubule disruption causes radial swelling without altering parallel cellulose microfibril deposition in Arabidopsis root cells. Plant Cell 15:1414–1429

Sugimoto K, Williamson RE, Wasteneys GO (2000) New techniques enable comparative analysis of microtubule orientation, wall texture, and growth rate in intact roots of Arabidopsis. Plant Physiol 124:1493–1506

Sugimoto K, Williamson RE, Wasteneys GO (2001) Wall architecture in the cellulose-deficient rsw1 mutant of Arabidopsis thaliana: microfibrils but not microtubules lose their transverse alignment before microfibrils become unrecognizable in the mitotic and elongation zones of roots. Protoplasma 215:172–183

Swarup R, Friml J, Marchant A, Ljung K, Sandberg G, Palme K, Bennett M (2001) Localization of the auxin permease AUX1 suggests two functionally distinct hormone transport pathways operate in the Arabidopsis root apex. Genes Dev 15:2648–2653

Szymanski DB (2005) Breaking the WAVE complex: the point of *Arabidopsis* trichomes. Curr Opin Plant Biol 8:103–112

Takahashi T, Gasch A, Nishizawa N, Chua NH (1995) The *DIMINUTO* gene of *Arabidopsis* is involved in regulating cell elongation. Genes Dev 9:97–107

Takeda T, Furuta Y, Awano T, Mizuno K, Mitsuishi Y, Hayashi T (2002) Suppression and acceleration of cell elongation by integration of xyloglucans in pea stem segments. Proc Natl Acad Sci USA 99:9055–9060

Thitamadee S, Tuchihara K, Hashimoto T (2002) Microtubule basis for left-handed helical growth in *Arabidopsis*. Nature 417:193–196

Trotochaud AE, Hao T, Yang Z, Clark SE (1999) The CLAVATA1 receptor-like kinase requires CLAVATA3 for its assembly into a signalling complex that includes KAPP and a Rho-related protein. Plant Cell 11:393–406

Tsuge T, Tsukaya H, Uchimiya H (1996) Two independent and polarized processes of cell elongation regulate leaf blade expansion in *Arabidopsis thaliana* (L.) Heynh. Development 122:1589–1600

Twell D, Park SK, Hawkins TJ, Schubert D, Schmidt R, Smertenko A, Hussey PJ (2002) MOR1/GEM1 has an essential role in the plant-specific cytokinetic phragmoplast. Nat Cell Biol 4:711–714

Van Damme D, Van Poucke K, Boutant E, Ritzenthaler C, Inzé D, Geelen D (2004) In Vivo Dynamics and Differential Microtubule-Binding Activities of MAP65 Proteins. Plant Physiol 136:3956–3967

Van der Straeten D, Zhou ZY, Prinsen E, Van Onckelen HA, Van Montagu MC (2001) A comparative molecular-physiological study of submergence response in lowland and deepwater rice. Plant Physiol 125:955–968

Voigt B, Timmers AC, Šamaj J, Müller J, Baluška F, Menzel D (2005) GFP-FABD2 fusion construct allows in vivo visualization of the dynamic actin cytoskeleton in all cells of *Arabidopsis* seedlings. Eur J Cell Biol 84:595–608

Vos JW, Dogterom M, Emons AM (2004) Microtubules become more dynamic but not shorter during preprophase band formation: a possible "search-and-capture" mechanism for microtubule translocation. Cell Motil Cytoskeleton 57:246–258

Wasteneys GO (1992) The characean cytoskeleton: spatial control in the cortical cytoplasm. In: Menzel D (ed) The cytoskeleton of the algae. CRC Press, Boca Raton, Florida, p 273–295

Wasteneys GO (2000) The cytoskeleton and growth polarity. Curr Opin Plant Biol 3:503–511

Wasteneys GO (2002) Microtubule organization in the green kingdom: chaos or self-order? J Cell Sci 115:1345–1354

Wasteneys GO (2003) Microtubules show their sensitive nature. Plant Cell Physiol 44:653–654

Wasteneys GO (2004) Progress in understanding the role of microtubules in plant cells. Curr Opin Plant Biol 7:651–660

Wasteneys GO, Collings DA (2004) Expanding beyond the great divide: the cytoskeleton and axial growth. In: Hussey PJ (ed) The plant cytoskeleton in cell differentiation and development. Blackwell, Oxford, p 83–115

Wasteneys GO, Fujita M (2006) Establishing and maintaining axial growth: wall mechanical properties and the cytoskeleton. J Plant Res 119:5–10

Wasteneys GO, Galway ME (2003) Remodeling the cytoskeleton for growth and form: an overview with some new views. Annu Rev Plant Biol 54:691–722

Wasteneys GO, Gunning BES, Hepler PK (1993) Microinjection of fluorescent brain tubulin reveals dynamic properties of cortical microtubules in living plant-cells. Cell Motil Cytoskel 24:205–213

Wasteneys GO, Jablonsky PP, Williamson RE (1989) Assembly of purified brain tubulin at cortical and endoplasmic sites in perfused internodal cells of the alga *Nitella tasmanica*. Cell Biol Int Rep 13:513–528
Wasteneys GO, Williamson RE (1989a) Reassembly of microtubules in *Nitella tasmanica*: assembly of cortical microtubules in branching clusters and its relevance to steady-state microtubule assembly. J Cell Sci 93:705–714
Wasteneys GO, Williamson RE (1989b) Reassembly of microtubules in *Nitella tasmanica*: quantitative analysis of assembly and orientation. Eur J Cell Biol 50:76–83
Webb M, Jouannic S, Foreman J, Linstead P, Dolan L (2002) Cell specification in the *Arabidopsis* root epidermis requires the activity of *ECTOPIC ROOT HAIR 3*—a katanin-p60 protein. Development 129:123–131
Wenzel CL, Williamson RE, Wasteneys GO (2000) Gibberellin-induced changes in growth anisotropy precede gibberellin-dependent changes in cortical microtubule orientation in developing epidermal cells of barley leaves. Kinematic and cytological studies on a gibberellin-responsive dwarf mutant, M489. Plant Physiol 124:813–822
Whittington AT, Vugrek O, Wei KJ, Hasenbein NG, Sugimoto K, Rashbrooke MC, Wasteneys GO (2001) MOR1 is essential for organizing cortical microtubules in plants. Nature 411:610–613
Wiesler B, Wang QY, Nick P (2002) The stability of cortical microtubules depends on their orientation. Plant J 32:1023–1032
Williamson RE (1990) Alignment of cortical microtubules by anisotropic wall stresses. Aust J Plant Physiol 17:601–613
Wiœniewska J, Xu J, Seifertová D, Brewer PB, Rùžièka K, Blilou I, Rouquié D, Benková E, Scheres B, Friml J (2006) Polar PIN localization directs auxin flow in plants. Science 312:883
Wu G, Gu Y, Li S, Yang Z (2001) A genome-wide analysis of Arabidopsis Rop-interactive CRIB motif-containing proteins that act as Rop GTPase targets. Plant Cell 13:2841–2856
Yang Z (2002) Small GTPases: versatile signaling switches in plants. Plant Cell 14(Suppl):S375–388
Yuan M, Shaw PJ, Warn RM, Lloyd CW (1994) Dynamic reorientation of cortical microtubules, from transverse to longitudinal, in living plant cells. Proc Natl Acad Sci USA 91:6050–6053
Yuen CY, Pearlman RS, Silo-Suh L, Hilson P, Carroll KL, Masson PH (2003) WVD2 and WDL1 modulate helical organ growth and anisotropic cell expansion in Arabidopsis. Plant Physiol 131:493–506
Zhong R, Burk DH, Morrison WH 3rd, Ye ZH (2002) A kinesin-like protein is essential for oriented deposition of cellulose microfibrils and cell wall strength. Plant Cell 14:3101–317

Plant Cell Monogr (5)
J.-P. Verbelen · K. Vissenberg: The Expanding Cell
DOI 10.1007/7089_2006_078/Published online: 7 October 2006
© Springer-Verlag Berlin Heidelberg 2006

The Control of Cell Size and Rate of Elongation in the *Arabidopsis* Root

Tinne De Cnodder · Jean-Pierre Verbelen · Kris Vissenberg (✉)

University of Antwerp, Dept. Biology, Plant Physiology and Morphology, Groenenborgerlaan 171, 2020 Antwerpen, Belgium
kris.vissenberg@ua.ac.be

Abstract The control of cell elongation is studied in the root of *Arabidopsis* by focusing on reduction of cell size. Reduced organ size is a common response of a plant to different exogenous and endogenous signals. There is not a single specific event/actor that is solely responsible for the fast inhibition of cell elongation, but a battery of different actors at different levels work together to result in cell elongation arrest.

A reduction of cell wall loosening seems to be achieved by cross-linking of structural proteins, by creating a suboptimal pH-environment for loosening enzymes by modulation of H^+-ATPase activity and by structural changes in the composition and architecture of the cell walls. Microtubules and the reorientation of cellulose microfibrils are not involved in the control of cell elongation in the *Arabidopsis* root. Significant alterations in cellular symplast exchange, brought about by modulation of plasmodesmal transport, certainly influence the cell's general metabolism. In combination with changes that directly influence cell wall properties this may lead to the observed cell elongation arrest in the *Arabidopsis* root.

1 The *Arabidopsis* Root and Cell Elongation

In many 19th century publications it was made clear that both cell division and cell expansion were the two fundamental processes governing plant size and shape. The localization and the importance of cell elongation for root development was documented quantitatively by Sachs (1874). His figure of the *Vicia faba* primary root was in continuous use in handbooks during one hundred years. In the 20th century, research on root elongation was refined and more related to hormone signaling (Erickson and Sax 1956; Avery and Burkholder 1936). Rapidly growing roots of common crop species were mostly used as model objects. It is, however, striking that the process of cell elongation in the root received relatively little attention, especially compared to the processes of cell division and cell patterning in the apical meristem. Even in a very recent handbook it is just absent (Beck 2005)!

The arrival of *Arabidopsis thaliana* as a model species has also boosted the research on cell elongation. Today, the hypocotyls and the root of *Arabidopsis thaliana* are often used as model systems to study cell elongation, or anisotropic cell expansion, at a physiological and molecular level. Under normal conditions, all three developmental stages that a plant cell goes through, division, elongation and differentiation are continuously present in the *Arabidopsis* root apex. These developmental processes are both spatially and temporally highly organized and occur in a predictable manner (Fig. 1, Dolan et al. 1993; Scheres et al. 1994, 2002; Kidner et al. 2000). Under standard growing conditions every 30 minutes a new cell enters the differentiation zone in the trichoblast cell files (http://webhost.ua.ac.be/fymo/Root.avi). Division and differentiation of *Arabidopsis* cells, however, fall out of the scope of this book; reviews can be found elsewhere (Dewitte and Murray 2003; De Veylder et al. 2003; Larkin et al. 2003; Serna 2005, respectively).

The introduction of in vivo methods for monitoring and measuring cell elongation with a high resolution was instrumental for a better definition of the process at the cellular level. Although authors differed in interpretation and in terminology, they essentially came to the same conclusion that the rate of cell elongation is not homogeneous throughout the whole elongation zone of the *Arabidopsis* root. It starts at a slow pace in the distal part and reaches its maximum rate in the proximal part, just prior to the onset of root hair formation (Beemster and Baskin 1998; Ishikawa and Evans 1995; Mullen et al. 1998; Le et al. 2001; van der Weele 2003). A similar observation was made before in roots of other species (Ishikawa and Evans 1993).

Thus the elongation zone, situated in between the division zone and the differentiation zone, can be subdivided into two zones with a substantially different elongation rate. The first zone, in which cell length increases marginally, was called "the transition zone" in maize roots (Baluška et al. 1996), a term that will be used throughout this review. Cells leaving the meristem are not capable yet of performing fast cell elongation. In the transition zone, these post mitotic cells are progressively prepared for fast cell elongation. In the adjacent "fast elongation zone", cells elongate intensively realizing a volume increase of 300% in about two hours in the Columbia ecotype (Le et al. 2001).

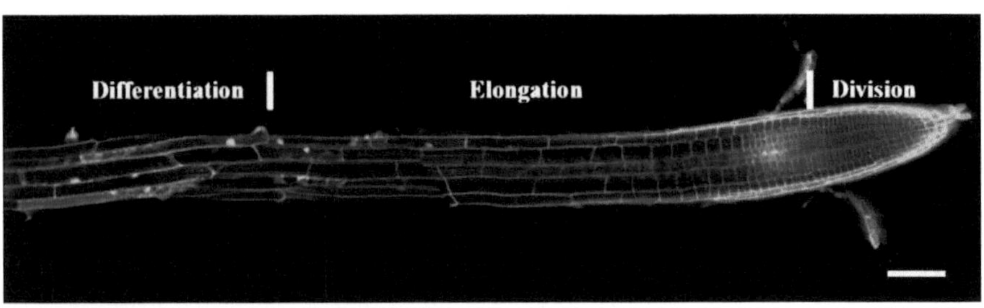

Fig. 1 Confocal micrograph of a 5-day-old *Arabidopsis* root (Col-0) stained with propidium iodide. The three developmental stages, division, elongation and differentiation are indicated. The *scale bar* represents 100 μm

Research on the developmental biology of *Arabidopsis* roots predominantly uses constitutive mutants with clear and severe root phenotypes (Arioli et al. 1998; Nicol et al. 1998; Sato et al. 2001), and therefore mainly focuses on the onset or the maintenance of cell elongation. Many of the actors that (are believed to) bring about or maintain cell elongation are reviewed in other parts of this volume. In contrast to this detailed knowledge, not much is known on the control itself of cell elongation. Which factors influence the final size of a root epidermal cell and how do they achieve this? The provision of answers to these questions is the goal of this work.

2 Ethylene as a Tool to Study the Control of Cell Elongation

Some time ago ethylene was recognized as a potential growth regulating factor (Neljubow 1901). The classic response of the etiolated seedling to ethylene is known as the triple response of the above ground parts of the plant (Guzman and Ecker 1990). The dramatic effect of exogenous ethylene on the whole plant is exemplified by the constitutive mutants *ctr* (Kieber et al. 1993) and *ein* or *etr* (Alonso et al. 1999; Bleecker et al. 1988; Chang et al. 1993). Ethylene is administered either as a gas or as its precursor 1-aminocyclopropane-1-carboxylic acid (ACC) (Kende 1993). Addition of ACC indeed effectively increases the endogenous ethylene concentration in plant tissues (Adams and Yang 1979; Lürssen et al. 1979). High concentrations of ethylene lead to inhibition of root growth within minutes. Treatments longer than 6 hours severely affect root development as roots start swelling and ectopic root hairs develop (Tanimoto et al. 1995; Dolan 1996). The first, fast response is relevant as a tool to study the control of elongation. Careful dosage and detailed analysis has shown that ethylene and ACC specifically control the fast cell elongation in a dose-dependent way (Le et al. 2001). This could well be a process used by the plant to continuously adjust root elongation rate in natural growing conditions.

The size reached by trichoblasts at the moment of root hair bulging has been proven to be a good tool for quantifying fast changes in cell elongation. The concept was introduced as the LEH (the Length of the first Epidermal cell with a visible root Hair bulge) (Fig. 2), (Le et al. 2001) and is used here to illustrate and quantify effects on cell elongation.

The availability of nutrients in the soil solution, such as nitrogen, phosphorus, iron and sulfur, can have profound impacts on root growth and root system architecture (López-Bucio et al. 2003). Changes in nitrate and phosphate concentrations have contrasting effects on the root elongation. In *Arabidopsis*, increasing nitrate availability reduces primary root elongation, whereas an increase in phosphate has the opposite effect (Linkohr et al. 2002).

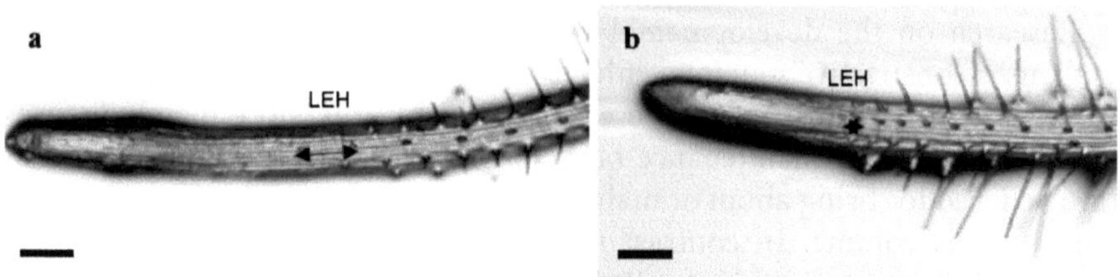

Fig. 2 Definition of LEH (Length of the first Epidermal cell with visible root Hair bulge) and the effect of 3 hrs 5 μM ACC. **a** The LEH is measured as the distance between the two first consecutive root hair bulges, as indicated with a *double-headed arrow*. In a wild-type root the LEH typically has a value of 128 μm. **b** In a root treated for 3 hrs with ACC (5 μM) the LEH is drastically reduced to 50 μm. The *scale bar* represents 100 μm

Besides the influence of nutrient availability, several common environmental stresses, such as osmotic- (Kreps et al. 2002), salt- (Sun et al. 2001) and metal stress (Sanita di Toppi and Gabbrielli 1999), and the application of hormones result in a clear reduction of epidermal cell size (Table 1).

So taken together, several environmental stresses and increased hormone levels evoke similar effects on the LEH value, and act in the same time frame as ACC. The use of ACC as an experimental tool to study cell elongation and its control is therefore justified. In what follows, the mechanism of cell elongation control is thus studied by unraveling the response of the

Table 1

Effector	Concentration	LEH (μm ± SE)	
No effector		128 ± 6	Control
Mannitol	200 mM	65 ± 3	Osmotic stress
PEG	10%	63 ± 3	
NaCl	200 mM	99 ± 4	Salt stress
KCl	150 mM	60 ± 3	
CsCl	5 mM	128 ± 3	
Cu^{2+}	200 mM	75 ± 3	Metal stress
Cd^{2+}	1 mM	93 ± 7	
Ag^{2+}	1 mM	104 ± 6	
ACC	5 μM	50 ± 4	Hormones
IAA	10 μM	65 ± 3	
ABA	50 μM	80 ± 3	
Methyl jasmonate	5 μM	36 ± 10	
Zeatin	1 μM	51 ± 14	

root to 3 hrs of treatment with the plant hormone ethylene, given as its precursor 1-aminocyclopropane-1-carboxylic acid (ACC) at 5 μM concentration. Treatment with ACC immediately and irreversibly reduces cell length in the *Arabidopsis* root by blocking the fast cell elongation as the LEH decreases from 128 μm (Fig. 2a), the value in control roots, to 50 μm (Fig. 2b) within 30 minutes. Both cell division and the rate of differentiation remain unaffected in time intervals < 6 hrs of treatment with ACC (Le et al. 2001). Besides experimentally increasing ethylene concentrations in the plant with ACC, or decreasing it with the ethylene biosynthesis inhibitor aminoethoxyvinylglycine (AVG) (Schaller and Kieber 2002), also constitutive mutants in the ethylene pathway (*eto2*: ethylene overproducer; *ctr1-1*: constitutive ethylene response; *ein2-1* and *etr1-3*: ethylene insensitive) were used as useful tools to monitor effects of changed ethylene conditions.

3 Where to Search

In theory, elongation rate is described by the "Lockhart equation" (Lockhart 1965):

$$r = \Phi(P - Y) \tag{1}$$

r = the rate or relative size of elongation,
Φ = the extensibility of the cell wall,
P = the turgor pressure (i.e. the source of cell wall stress),
Y = the yield threshold (i.e. the minimum pressure required for growth).

In theory, an inhibition or reduction of cell elongation (r) can result from a drop in turgor pressure (P), or from changes in cell wall architecture decreasing the extensibility (Φ) or increasing the yield threshold (Y). In reality, however, hormones and other effectors seem to modulate the rate of cell expansion only by influencing cell wall properties (Cosgrove 1999). The turgor pressure remains rather constant and recovers very quickly after the onset of a stress (Shabala and Lew 2002; Spollen and Sharp 1991; Tomos and Pritchard 1994).

The most obvious regulatory processes are indeed localized in the cell wall and will be discussed in detail. Besides the cell wall, the involvement of the cytoskeleton and functional cell-to-cell communication will also be highlighted. Ultimately, a CATMA (Complete Arabidopsis Transcriptome MicroArray) analysis indicates that the ACC-induced inhibition of cell elongation involves many levels of control, as the expression of genes with various functions in the cell's metabolism is altered significantly.

4 Microtubule-Microfibril Orientation and the Inhibition of Fast Cell Elongation

Elongation is a form of anisotropic cell expansion. During the fast cell elongation in the *Arabidopsis* root, cells grow rapidly in length but not in width or thickness. In primary cell walls the orientation of cellulose microfibrils determines the anisotropic mechanical properties of the wall (Kerstens et al. 2001) and cortical microtubules are believed to be instrumental in their orientation. The predominant hypothesis, known as the "cellulose synthase constraint model", states that cortical microtubules regulate cell expansion by directing the movement of cellulose synthase complexes during the deposition of microfibrils in the wall (Giddings and Staehelin 1991; Hable et al. 1998). In some cases, a straightforward causal effect between the microtubule orientation and the cellulose microfibril orientation indeed exists in *Arabidopsis* (see Wasteneys and Collings 2007, in this volume). The fact that perturbing microtubule dynamics or eliminating microtubules impairs growth anisotropy is also in favor of the model (Baskin et al. 1994). However, several more recent studies described a role for cortical microtubules in anisotropic growth without changing the direction of cellulose microfibrils (Himmelspach et al. 2003; Baskin et al. 2004; Wasteneys 2004). For example, the disruption of cortical microtubules in the *Arabidopsis mor1-1* mutant, leads to a severe root phenotype with left-handed root twisting and radial swelling but does not alter the transverse orientation of cellulose microfibrils (Sugimoto et al. 2003). The precise role of cortical microtubules and the cross-talk between microtubules and cellulose microfibrils in the control of cell expansion therefore remains controversial to date (Wasteneys 2004).

The orientation of microtubules in the *Arabidopsis* root was imaged by classical immunolocalization and by confocal microscopy of GFP-MAP4 transgenic lines (Granger and Cyr 2001). Both in control roots and in roots impaired in cell elongation by ACC, the orientation of microtubules is transverse to the root axis and switches to an oblique or longitudinal orientation in the cells that have arrested their elongation in the differentiation zone. In atrichoblasts of ACC-treated roots, this reorientation occurs immediately after the elongation stops, whereas in control roots this transition is more gradual (Fig. 3). After the disruption of microtubules by oryzalin, both elongation and the ACC-induced inhibition of elongation still occur (Fig. 4). Thus, neither the reorientation nor the absence of microtubules cause the elongation arrest.

Moreover, using Field Emission Scanning Electron Microscopy (FESEM) the cellulose microfibrils in the cell walls of the elongation zone were found to be orthogonal to the root axis both in untreated and in ACC-treated roots (Fig. 5). Thus, the stop in elongation is not caused by a switch in orientation of cellulose microfibrils.

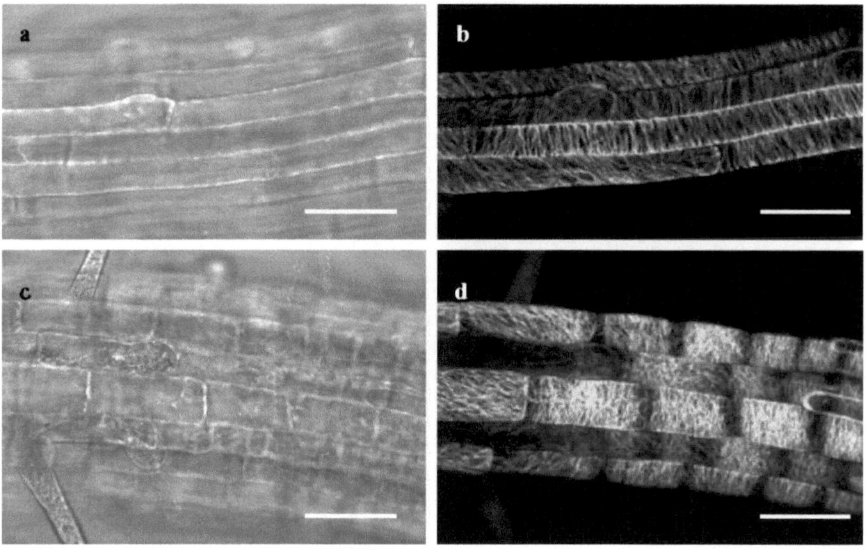

Fig. 3 Microtubules in roots of the *Arabidopsis* GFP-MAP4 line. Bright-field image of an untreated root (**a**) and a root treated for 3 hrs with ACC (**c**). **b,d** Confocal image of GFP-MAP4 marked microtubules in the same roots as in **a** and **c** respectively. In normal conditions, microtubules in the elongation zone are transverse to the main axis of the cell. Upon ACC addition, microtubules change to directions oblique or longitudinal in cells ≥ the ACC-imposed LEH. The *scale bar* represents 30 μm

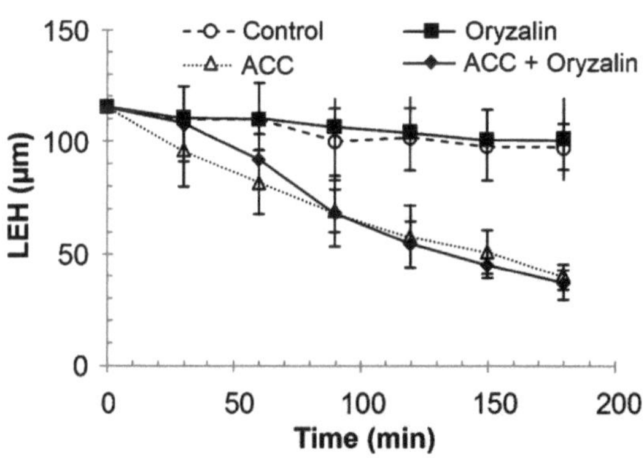

Fig. 4 Effects of the anti-microtubule drug oryzalin and ACC on the mean LEH values (± SD). The absence (oryzalin) of microtubules has no effect on the LEH value in "control" roots, nor on the characteristic ACC response where the LEH decreases

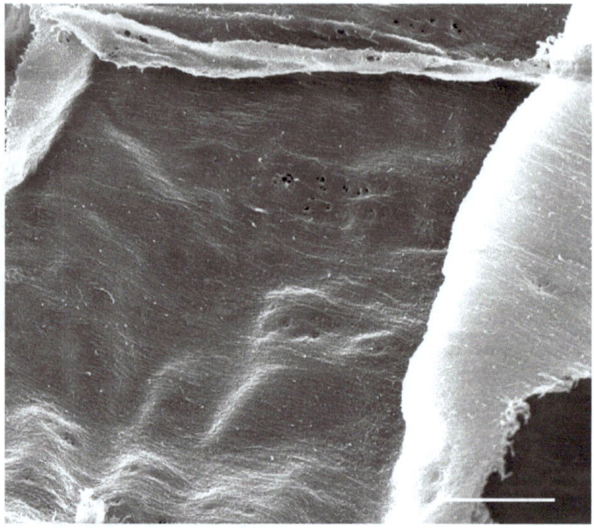

Fig. 5 FESEM image of the cellulose microfibrils in the innermost cell wall layer of part of a cell in the root elongation zone. Individual cellulose microfibrils are aligned parallel and perpendicular to the long axis of the cell. On the picture, several pitfields can be detected as well. The *scale bar* represents 1 μm

5 Apoplastic Events and Inhibition of Fast Cell Elongation

In the CATMA microarray comparison of control and ACC-treated roots, 18% of the genes with a significantly altered expression level have a putative func-

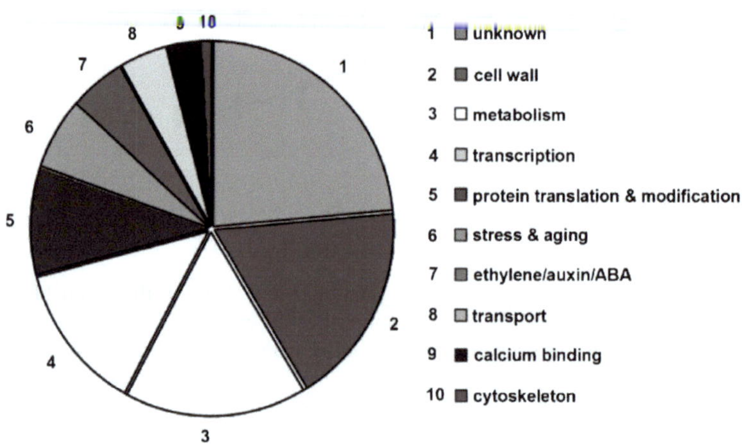

Fig. 6 Groups of genes differentially expressed between control and ACC-treated roots subdivided according to their putative function(s) in the cell

tion in cell wall metabolism, which stresses the importance of the apoplast in the control of cell elongation (Fig. 6). Theoretically cell elongation arrest could be caused by an increase in cross-linking events in the cell wall, a decrease in acid-induced wall loosening and/or changes in the deposition and structure of wall material. These different processes were therefore investigated in detail.

5.1 Cross-Linking of Structural Cell Wall Proteins

The extent of cross-linking between hydroxyproline-rich glycoproteins (HRGPs) can be an important mechanism to restrict cell growth during development (Brownleader et al. 2000; Knox 1995). HRGPs (Sommer-Knudsen et al. 1998), like arabinogalactan proteins (AGPs; Schultz et al. 2000; Showalter 2001) and extensins (Kieliszewski and Lamport 1994) are well-characterized structural cell wall proteins. In the process of HRGP-formation many Pro-residues are hydroxylated to form Hyp (Showalter 1993), hence its name. Cross-linking occurs by the formation of intramolecular isodityrosine (Idt) bridges (Epstein and Lamport 1984; Zhou et al. 1992) or tetrameric derivatives of tyrosine, di-isodityrosine (Di-Idt). This process is driven by peroxidase-action and catalyzed by hydrogen peroxide (Brady and Fry 1997). The Di-Idt-formation was even suggested to form intermolecular cross-links between extensins (Brady et al. 1996).

The involvement of HRGP cross-linking was studied in three steps. Interference in the hydroxylation of Pro to Hyp by the inhibitor 3,4-DL-dehydroproline (DP) significantly increased the size of control and ACC-treated cells (Fig. 7). Thus, fully functional HRGPs seem therefore necessary to stop cell elongation. As a next step, in a competition experiment for Di-Idt-cross-linking, the application of exogenous tyrosine significantly reduced the inhibitory effect of ACC on root cell elongation, but did not affect cell size in control plants (Fig. 7). Di-Idt-cross-linking thus seems instrumental for the ACC-induced arrest of cell elongation. Finally, the catalysators of the cross-linking events, extracellular "reactive oxygen species (ROS)" such as hydrogen peroxide, were detected in the short elongation zone by the specific fluorochrome Oxyburst Green H_2HFF-BSA only after ACC treatment (De Cnodder et al. 2005).

Moreover, in the CATMA analysis several genes coding for structural proteins and peroxidases were found to be up-regulated by ACC (unpublished results).

These evidences strongly suggest that HRGPs are involved in the control of cell elongation and that their cross-linking is necessary to limit the final cell length (De Cnodder et al. 2005). The involvement of HRGPs in cell elongation is further strengthened by the findings that fucosylated root AGPs (Van Hengel and Roberts 2002) and extensins LRX1 and LRX2 (Baumberger et al. 2003) are indispensable for normal cell elongation in *Arabidopsis*. Moreover, similar

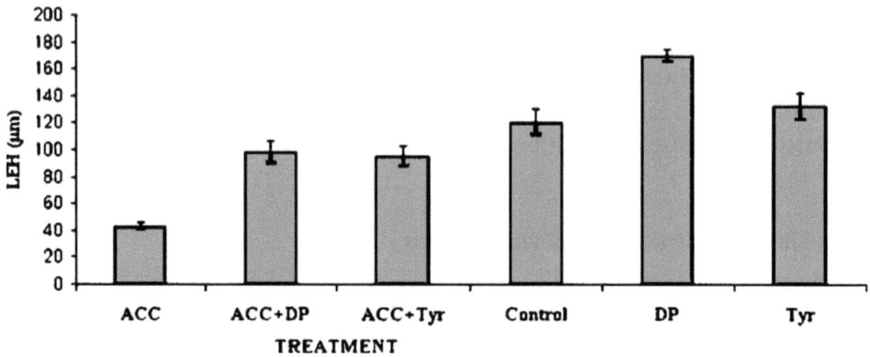

Fig. 7 Effect of 3,4-DL-dehydroproline (DP) and exogenous tyrosine on cell elongation in the *Arabidopsis* root. The LEH (± SE) is increased in control and ACC-treated roots after 3 hrs of treatment with 500 μM DP, whereas only the ACC-treated seedlings are affected by 3 hrs of treatment with 500 μM tyrosine

structural proteins are deposited in response to several elicitor-preparations (Bolwell et al. 2002; Bradley et al. 1992; Brownleader et al. 1995; El-Gendy et al. 2001; Wojtaszek et al. 1995).

5.2 Cell Wall Loosening

The acid-growth theory indicates protons as the primary wall loosening factor and states that the apoplastic acidification causes the cleavage of load-bearing bonds in the cell wall which results in cell elongation (Rayle and Cleland 1970, 1992; Cosgrove 1989). A low apoplastic pH (< 5) activates expansins in the wall, which probably break the hydrogen bonds between the cellulose chains and the cross-linking glycans (McQueen-Mason et al. 1992; Cosgrove 1999, 2000). Using the MIFE (Microelectrode Ion Flux Estimation) technique (Newman 2001; Shabala et al. 1997), the pH profile along the root surface was measured.

In control conditions, the surface pH is highest in the transition zone (TZ), and lowest in the adjacent fast elongation zone (EZ) (Fig. 8). In maize seedlings a similar spatial profile of root-surface acidification coincides with the spatial profile of growth along the root (Fan and Neumann 2004; Peters and Felle 1999; Pilet et al. 1983). The drop in pH at the onset of the elongation zone can therefore serve as an indicator of the transition from a slow to an explosive growth. Between 20 and 60 min after ACC application, however, the surface pH at this site increased steeply with 0.2–0.25 pH units, and remained fairly constant afterwards (Fig. 8, arrow and asterisk). Thus, ACC application causes an extracellular alkalinization that coincides both in time and space with the inhibition of cell elongation.

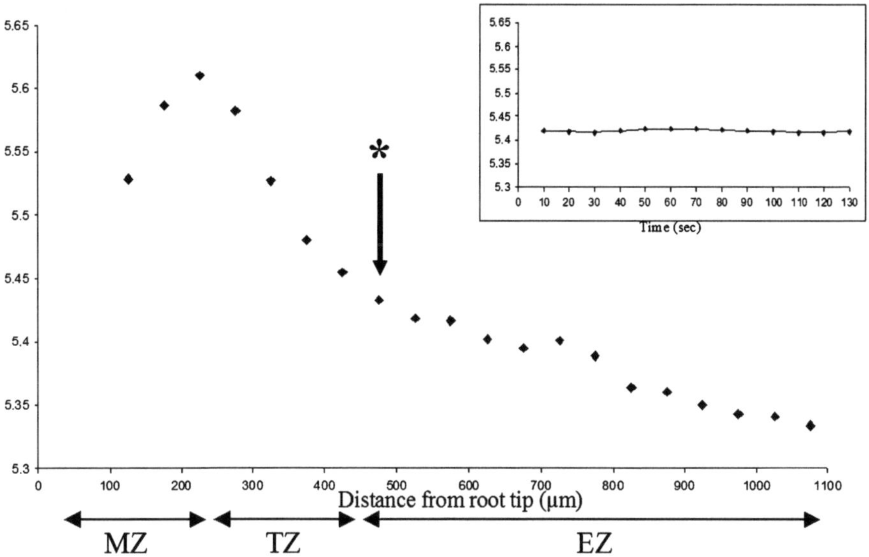

Fig. 8 Plot of the surface pH along the *Arabidopsis* root. The surface pH along the *Arabidopsis* root was recorded using the MIFE system and sampled every 50 μm from 150 μm of the root tip off to the root hair zone. The pH values are stable in one point (see *inset*). Each single point on the plot is the average of at least 10 measurements. ACC causes a rise of 0.2–0.25 pH units (ΔpH) at a distance of 400–500 μm from the *Arabidopsis* root tip (see *arrow* and *asterisk*). MZ = meristematic zone, TZ = transition zone, EZ = elongation zone

Extracellular acidification is mainly the result of plasma membrane H^+-ATPase activity, which governs the efflux of protons across the plasma membrane (Palmgren 1998, 2001). The possible link between H^+-ATPase activity and the alkalinization of the cell wall in the ACC-growth response was investigated by modulation of the enzyme's activity. Fusicoccin (FC) has been shown to bind to a plasma membrane receptor complex that includes both an H^+-ATPase and a 14-3-3 protein (Alsterfjord et al. 2004; Baunsgaard et al. 1998), and to promote the activation of H^+-ATPases (De Boer 1997; Malerba et al. 2004; Olivari et al. 1998). On the other hand *N*,*N*′-dicyclohexylcarbodiimide (DCCD) inhibits H^+-ATPases. Experiments with FC and DCCD have proven that ACC blocks the final cell length partially by locking the H^+-ATPases in their low activity state. This then results in extracellular alkalinization, rendering the apoplast unfavorable for expansin action.

Moreover, in ACC-treated roots, the expression of expansins, the primary actors during acid-growth, is reduced in the CATMA data (unpublished results). Besides expansins, xyloglucan endotransglycosylase/hydrolases (XTHs) are suggested to be involved in the loosening of cell walls (Fry et al. 1992). These enzymes cut and rejoin xyloglucan chains, the most dominant hemicelluloses tethering adjacent microfibrils in the primary cell wall of dicotyledons (Carpita and Gibeaut 1993; Brett and Waldron 1996). However, using a fluorescent in vivo technique (Vissenberg et al. 2000), it was observed that XTH action was not altered upon ACC treatment (Verbelen et al. 2006).

5.3 Cell Wall Composition and Architecture

The primary cell wall is composed of three independent but interacting networks: cellulose-hemicelluloses, pectins and structural cell wall proteins with a minority of phenolic esters and minerals built in (Carpita and Gibeaut 1993; Brett and Waldron 1996). To detect possible changes in the cell wall composition along the root upon ACC treatment, "Fourier-Transform Infrared" (FT-IR) spectroscopy was used (see box in Hématy and Höfte 2007, in this volume). This technique is based on the fact that different functional bonds in cell wall components absorb at characteristic frequencies in the infrared range (McCann et al. 1992). In particular, FT-IR can quantitatively detect saturated and unsaturated esters (Morikawa et al. 1978), amides of proteins, carboxylic acids (McCann et al. 1992), and it can provide a complex "fingerprint" of carbohydrates such as cellulose and xyloglucan (Kačuráková et al. 2000). FT-IR spectra were collected along untreated and ACC-treated roots by sampling in adjacent points from the root tip until the appearance of the first root hair bulge.

A color-coded representation of the absorbances at different wavenumbers reveals clear differences between control and ACC-treated roots (Fig. 9). In the control, the border between transition- and elongation zone is marked by a "switch" to a higher level of cellulose, xyloglucan and methyl-esterified pectins and to a lower protein content. ACC-treated roots lack this "switch" in the cell wall components, and consequently the spectral characteristics of the transition zone are extended. This observation fits with the findings that

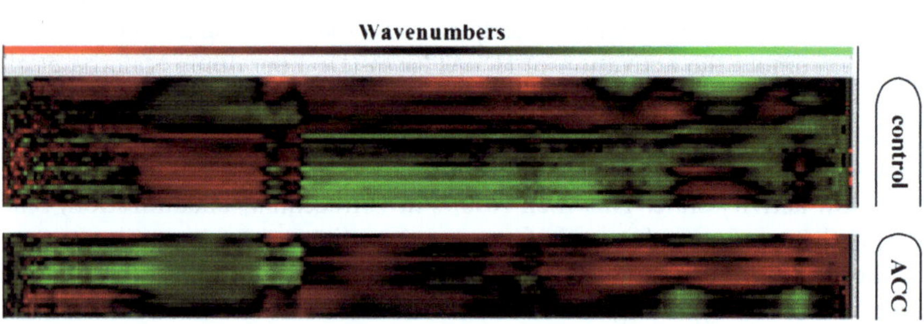

Fig. 9 Color-coded representation of average FT-IR spectra along the control and ACC-treated root. *Red* means a high absorbance at the wavenumber, *green* a low absorbance. Clear differences can be seen along a single root and between the two kinds of roots, reflecting alterations in cell wall components/composition

in mutants impaired in cellulose synthesis a reduced cellulose content causes a pronounced shortening of the epidermal cells (Fagard et al. 2000; Pagant et al. 2002). The decrease in methyl-esterification of pectins in ACC-treated seedlings opens up the possibility for more calcium cross-links at the free carboxylic acid residues (McCann et al. 1997). Calcium cross-linking of pectins can increase the stiffness of the wall and limit the movement of cell wall modifying enzymes, contributing to the inhibition of cell elongation.

Also a rise in phenolic ring absorbance, present in the cell wall components lignin, lignan and flavonoids, was detected in the FT-IR spectral map of ACC-treated roots. The mRNA levels of genes involved in the biosynthesis of these compounds were substantially up-regulated in response to ACC, as found in the CATMA data (unpublished results).

To summarize: the apoplastic events related to the definition of cell size involve actions as well on the level of protein content and cross-linking, as on the composition and the loosening of the cellulose-hemicellulose network and the pectin component of the cell wall.

6
Symplastic Events and Fast Cell Elongation

Besides the clear involvement of the apoplast in the response to ACC, changes in symplastic structures are also obviously linked with the ACC-induced inhibition of cell elongation.

Plasmodesmata (PD) are specialized channels that allow the movement of water, various nutrients, and other molecules, including signaling molecules, between adjacent plant cells (Lucas and Lee 2004). They are keystone components in the integration of development and may help to determine the program of cell differentiation. Depending on endogenous and exogenous cues they can either be open or closed and in this way determine symplastic fields in a tissue (Rinne and van der Schoot 1998). Modifications in symplastic transport through PD between cells have been considered in the control of elongation in the root (Roberts and Oparka 2003). They were shown to symplastically isolate epidermis cells in the *Arabidopsis* root at the onset of differentiation (Oparka et al. 1994; Ducket 1994). Recent studies have shown that callose ($\beta(1\rightarrow3)$-glucan) participates in the gating of plasmodesmata in vivo. Callose deposition inhibited symplasmic transport in wheat (*Triticum aestivum*) root apices exposed to aluminum toxicity (Sivaguru et al. 2000), while intensive depositions of callose in transgenic tobacco (*Nicotiana tabacum*) plants deficient in $\beta(1\rightarrow3)$-glucanase reduced the size exclusion limit of the plasmodesmata (Iglesias and Meins 2000).

The plant-specific unconventional myosin VIII is considered to be a linker between the cytoskeleton and the cell wall (Baluška et al. 2003). In maize, the myosin VIII protein accumulates at plasma membrane sites of high callose deposition, such as the cell plate and plasmodesmata (Baluška et al. 2001; Reichelt et al. 1999). It was suggested that myosin VIII binds either directly or via some further adaptor proteins to the callose synthase complexes (Verma and Hong 2001; Østergaard et al. 2003).

Factors that affect the permeability of plasmodesmata, such as callose and myosin (Cilia et al. 2002), could thus be involved in the control of symplastic transport and as mentioned before also in the control mechanism of elongation.

Callose was detected using aniline blue. This is a commonly used stain to detect (Nickle and Meinke 1998) and even to quantify callose (Beffa et al. 1996). In control *Arabidopsis* roots callose deposition was only seen in the cell plate of dividing cells and in the sieve plates of the phloem (both indicated with an arrow in Fig. 10a). It was clearly absent in cell walls of the elongation and differentiation zone. Treatment with ACC caused callose deposition in atrichoblast and trichoblast cell files of the epidermis and in the underlying cortical cells of the elongation and differentiation zone (Fig. 10b) (De Cnodder et al. 2005).

Via immunolocalization myosin VIII is detected in the meristem near the cell plate of dividing cells (Fig. 11). The protein is absent in the elongation

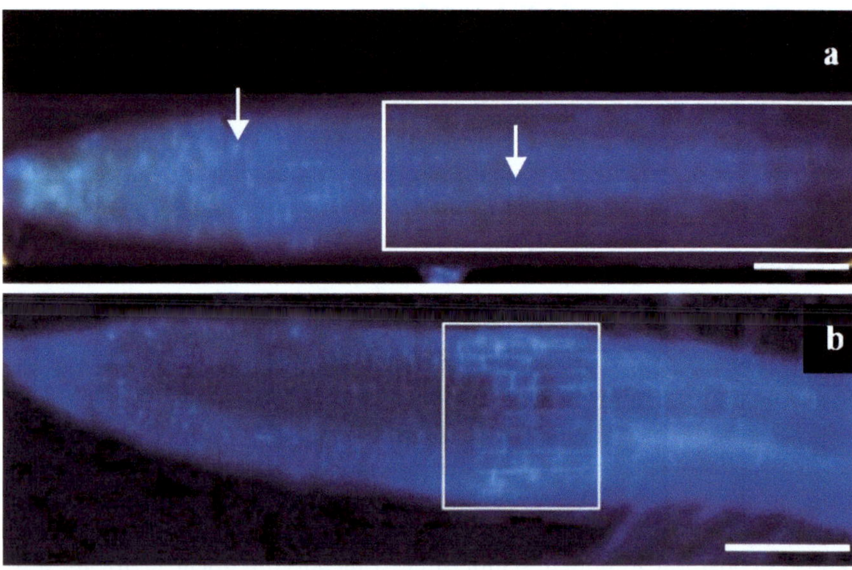

Fig. 10 Callose deposition in *Arabidopsis* roots. Roots were stained with aniline blue and observed under UV light. The *rectangles* on the figures indicate the elongation zones. **a** Callose is seen in the cell plates (*left arrow*) of dividing cells and the sieve plates of the phloem (*right arrow*). Control roots have no callose deposited in the epidermis and cortex of the elongation and differentiation zone (see absence of a clear signal in the *rectangle*). **b** Roots treated for 3 hrs with ACC accumulate callose in the elongation zone and differentiation zone in cell walls of epidermal and cortex cells (see labeling in the *rectangle*). The *scale bar* represents 100 μm. (Reprinted with permission of The New Phytologist)

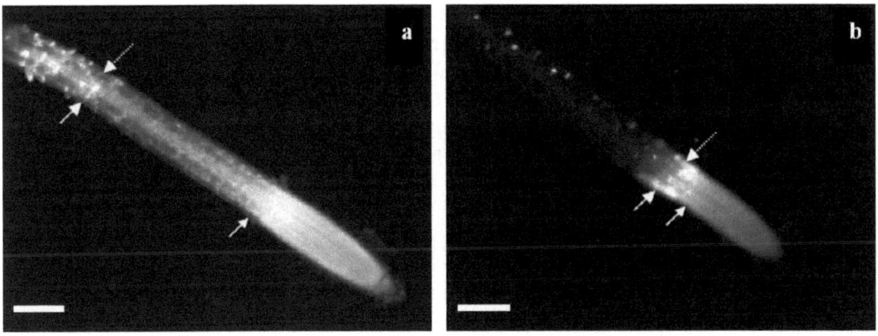

Fig. 11 Myosin VIII immunolocalization in *Arabidopsis* roots. **a** In control roots myosin VIII is located in the transverse walls of the elongation zone. **b** 3 hrs of ACC treatment reduces the elongation zone and shifts the labeling with the anti-myosin VIII antibody towards the root tip. The *lower arrows* in **a** and **b** mark the beginning and the end of the elongation zone, the *dotted (upper) arrows* point to myosin VIII occurrence. The *scale bar* represents 100 μm

zone (marked by two white arrows at the underside of the root) but is again very typically present near the transverse cell walls at the end of the elongation zone where differentiation starts (Fig. 11a,b, dotted arrows). There myosin VIII might be involved in setting up a symplastic limit between the elongation and differentiation zone (Oparka et al. 1994). Its localization shifts along with this border towards the root tip in ACC-treated roots. In parallel, a putative myosin heavy chain gene was up-regulated in response to ACC, suggesting a role for myosin in elongation control (unpublished results).

Taken together, the ectopic deposition of callose and the localization of the unconventional plant-specific myosin VIII point to possible adjustments of plasmodesmal transport in the set of mechanisms used for cell elongation control.

7 Conclusions and Further Perspectives

Cell elongation is controlled by the apoplast as well as the symplast, but many aspects remain to be uncovered. It is evident that changes in cell wall composition and structure, in cross-linking and on the level of cell wall loosening are instrumental in controlling cell wall extensibility. The link between symplastic events and the changes observed in the apoplast is certainly not less important. Actual research shows that cell–cell communication is indeed crucial for long distance signaling and whole plant development.

Future research on cell elongation using integrative approaches including transcriptome and proteome analysis will expand our knowledge on the general regulation of this process.

Acknowledgements Authors acknowledge the financial support of the Fund for Scientific Research Flanders (FWO—Belgium), grant G0345.02. K.V. is a postdoctoral fellow of the Fund for Scientific Research Flanders (FWO—Belgium). Authors greatly acknowledge the help of B. Van Loock, S. Foubert, J. Van Orden, M. Liao, J. Eysermans—EMAT (Univ. Antwerpen), Dr. J. Le (Purdue Univ.), Dr. T. Elzenga and M. Staal (Univ. Groningen), Dr. F. Baluška (Univ. Bonn), Dr. H. Höfte and Dr. G. Mouille (INRA, Versailles), Dr. S.C. Fry (Univ. Edinburgh).

References

Adams DO, Yang SF (1979) Ethylene biosynthesis: identification of 1-aminocyclopropane-1-carboxylic acid as an intermediate in the conversion of methionine to ethylene. Proc Natl Acad Sci USA 76:170–174

Alonso JM, Hirayma T, Roman G, Nourizodeh S, Ecker JR (1999) A multi-responsive gene encoding 1-aminocyclopropane-1-carboxylate synthase (ACS6) in mature *Arabidopsis* leaves. Plant Mol Biol 39:209–219

Alsterfjord M, Sehnke PC, Arkell A, Larsson H, Svennelid F, Rosenquist M, Ferl RJ, Sommarin M, Larsson C (2004) Plasma membrane H^+-ATPase and 14-3-3 isoforms of *Arabidopsis* leaves: evidence for isoform specificity in the 14-3-3/H^+-ATPase interaction. Plant Cell Physiol 45:1202–1210

Arioli T, Peng L, Betzner AS (1998) Molecular analysis of cellulose biosynthesis in *Arabidopsis*. Science 279:717–720

Avery GS, Burkholder PR (1936) Polarized growth and cell studies on the Avena coleoptile, phytohormone test object. Bull Torrey bot Club 63:1–15

Baluška F, Volkmann D, Barlow RW (1996) Specialized zones of development in roots. View from the cellular level. Plant Physiol 112:3–4

Baluška F, Cvrclova F, Kendrick-Jones J, Volkmann D (2001) Sink plasmodesmata as gateways for phloem unloading. Myosin VIII and calreticulin as molecular determinants of sink strength? Plant Physiol 126:39–46

Baluška F, Samaj J, Wojtaszek P, Volkmann D, Menzel D (2003) Cytoskeleton-plasma-membrane-cell wall continuum in plants. Emerging links revisited. Plant Physiol 133:482–491

Baskin TI, Wilson JE, Cork A, Williamson RE (1994) Morphology and microtubule organization in *Arabidopsis* roots exposed to oryzalin or taxol. Plant Cell Physiol 35:935–942

Baskin TI, Beemster GT, Judy-March JE, Marga F (2004) Disorganization of cortical microtubules stimulates tangential expansion and reduces the uniformity of cellulose microfibril alignment among cells in the root of *Arabidopsis*. Plant Physiol 135:2279–2290

Baumberger N, Steiner M, Ryser U, Keller B, Ringli C (2003) Synergistic interaction of the two paralogous *Arabidopsis* genes *LRX1* and *LRX2* in cell wall formation during root hair development. Plant J 35:71–81

Beck CB (2005) An introduction to plant structure and development. Cambridge University Press, Cambridge

Beemster GTS, Baskin TI (1998) Analysis of cell division and elongation underlying the developmental acceleration of root growth in Arabidopsis thaliana. Plant Physiol 116:1515–1526

Beffa RS, Hofer RM, Thomas M, Meins F (1996) Decreased susceptibility to viral disease of beta-1,3-glucanase-deficient plants generated by antisense transformation. Plant Cell 8:1001–1011

Bleecker AB, Estelle MA, Somerville C, Kende H (1988) Insensitivity to ethylene conferred by a dominant mutation in *Arabidopsis thaliana.* Science 241:1086–1089

Bolwell GP, Bindschedler LV, Blee KA, Butt VS, Davies DR, Gardner SL, Gerrish C, Minibayeva F (2002) The apoplastic oxidative burst in response to biotic stress in plants: a three-component system. J Exp Bot 53:1367–1376

Bradley DJ, Kjellbom P, Lamb CJ (1992) Elicitor- and wound-induced oxidative cross-linking of a proline-rich plant cell wall protein: a novel and rapid response. Cell 70:21–30

Brady DJ, Fry SC (1997) Formation of di-isodityrosine and loss of isodityrosine in the cell walls of tomato cell-supension cultures treated with fungal elicitors and H_2O_2. Plant Physiol 115:87–92

Brady DJ, Sadler IH, Fry SC (1996) Di-isodityrosine, a novel tetrameric derivative of tyrosine in plant cell wall proteins: a new potential cross-links. J Biochem 315:323–327

Brett CT, Waldron KW (1996) Physiology and biochemistry of plant cell walls, 2nd edn. Chapman and Hall, London

Brownleader MD, Ahmed N, Trevan M, Chaplin MF, Dey PM (1995) Purification and partial characterization of Tomato extensin peroxidase. Plant Physiol 109:1115–1123

Brownleader MD, Hopkins J, Mobasheri A, Dey PM, Jackson P, Trevan M (2000) Role of extensin peroxidase in tomato (*Lycopersicon esculentum* Mill.) seedling growth. Planta 210:668–676

Carpita NC, Gibeaut DM (1993) Structural models of primary cell walls in flowering plants: consistency of molecular structure with the physical properties of the walls during growth. Plant J 3:1–30

Chang C, Kwok SF, Bleecker AB, Meyerowitz EM (1993) *Arabidopsis* ethylene-response gene ETR1: Similarity of product to two-component regulators. Science 262:539–544

Cilia M, Cantrill L, van Bel A (2002) Plasmodesmata 2001: On safari through the symplast. Plant Cell 14:7–10

Cosgrove DJ (1989) Characterization of long-term extension of isolated cell walls from growing cucumber hypocotyls. Plant 177:121–133

Cosgrove DJ (1999) Enzymes and other agents that enhance cell wall extensibility. Ann Rev Plant Mol Biol 50:391–417

Cosgrove DJ (2000) Loosening of plant cell walls by expansins. Nature 407:321–326

De Boer AH (1997) Fusicoccin – a key to multiple 14-3-3 locks? Trends Plant Sci 2:60–66

De Cnodder T, Vissenberg K, Van Der Straeten D, Verbelen J-P (2005) Regulation of cell length in the *Arabidopsis thaliana* root by the ethylene precursor 1-aminocyclopropane-1-carboxylic acid: A matter of apoplastic reactions. New Phytol 168:541–550

De Veylder L, Joubes J, Inzé D (2003) Plant cell cycle transitions. Curr Opin Plant Biol 6:536–543

Dewitte W, Murray JAH (2003) The plant cell cycle. Ann Rev Plant Biol 54:235–264

Dolan L, Janmaat K, Willemsen V, Linstead P, Poethig S, Roberts K, Scheres B (1993) Cellular organisation of the Arabidopsis root. Development 119:71–84

Dolan L (1996) Pattern in the root epidermis: an interplay of diffusible signals and cellular geometry. Ann Bot 77:547–553

Duckett CM, Oparka KJ, Prior DAM, Dolan L, Roberts K (1994) Dye-coupling in the root epidermis is progessively reduced during development. Development 120:3247–3255

El-Gendy W, Brownleader MD, Ismail H, Clarke PJ, Gilbert J, El-Bordiny F, Trevan M, Hopkins J, Naldrett M, Jackson P (2001) Rapid deposition of wheat cell wall structural proteins in response to Fusarium-derived elicitors. J Exp Bot 52:85–90

Epstein L, Lamport DTA (1984) An intramolecular linkage involving isodityrosine in extensin. Phytochemistry 23:1241–1246

Erickson RO, Sax KB (1956) Elemental growth rate of the primary root of Zea mays. Proc Amer Philos Soc 100:487–498

Fagard M, Desnos T, Desprez T, Goubet F, Refregier G, Mouille G, McCann M, Rayon C, Vernhettes S, Höfte H (2000) *PROCUSTE1* encodes a cellulose synthase required for normal cell elongation specifically in roots and dark-grown hypocotyls of *Arabidopsis*. Plant Cell 12:2409–2423

Fan N, Neumann PM (2004) The spatially variable inhibition by water deficit of maize root growth correlated with altered profiles of proton flux and cell wall pH. Plant Physiol 135:2291–2300

Fry SC, Smith RC, Renwick KF, Martin DJ, Hodge SK, Matthews KJ (1992) Xyloglucan endotransglycosylase, a new wall-loosening enzyme activity from plants. Biochem J 282:821–828

Giddings TH Jr, Staehelin LA (1991) Microtubule-mediated control of microfibril deposition: a re-examination of the hypothesis. In: Lloyd CW (ed) The cytoskeletal basis of plant growth and form. Academic Press, San Diego, p 85–100

Granger CL, Cyr RJ (2001) Spatiotemporal relationships between growth and microtubule orientation as revealed in living root cells of *Arabidopsis thaliana* transformed with green-fluorescent-protein gene construct GFP-MBD. Protoplasma 216:201–214

Guzman P, Ecker JR (1990) Exploiting the triple response of Arabidopsis to identify ethylene-related mutants. Plant Cell 2:513–523

Hable WE, Bisgrove SR, Kropf DL (1998) To shape a plant – The cytoskeleton in plant morphogenesis. Plant Cell 10:1772–1774

Hématy K, Höfte H (2007) Cellulose and cell elongation. In: Verbelen JP, Vissenberg K (eds) The expanding cell. Plant Cell Monographs, vol 5. Springer, Berlin Heidelberg New York (in press)

Himmelspach R, Williamson RE, Wasteneys GO (2003) Cellulose microfibril alignment recovers from DCB-induced disruption despite microtubule disorganization. Plant J 36:565–575

Iglesias VA, Meins F Jr (2000) Movement of plant viruses is delayed in a β-1,3-glucanase-deficient mutant showing a reduced plasmodesmatal size exclusion limit and enhanced callose deposition. Plant J 21:157–166

Ishikawa H, Evans ML (1993) The role of the distal elongation zone in the response of maize roots to auxin and gravity. Plant Physiol 102:1203–1210

Ishikawa H, Evans ML (1995) Specialized zones of development in roots. Plant Physiol 109:725–727

Kačuráková M, Capek P, Sasinkova V, Wellner N, Ebringerova A (2000) FT-IR study of plant cell wall model compounds: pectic polysaccharides and hemicelluloses. Carbohydr Res 43:195–203

Kende H (1993) Ethylene biosynthesis. Annu Rev Plant Physiol Plant Mol Biol 44:283–307

Kerstens S, Decraemer FWS, Verbelen J-P (2001) Cell walls at the surface behave mechanically like fiber-reinforced composite materials. Plant Physiol 127:281–285

Kidner C, Sundareson V, Roberts K, Dolan L (2000) Clonal analysis of the Arabidopsis root confirms that position, not lineage, determines cell fate. Planta 211:191–199

Kieber JJ, Rothenberg M, Roman G, Feldmann KA, Ecker JR (1993) CTR1, a negative regulator of the ethylene response pathway in Arabidopsis, encodes a member of the raf family of protein kinases. Cell 72:427–441

Kieliszewski MJ, Lamport DTA (1994) Extensin: repetitive motifs, functional sites, post-translational codes, and phylogeny. Plant J 5:157–172

Knox JP (1995) The extracellular matrix in higher plants. Developmentally regulated proteoglycans and glycoproteins of the plant cell surface. FASEB J 9:1004–1012

Kreps JA, Wu Y, Chang H-S, Zhu T, Wang X, Harper JF (2002) Transcriptome changes for *Arabidopsis* in response to salt, osmotic and cold stress. Plant Physiol 130:2129–2141

Larkin JC, Brown ML, Schiefelbein J (2003) How cells know what they want to be when they grow up? Lessons from epidermal patterning in *Arabidopsis*. Ann Rev Plant Biol 54:403–430

Le J, Vandenbussche F, Van Der Straeten D, Verbelen J-P (2001) In the early response of Arabidopsis roots to ethylene, cell elongation is up- and down regulated and uncoupled from differentiation. Plant Physiol 125:519–522

Linkohr BI, Williamson LC, Fitter AH, Leyser O (2002) Nitrate and phosphate availibility and distribution have different effects on root system architecture of Arabidopsis. Plant J 29:751–760

Lockhart JA (1965) An analysis of irreversible plant cell elongation. J Theoret Biol 8:264–275

López-Bucio J, Cruz-Ramírez A, Herrera-Estrella L (2003) The role of nutrient availibility in regulationg root architecture. Curr Opin Plant Biol 6:280–287

Lucas WJ, Lee JY (2004) Plant cell biology - Plasmodesmata as supracellular control network in plants. Nature Rev Mol Cell Biol 5:712–726

Lürssen K, Naumann K, Schröder R (1979) 1-Aminocyclopropane-1-carboxylic acid - an intermediate of the ethylene biosynthesis in higher plants. Z Pflanzenphysiol 92:285–294

Malerba M, Cerana A, Crosti P (2004) Comparison between the effect of fusicoccin, tunicamycin, and brefeldin A on programmed cell death of cultured sycamore (*Acer pseudoplatanus* L.) cells. Protoplasma 224:61–70

McCann MC, Hammouri M, Wilson R, Belton P, Roberts K (1992) Fourier transform infrared microspectroscopy is a new way to look at plant cell walls. Plant Physiol 100:1940–1947

McCann MC, Chen L, Roberts K, Kemsley EK, Sene C, Carpita NC, Stacey NJ, Wilson RH (1997) Infrared microspectroscopy: Sampling heterogeneity in plant cell wall composition and architecture. Physiol Plant 100:729–738

McQueen-Mason SJ, Durachko DM, Cosgrove DJ (1992) Two endogenous proteins that induce cell wall extension in plants. Plant Cell 4:1425–1433

Morikawa H, Hayashi R, Senda M (1978) Infrared analysis of pea stem cell walls and oriented structure of matrix polysaccharides in them. Plant Cell Physiol 19:1151–1159

Mullen JL, Ishikawa H, Evans ML (1998) Analysis of changes in relative elemental growth rate in the elongation zone of the Arabidopsis roots upon gravistimulation. Planta 206:598–603

Neljubow DN (1901) Über die horizontale Nutation der Stengel von Pisum sativum und einiger anderen. Pflanzen Beiträge und Botanik Zentralblatt 10:128–139

Newman IA (2001) Ion transport in roots: measurements of fluxes using ion-selective microelectrodes to characterize transporter function. Plant Cell Environ 24:1–14

Nickle TC, Meinke DW (1998) A cytokinesis-defective mutant of *Arabidopsis* (cytl) characterized by embryonic lethality, incomplete cell walls, and excessive callose accumulation. Plant J 15:321–332

Nicol R, His I, Jauneau A, Vernhettes S, Canut H, Höfte H (1998) A plasma membrane-bound putative endo-1,4-beta-D-glucanase is required for normal wall assembly and cell elongation in Arabidopsis. EMBO J 17:5563–5576

Olivari C, Meanti C, De Michelis MI, Rasi-Caldogno F (1998) Fusicoccin induces the association between the plasma membrane H^+-ATPase and the fusicoccin receptor. Plant Physiol 116:529–537

Oparka KJ, Duckett CM, Prior DAM, Fisher DB (1994) Real-time imaging of phloem unloading in the root tip of *Arabidopsis*. Plant J 6:759–766

Østergaard L, Petersen M, Mattson O, Mundy J (2003) An Arabidopsis callose synthase. Plant Mol Biol 49:559–566

Pagant S, Bichet A, Sugimoto K, Lerouxel O, Desprez T, McCann M, Lerouge P, Vernhettes S, Höfte H (2002) *KOBITO1* encodes a novel plasma membrane protein necessary for normal synthesis of cellulose during cell expansion in *Arabidopsis*. Plant Cell 14:2001–2013

Palmgren MG (1998) Proton gradients and plant growth: Role of the plasma membrane H^+-ATPase. Adv Bot Res 28:1–70

Palmgren MG (2001) Plasma membrane H^+-ATPases: Powerhouses for nutrient uptake. Annu Rev Plant Physiol Plant Mol Biol 52:817–845

Peters WS, Felle HH (1999) The correlation of profiles of surface pH and elongation growth in maize roots. Plant Physiol 121:905–912

Pilet P-E, Versel J-M, Mayor G (1983) Growth distribution and surface pH paterns along maize roots. Planta 158:398–402

Rayle DL, Cleland RE (1970) Enhancement of wall loosening and elongation by acid solutions. Plant Physiol 46:250–253

Rayle DL, Cleland RE (1992) The acid growth theory of auxin-induced cell elongation is alive and well. Plant Physiol 99:1271–1274

Reichelt S, Knight AE, Hodge TP, Baluška F, Šamaj J, Volkmann D, Kendrick-Jones J (1999) Characterization of the unconventional myosin VIII in plant cells and its localization at the post-cytokinetic cell wall. Plant J 19:555–569

Rinne PLH, van der Schoot C (1998) Symplasmic fields in the tunica of the shoot apical meristem coordinate morphogenetic events. Development 125:1477–1485

Roberts AG, Oparka KJ (2003) Plasmodesmata and the control of symplastic transport. Plant Cell Environ 26:103–124

Sachs J (1874) Ueber das Wachstum der Haupt- und Nebenwurzeln. Arb D Bot Intst Würzb, I

Sanita di Toppi L, Gabbrielli R (1999) Response to cadmium in higher plants. Environ Exp Bot 41:105–130

Sato S, Kato T, Kakegawa K, Ishii T, Liu YG, Awano T, Takabe K, Nishiyama Y, Kuga S, Sato S, Nakamura Y, Tabata S, Shibata D (2001) Role of the putative membrane-bound endo-1,4-β glucanase KORRIGAN in cell elongation and cellulse synthesis in Arabidopsis thaliana. Plant Cell Physiol 42:251–263

Schaller GE, Kieber JJ (2002) Ethylene. doi: 10.1199/tab.0071, The Arabidopsis book, p 1–18

Scheres B, Wolkenfelt H, Willemsen V, Terlouw M, Lawson E, Dean C, Weisbeek P (1994) Embryonic origin of the Arabidopsis primary root and root meristem initials. Development 120:2475–2487

Scheres B, Benfey P, Dolan L (2002) Root development. doi: 10.1199/tab.0101, The Arabidopsis book, p 1–18

Schultz CJ, Johnson KL, Currie G, Bacic A (2000) The classical arabinogalactan protein gene family of *Arabidopsis*. Plant Cell 12:1751–1767

Serna L (2005) Epidermal cell patterning and differentiation throughout he apical-basal axis of the seedling. J Exp Bot 56:1983–1989

Shabala NS, Lew RR (2002) Turgor regulation in osmotically stressed *Arabidopsis* epidermal root cells. Direct support for the role of inorganic ion uptake as revealed by concurrent flux and cell turgor measurements. Plant Physiol 129:290–299

Shabala S, Newman IA, Morris J (1997) Oscillations in H^+ and Ca^{2+} ion fluxes around the elongation region of corn roots and effects of external pH. Plant Physiol 113:111–118

Showalter AM (1993) Structure and function of plant cell wall proteins. Plant Cell 5:9–23
Showalter AM (2001) Arabinogalactan proteins: structure, expression and function. Cell Mol Life Sci 58:1399–1417
Sivaguru M, Fujiwara T, Samaj J, Baluška F, Yang Z, Osawa H, Maeda T, Mori T, Volkmann D, Matsumoto H (2000) Aluminium-induced 1-3-β-D-glucan inhibits cell-to-cell trafficking of molecules through plasmodesmata. A new mechanism of aluminium toxicity in plants. Plant Physiol 124:991–1005
Sommer-Knudsen J, Bacic A, Clarke AE (1998) Hydroxyproline-rich plant glycoproteins. Phytochemistry 47:483–497
Spollen WG, Sharp RE (1991) Spatial distribution of turgor and root growth at low water potentials. Plant Physiol 96:438–443
Sugimoto K, Himmelspach R, Williamson RE, Wasteneys GO (2003) Mutation or drug-dependent microtubule disruption causes radial swelling without altering parallel cellulose microfibril deposition in *Arabidopsis* root cells. Plant Cell 15:1414–1429
Sun W, Bernard C, van de Cotte B, Van Montagu M, Verbruggen N (2001) At-HSP17.6A, encoding a small heat-shock protein in *Arabidopsis*, can enhance osmotolerance upon overexpression. Plant J 27:407–417
Tanimoto M, Roberts K, Dolan L (1995) Ethylene is a positive regulator of root hair development in *Arabidopsis thaliana*. Plant J 8:943–948
Tomos AD, Pritchard J (1994) Biophysical and biochemical control of cell expansion in roots and leaves. J Exp Bot 45:1721–1731
van der Weele CM, Jiang H, Palaniappan KK, Ivanov VB, Palaniappan K, Baskin TI (2003) A new algorithm for computational image analysis of deformable motion at high spatial and temporal resolution applied to root growth. Rougly uniform elongation in the meristem and also, after an abrupt acceleration, in the elongation zone. Plant Physiol 132:1–11
Van Hengel AJ, Roberts K (2002) Fucosylated arabinogalactan proteins are required for full root cell elongation in *Arabidopsis*. Plant J 32:105–113
Verbelen J-P, Le J, Vissenberg K, De Cnodder T, Vandenbussche F, Sugimoto K, Van Der Straeten D (2006) Microtubules and the control of cell elongation in *Arabidopsis* roots. *NATO monograph: The Plant Cytoskeleton: Functional diversity and biotechnological implications* (in press)
Verma DPS, Hong Z (2001) Plant callose synthase complexes. Plant Mol Biol 47:693–701
Vissenberg K, Martinez-Vilchez IM, Verbelen J-P, Miller JG, Fry SC (2000) In vivo colocalizatin of xyloglucan endotransglycosylase activity and its donor substrate in the elongation zone of Arabidopsis roots. Plant Cell 12:1229–1238
Wasteneys GO (2004) Progress in understanding the role of microtubules in plant cells. Curr Opin Plant Biol 7:651–660
Wasteneys GO, Collings DA (2007) The cytoskeleton and co-ordination of directional expansion in a multicellular context. In: Verbelen JP, Vissenberg K (eds) The expanding cell. Plant Cell Monographs, vol 5. Springer, Berlin Heidelberg New York (in press)
Wojtaszek P, Trethowan J, Bolwell GP (1995) Specificity in the immobilization of cell wall proteins in response to different elicitor molecules in suspenson-cultured cells of French bean (*Phaseolus vulgaris* L.). Plant Mol Biol 28:1075–1087
Zhou J, Rumeau D, Showalter AM (1992) Isolation and characterization of two wound-regulated tomato extensin genes. Plant Mol Biol 20:5–17

Plant Cell Monogr (5)
J.-P. Verbelen · K. Vissenberg: The Expanding Cell
DOI 10.1007/7089_2006_079/Published online: 7 October 2006
© Springer-Verlag Berlin Heidelberg 2006

Signal Crosstalk in the Control of Hypocotyl Elongation in *Arabidopsis*

Liesbeth De Grauwe · Filip Vandenbussche · Dominique Van Der Straeten (✉)

Unit Plant Hormone Signalling and Bio-imaging, Department of Molecular Genetics, Ghent University, Ledeganckstraat 35, 9000 Ghent, Belgium
Dominique.VanDerStraeten@ugent.be

Abstract Since plants are sessile organisms, environmental factors have a strong influence on their development. Hence, there is a need for adaptive mechanisms which integrate the multiple endogenous and environmental signals resulting in the most adequate biological response. Because of its morphological simplicity, the hypocotyl is often used as a model to study signal integration. Hypocotyl extension is a process regulated by a network of interacting signals, including light and hormones. Both are subject to reciprocal regulation, with hormonal factors influencing the biosynthesis and/or signalling of other hormones. In the following chapter, the current state of knowledge on the regulation of hypocotyl growth will be reviewed.

Abbreviations

ACC	1-Aminocyclopropane-1-carboxylic acid
ACO	ACC oxidase
AMP	Altered meristem program
ARF	Auxin response factor
ARR	*Arabidopsis* response regulator
BA	Benzyladenine
BAS1-D	phyB activation-tagged suppressor1-dominant
BR	Brassinosteroid
CAB	Chlorophyll A/B binding protein
CBB	Cabbage
CCA	Circadian clock-associated
CDD	cop10/det1/dbb1
COP	Constitutive photomorphogenesis
CPD	Constitutive photomorphogenesis and dwarfism
CRY	Cryptochrome
CSN	COP9-signalosome
DET	De-etiolated
EBR	Epibrassinolide
EIN	Ethylene insensitive
FUS	Fusca
GA	Gibberellin
GUS	β-Glucuronidase
HLS	Hookless
HPT	Hauptling
HSS	Hookless suppressor

HY	Long hypocotyl
IAA	Indol-3-acetic acid
LHY	Late elongated hypocotyl
LNM	Low-nutrient medium
LRE	Light-responsive element
MS	Murashige and Skoog
NPA	α-Naphthylphthalamic acid
NPH	Nonphototropic hypocotyl
PCK	Phosphoenolpyruvate carboxykinase
PHOT	Phototropin
PHY	Phytochrome
PIN	Pin-formed
PT	Primordial timing
SAX	Hypersensitive to abscisic acid and auxin
SCF	SKP1/Cullin/F-box
SPY	Spindly
TOC	Timing of CAB-expression

1 Introduction

The hypocotyl of dicotyledonous plants forms the connection between the embryonic leaves (also referred to as cotyledons) and the seedling root. The hypocotyl is a plastic organ, strongly influenced by different factors regulating cell elongation such as light, temperature and phytohormones. Because of its morphological simplicity, the hypocotyl can be used as a model to study the integration of signals that control plant growth. Both physiological and genetic studies have revealed a complex regulation of hypocotyl growth.

Both light- and dark-grown *Arabidopsis thaliana* hypocotyls contain parallel longitudinal files of approximately 20 epidermal cells of embryonic origin. Since no obvious increase in cell number could be observed during seedling growth, it can be assumed that epidermal cell divisions are absent in this tissue. Therefore, hypocotyl elongation is only due to longitudinal cell expansion (Gendreau et al. 1997) and cell division only occurs during the formation of stomata (Berger et al. 1998). However, the absence of cell division during hypocotyl growth is not general for all species, as cell division is correlated with hypocotyl elongation in *Lupinus* (Nicolas et al. 2001).

The hypocotyl has a simple organization, comparable to that of the root: one endodermal cell layer surrounds the central cylinder; furthermore, there are two cortical layers and one epidermal layer. The epidermis contains two types of longitudinal cells, protruding and non-protruding, of which only the latter develop stomata (Gendreau et al. 1997; Berger et al. 1998). After completion of the extension growth, which takes around 4 weeks, secondary thickening begins: epidermal, cortical and endodermal cell layers gradually disintegrate, while cell divisions in the central cylinder induce the formation of secondary xylem, vascular cambium and secondary phloem (Gendreau et al. 1997).

2 Hormonal Influences in Light and Darkness

The hypocotyl has distinct growth patterns in light and darkness (designated photomorphogenesis and skotomorphogenesis, respectively). In darkness, the hypocotyl of *Arabidopsis thaliana* seedlings elongates up to ten times its size in the light. In addition, the etiolated seedlings form an apical hook (a curvature of the upper part of the hypocotyl), have closed cotyledons (folded along the midrib) and a reduced root growth (Cosgrove 1994; Quail et al. 1995). In the light, the growth of the hypocotyl is reduced, and cotyledons open and expand. Throughout development plants establish the photosynthetic apparatus (Cosgrove 1994; Quail et al. 1995) (Fig. 1).

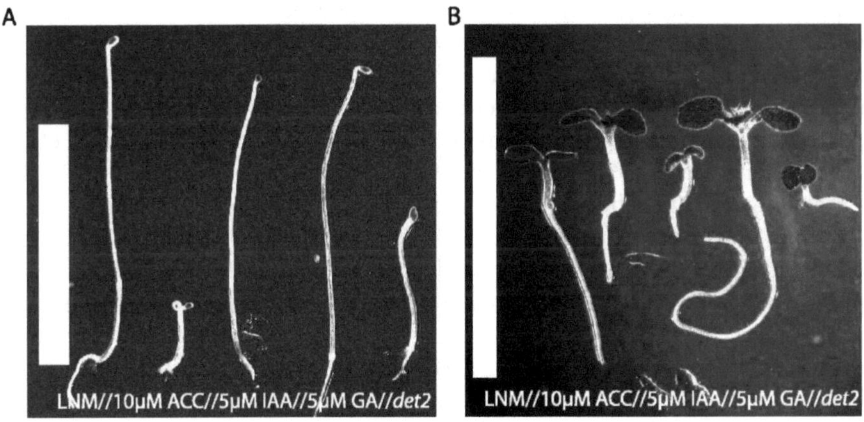

Fig. 1 Representative pictures of the influence of various plant hormones on *Arabidopsis* hypocotyl growth. From *left* to *right*: A wild-type plant grown on low-nutrient medium without exogenously added hormones; a wild-type plant grown on a medium supplemented with 10 μM ACC, 5 μM IAA, 5 μM GA_3 and a mutant deficient in BR synthesis (*det2*). **A** In darkness; **B** in the presence of light. *White bars* represent 1 cm

2.1 Hormonal Interactions During Skotomorphogenesis

The establishment of an elongated hypocotyl in wild-type etiolated *Arabidopsis* seedlings is a process that can be strongly affected by exposure to hormones, a typical example being the triple response as a consequence of ethylene treatment (reviewed by Ecker 1995). In *Arabidopsis*, the triple response is manifested by an inhibition of root and hypocotyl elongation, radial swelling of the hypocotyl and an exaggerated apical hook (Fig. 1; Guzman and Ecker 1990). It was first found as a consequence of treatment with exogenous ethylene (Bleecker et al. 1988). However, other hormones (auxin, BRs, GA and cytokinins) are necessary for the correct development of the apical hook (Lehman et al. 1996; Friml et al. 2002; Alabadí et al. 2004; Achard et al. 2003; Vriezen et al. 2004; Helliwell et al. 2001; De Grauwe et al. 2005).

An asymmetric elongation of cells in the outer compared to the inner cell layers is essential for hook exaggeration. Given the conspicuous effect of ethylene on hook morphology, the influence of this hormone was analysed in much detail. It was shown that accumulation of ACO2, an enzyme active in ethylene biosynthesis, was tightly correlated with the formation of the hook in both wild-type seedlings and in ethylene-treated plants. The differential distribution of this enzyme suggests that the bending process resulted from differential ethylene biosynthesis. Alternatively, hook formation might be a result of altered ethylene sensitivity (Raz and Ecker 1999).

Since auxin is able to induce the expression of ACC synthases (all except ACS7 and ACS9), a regulatory feedback between auxins and ethylene in hook formation was suggested (Yamagami et al. 2003). Although transport of endogenous auxins does not alter hypocotyl length in etiolated seedlings (Jensen et al. 1998), the characterization of several hookless mutants supported the essential role of auxins in hook formation (Fig. 1; Lehman et al. 1996; Celenza et al. 1995; King et al. 1995; Boerjan et al. 1996; Li et al. 2004). An asymmetric distribution of the auxin signal is necessary for this process (Lehman et al. 1996; Friml et al. 2002; De Grauwe et al. 2005). PIN1 auxin efflux carriers position asymmetrically in differential growth, as shown for phototropism, with less PIN1 on the basal side of the cells in the elongating region. They thereby cause differential auxin transport and asymmetric cell elongation (Blakeslee et al. 2004). This asymmetric auxin distribution is found to result from crosstalk between auxin and ethylene. The hookless phenotype of the *hls1* mutant, which is affected in an *N*-acetyltransferase, could be phenocopied by application of auxin transport inhibitors. Since blocking of auxin transport or *hls1* both caused similar effects on auxin primary response genes, it can be concluded that HLS1 is affecting auxin localization (Lehman et al. 1996). While transcription of *HLS1* in etiolated seedlings was induced by ethylene treatment, it was decreased upon light exposure. Moreover, HLS1 was able to promote degradation of the auxin-responsive ARF2/HSS protein, thereby altering the auxin responses. These data suggest that HLS1 could act as a key integrator of the auxin, ethylene and light signalling pathways that control hypocotyl bending (Li et al. 2004). This is consistent with the finding that ethylene could rescue the phenotype of null mutants in NPH4/ARF7, an auxin-responsive transcriptional regulator (Harper et al. 2000). The interaction could be explained by functional redundancy of ARF transcription factors; NPH4/ARF7 is the predominant factor in the absence of ethylene, whereas other ARFs may also be functional in the presence of ethylene, possibly through HLS1 activity (Harper et al. 2000). The unique and overlapping function of related *ARF* gene family members was demonstrated by Okushima et al. (2005) using loss-of-function mutants for all the *ARF* genes. Interestingly, the altered response of the *nph4/ARF7* mutant to unilateral blue light provides an additional link between hormones and light signalling (Sect. 3.3; Stowe-Evans et al. 1998).

More recently, the essential role of GAs in all aspects of skotomorphogenesis was demonstrated. GAs can mediate cell elongation, in some cases in cooperation with ethylene (Raskin and Kende 1984; Saibo et al. 2003) or auxin (Fu and Harberd 2003; Lehman et al. 1996; Saibo et al. 2003). In etiolated wild-type *Arabidopsis* seedlings, hypocotyl elongation requires GA. However, since the response is close to saturation, exogenous GA cannot further stimulate elongation (Fig. 1; Cowling and Harberd 1999). GAs basically work in two ways: they promote skotomorphogenic responses, such as hypocotyl elongation and formation/maintenance of the apical hook, and repress photomorphogenesis in the dark, as evidenced by the regulation of light-induced genes (Alabadí et al. 2004). The requirement of GAs in hook formation and hook exaggeration upon ethylene treatment has been demonstrated. DELLA proteins are destabilized by GA action through degradation by the 26S proteasome, a process that can be enhanced by auxins and counteracted by ethylene (Fu and Harberd 2003; Achard et al. 2003; Vriezen et al. 2004). A model for GA, auxin and ethylene interactions in hook formation was presented by Vriezen et al. (2004). Ethylene can inhibit the GA signal at the level of DELLA proteins stronger on the inner side of the curve than on the outer side due to differential ethylene sensitivity. Alternatively, it is possible that a differential GA sensitivity resulting in differential GA response plays a role in hook exaggeration, as suggested by the pattern of *GASA1:GUS* expression in ACC-treated seedlings (Vriezen et al. 2004).

The role of BRs in hook development is long-standing. BR receptor mutants and mutants such as *cbb1, det2* and *cpd*, which are defective in the synthesis of BRs, have a de-etiolated phenotype when grown in darkness, including a short hypocotyl and the lack of the characteristic hook (Fig. 1). It was recently discovered that disruption of BR synthesis as well as the exogenous application of BRs can alter auxin transport, resulting in the disappearance of the apical hook (De Grauwe et al. 2005). However, BRs cannot stimulate hypocotyl elongation in the wild-type which might point to a saturated response, as is the case with gibberellins (Chory et al. 1991; Kauschmann et al. 1996; Szekeres et al. 1996).

Interaction between GAs and BRs in hypocotyl elongation in *Arabidopsis* has been reported as well (Bouquin et al. 2001). In general, both hormones positively mediate the control of common processes (seed germination and hypocotyl elongation in darkness), with BRs acting downstream of GAs or in a parallel pathway (Alabadí et al. 2004).

Evidence for the involvement of cytokinins in hook exaggeration can be found in the *amp1/hpt/cop2/pt* mutant. Disruption of a carboxypeptidase in the mutant results in a hookless phenotype and leads to an accumulation of cytokinins (Chaudhury et al. 1993; Helliwell et al. 2001). The effect of cytokinin accumulation is comparable to that of ethylene, since application of BA leads to the characteristic triple response which can be blocked by ethylene inhibitors, or since the effect is absent in ethylene-resistant mutants treated with cytokinin (Chory et al. 1994; Cary et al. 1995). Moreover, cytokinins can induce ethylene biosynthesis by stabilizing ACC synthase proteins (ACS5) and consequently reduce hypocotyl elongation in darkness (Clark et al. 1998; Vogel et al. 1998; Chae et al. 2003). In plant species such as soybean, high amounts of jasmonic acid were discovered in the hook of the hypocotyl. However, no such data have been published for *Arabidopsis* until now (Creelman and Mullet 1995).

A model for the interaction between ethylene, auxins and BRs was proposed by De Grauwe et al. (2005). In Fig. 2, Cytokinin and GA were added to

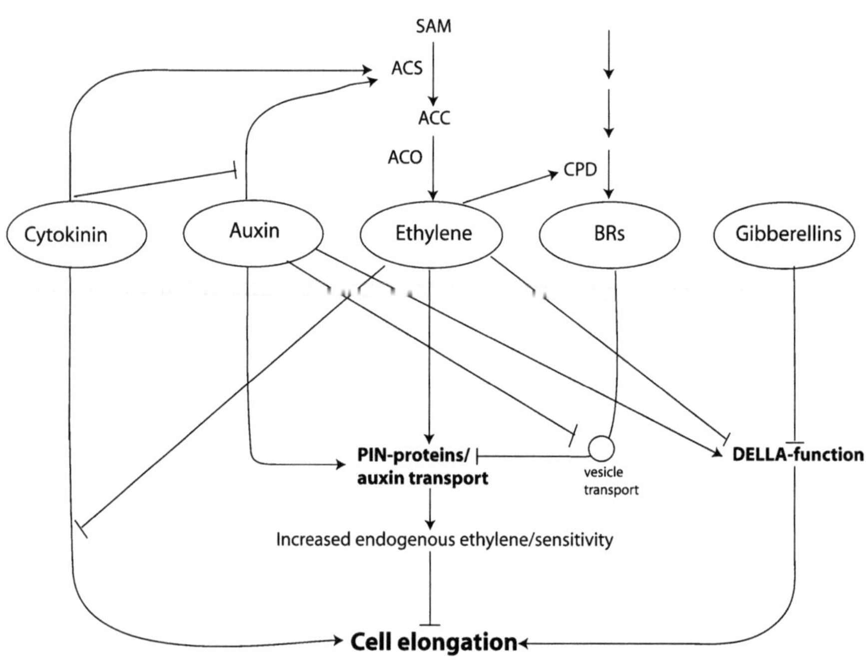

Fig. 2 Model of the interaction between ethylene, auxin, brassinosteroids, cytokinin and gibberellins in the apical hook, based on Geldner et al. 2001, Raz and Ecker 1999, Achard et al. 2003, Vriezen et al. 2004, Asami et al. 2004, De Grauwe et al. 2005 and Smets et al. 2005

form a complete picture of the proposed interactions. The figure shows that endogenous accumulation of ethylene or treatment with exogenous ethylene or its precursor ACC alters auxin distribution, either by directly influencing PIN proteins or through a change in BR biosynthesis. In turn, a differential auxin distribution might be responsible for an altered ethylene production or sensitivity at the inner side of the hook (Raz and Ecker 1999; Vriezen et al. 2004), thereby causing an inhibition of cell elongation and the appearance of an exaggerated apical hook (De Grauwe et al. 2005). It is not clarified yet how cytokinins come into play, but there is evidence that the effect is mediated by an enhanced production of ethylene (Cary et al. 1995; Vogel et al. 1998; Chae et al. 2003). Like auxin, cytokinins stimulate ethylene production at the level of ACC synthases (Mattoo and White 1991), a stimulation that appears more effective in conjunction with auxins (Fuchs and Liebermann 1968; Aharoni et al. 1979). The crosstalk with GA has been explained above.

Apart from the apical hook, an etiolated hypocotyl can be distinguished from a light-grown one by three main characteristics (Gendreau et al. 1997). First, epidermal cells of wild-type dark-grown hypocotyls reach lengths of more than 1 mm, about tenfold the length of light-grown hypocotyl cells. The latter, however, underwent a more pronounced lateral expansion. Longitudinal cell expansion is often related to a transverse orientation of cortical microtubules, together with an anisotropy in the extensibility of the cell wall (Refrégier et al. 2004; Le et al. 2005). This difference between light- and dark-grown seedlings results from the photomorphogenic developmental program and not from an altered photosynthesis. A second characteristic of dark-grown hypocotyls is that they undergo an extra round of endoreduplication (three rounds or 2C, 4C, 8C and 16C in darkness, only two rounds or up to 8C in the light). Application of ethylene, auxin or GA to light-grown seedlings stimulates endoreduplication. The combination of GA with either ACC or IAA provokes the strongest effect, reflected in an increase in 8C:4C ratio and in one extra round of DNA synthesis (Saibo et al. 2003). A third element specific for hypocotyl cells in the dark is that they elongate along a steep acropetal gradient, with the cells at the top reaching the longest final size (Gendreau et al. 1997; Le et al. 2005). This is brought about by a biphasic elongation pattern. At 24 h after imbibition, all cells in the hypocotyl start slow but synchronous elongation. At 48 h after imbibition, cells at the base start fast cell elongation. This zone with fast-elongating cells moves up the hypocotyl, to become restricted in a small area below the apical hook at advanced growth stages (Refrégier et al. 2004). In the light, all cells elongate during the entire growth period, but the area of fastest growth moves up from the base to the middle of the hypocotyl with time (Le et al. 2005). Light exerts the strongest inhibition on cell elongation at the top of the hypocotyl, a region covering the same cells that form the apical hook in darkness (Gendreau et al. 1997; Bleecker et al. 1988; Raz and Ecker 1999; Le et al. 2005).

2.2 Hormonal Interactions During Photomorphogenesis

There are some parallels between hook maintenance/exaggeration in the dark and hypocotyl elongation in the light. First, both depend upon ethylene and auxin (Lehman et al. 1996; Smalle et al. 1997; Vandenbussche et al. 2003a). Second, both have a time window of sensitivity (Raz and Ecker 1999; Saibo et al. 2003). Third, analysis of *hls* mutants shows that their respective products function not only in ethylene-mediated hook maintenance/exaggeration in the dark, but also in ethylene-mediated hypocotyl elongation in the light. Hence, these responses appear to be largely controlled by the same mechanisms, although differences can be observed (De Grauwe et al. 2005) (Fig. 2).

In the light, hypocotyls of *Arabidopsis* seedlings elongate after treatment with ethylene. This effect is most obvious on 2- to 5-day-old seedlings, is more pronounced on low-nutrient medium (LNM) as compared to a rich medium (MS/2), and results from an increase in cell elongation, rather than cell division (Smalle et al. 1997; Saibo et al. 2003). The carboxy terminus of EIN2, a positive regulator in the ethylene-signalling pathway, is sufficient to rescue the phenotype of *ein2-5* mutants in the light, but it can not induce the triple response in darkness, indicating the existence of two partially separated ethylene-signalling networks in *Arabidopsis* (Alonso et al. 1999).

It was reported that auxin transport plays a role in hypocotyl elongation in the light but not in the dark (Jensen et al. 1998). Application of the same concentration of auxin to light-grown plants can be inhibiting as well as stimulating hypocotyl elongation (Jensen et al. 1998; Smalle et al. 1997; Lincoln et al. 1990). It was suggested that the levels of auxin in seedlings are already optimal for elongation, resulting in supra-optimal levels upon addition of auxin to the wild type (Vandenbussche et al. 2003b; Collett et al. 2000). However, the contradiction in response to auxin seen by Jensen et al. (1998) compared to Smalle et al. (1997) can also be due to differences in temperature (Gray et al. 1998; Collett et al. 2000; Sect. 5.1). The interaction between ethylene and auxin is complex, since ethylene probably affects several auxin efflux carriers, mostly stimulating auxin transport (Fig. 2) (Benkova et al. 2003; Li et al. 2004; De Grauwe et al. 2005).

Gibberellins can stimulate hypocotyl elongation in both light and darkness, while exogenous GA only has an additive effect in the light (in darkness, GA levels are close to optimal) (Cowling and Harberd 1999; Saibo et al. 2003). When GAs were applied together with auxins and ethylene, a synergistic effect on hypocotyl elongation was observed, as well as an extra round of endoreduplication. Since the highest DNA content is found in those cells that elongated the most, there might be a direct correlation between higher DNA levels and the ability to elongate (Saibo et al. 2003). Also, a slight increase in cell number was observed and most new cells were related to an increased stomata formation (Saibo et al. 2003). The elevated cell number could not be explained by cell division, but rather by a change in cell specification (Saibo et al. 2003). Since cell fate is not defined at the early stages of embryogenesis (Scheres et al. 1994), it is possible that cells from the root/hypocotyl or from the hypocotyl/cotyledon junction might become hypocotyl cells upon germination (Saibo et al. 2003). Interestingly, the new cells remain in the apical part of the hypocotyl analogous with the formation of the apical hook, where cell division activity has also been demonstrated (Saibo et al. 2003; Vriezen et al. 2004).

Brassinosteroids also interact with GA in light-mediated hypocotyl growth. The *sax1* mutant, defective in a very early step in BR biosynthesis (catalysing the oxidation and isomerization of the steroid precursors), shows an almost normal phenotype in darkness, except that the hypocotyl is a little shorter. Although in the light *sax1* roots are hypersensitive to abscisic acid and auxin, its hypocotyl is insensitive to GA and ACC. Apart from the ACC insensitivity, all hormonal phenotypes can be reversed by addition of EBR (Ephritikhine et al. 1999). In contrast, BRs are able to compensate for the insensitivity to ethylene in light-grown *hls* mutants, suggesting a downstream role for BRs. Application of BRs can still induce hypocotyl elongation, possibly by altering auxin distribution (De Grauwe et al. 2005).

Cytokinins have no obvious effect in the light provided a minimal light intensity ($7\,\mu E\,m^{-2}\,s^{-1}$) is given. This could be explained by saturation of the hypocotyl inhibition response above threshold light levels (Su and Howell 1995). However, cytokinins promote hypocotyl elongation in the light when ethylene action is blocked. At the same time, the level of ACC is strongly increased, indicating that ethylene biosynthesis is up-regulated by treatment with cytokinin in the light as in darkness (Clark et al. 1998; Cary et al. 1995; Vogel et al. 1998; Smets et al. 2005). A similar phenotypic response to cytokinins could be observed when auxin transport is blocked by NPA. This supports the finding that, in the light, cytokinins interact with the ethylene-signalling pathway and conditionally up-regulate ethylene and auxin synthesis (Smets et al. 2005) (Fig. 2).

3 Influences of Light

3.1 Spectral Effects

Photomorphogenesis can be stimulated by different wavelengths (Koornneef et al. 1980; Reed et al. 1994; Nagatani et al. 1993; Parks and Quail 1993; Whitelam and Smith 1993). In *Arabidopsis*, four major classes of photoreceptors have been identified: five phytochromes (PHYA-E) acting predominantly in red/far-red wavelengths, two cryptochromes (CRY1 and CRY2) responding to blue light and UV-A, two phototropins (PHOT1 and PHOT2) only responding to blue light and, finally, the UV-B receptors (reviewed by Quail 2002). Specific receptors for green light have not yet been identified.

Upon exposure to light, even at very low fluence, hypocotyl elongation is inhibited, and cotelydons start to expand and become photosynthetically competent (Frankhauser and Chory 1997; von Arnim and Deng 1996). This very low fluence response (VLFR) is primarily controlled by PHYA, hence de-etiolation is mainly considered to be phytochrome-mediated (Reed et al. 1994). Once the seedling emerges from the soil, PHYA is rapidly degraded and the effects of photostable phytochromes and cryptochromes begin to dominate (Vierstra 1994). Because of its rate-limiting effect, CRY1 seems to be the most important factor in blue-light-mediated de-etiolation in plants (Lin et al. 1998). However, especially under low light intensities, there is also a role for CRY2 in the blue-light regulation of hypocotyl growth (Lin 2000). In addition, a small effect of PHOT1 was noticed at high-intensity broad band blue light, although phototropins are mainly necessary for detecting unilateral or asymmetric light, resulting in phototropic bending (Christie et al. 1998).

Furthermore, it was shown that red light is less effective than far-red or blue light in stimulating hook opening (Liscum et al. 1993). Red light stimulation of hook opening shows reciprocity and is far-red reversible, as is the stimulation of hook opening by low-fluence blue light. High-fluence blue light does not exhibit reciprocity (Liscum et al. 1993). Moreover, the fluence rate-responses for red, far-red and blue light stimulated hook opening are very similar to those generated for the inhibition of hypocotyl elongation in wild-type *Arabidopsis* (Young et al. 1992), pointing to a common signal transduction pathway for both processes. It has been suggested that apical hook opening is controlled by multiple signal transduction pathways, including low-fluence and high-irradiance phytochrome systems and a blue light sensitive high-irradiance photosensory system (Liscum et al. 1993). The effect of green light is different. Folta et al. (2004) showed that, in contrast to other wavelengths, green light can rapidly stimulate hypocotyl growth, a phenomenon that in white light conditions is masked by the inhibition through phytochrome action.

3.2
Mechanism of Light Signalling

Thus, several wavelengths can initiate similar processes (inhibition of hypocotyl elongation, cotyledon expansion and greening). This suggests that different signals are controlled by the same machinery (Fig. 3). The process has been elucidated by the use of constitutively photomorphogenic (COP), de-etiolated (DET) or fusca (FUS) mutants and subsequent cloning of the cor-

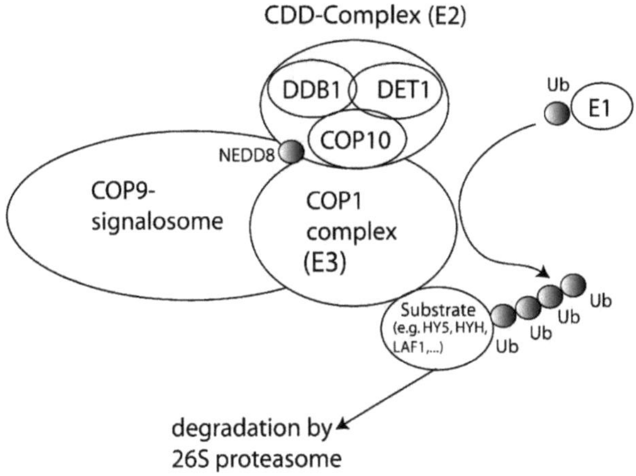

Fig. 3 Model of regulation of the COP/DET/FUS signalosome based on Deshaies 1999, Wei and Deng 1999, Lyapina et al. 2001, Schwechheimer and Deng 2001, Suzuki et al. 2002, Holm et al. 2002 and Yanagawa et al. 2004. The COP9-signalosome (CSN) acts downstream of the phytochromes and promotes neddylation of the cullin-containing E3 ubiquitin ligase (COP1), thereby facilitating its interaction with the CDD (COP10, DET1, DDB1) complex, which has an E2 activity through COP10

responding genes. Recently, it has been shown that the COP/DET/FUS gene products regulate protein degradation of transcription factors involved in light signalling.

Targeted protein degradation by the 26S proteasome is one of the most important means to control transcription, signal transduction, cell cycle progression and metabolic activities (Weissman 2001; Suzuki et al. 2002). For many substrates, becoming tagged by ubiquitin attachment is an important signal for proteasome-mediated degradation. Mostly, three sequential enzymatic steps are involved in ubiquitylation of substrates, catalysed by the following enzymes: the ubiquitin activating enzyme (E1), ubiquitin conjugating enzyme (E2) and the ubiquitin protein ligase (E3) (Deshaies 1999; Weissman 2001).

The COP9-signalosome (CSN) has been identified as a repressor of photomorphogenesis. CSN is a nuclear-enriched protein complex showing architectural and functional homology to the lid subcomplex of the 26S proteasome (Osterlund et al. 1999; Wei and Deng 1999; Peng et al. 2001). The CSN complex is supposed to act downstream of the phytochromes and regulates a light signalling specific E3 complex. It promotes neddylation of the cullin-containing E3 ubiquitin ligase (COP1), thereby facilitating its interaction with the CDD (COP10, DET1, DDB1) complex, which has an E2 activity through COP10 (Deshaies 1999; Wei and Deng 1999; Lyapina et al. 2001; Schwechheimer and Deng 2001; Suzuki et al. 2002; Yanagawa et al. 2004) (Fig. 3).

In darkness, COP1 accumulates in the nucleus and represses photomorphogenic development, partly through regulation of degradation of transcription factors such as HY5 (Holm et al. 2002), which bind to light-responsive elements (LREs) in the promoters of light-responsive genes (Millar and Kay 1996; Terzaghi and Cashmore 1995; Tobin and Kehoe 1994; Chattopadhyay et al. 1998). The amount of transcription factor (e.g. HY5) is limiting in the dark as long as COP1 is localized in the nucleus. Light induces the redistribution of COP1 to the cytoplasm and abrogates its repressor activity on photomorphogenesis. Consequently the transcriptional activators, including HY5, accumulate, resulting in the induction of light-regulated genes (Osterlund et al. 2000).

Interaction of the CSN with other pathways has been reported, e.g. with the SCF^{SLY1} complex, necessary for the degradation of the DELLA proteins in the GA signalling pathway (Fu et al. 2002), as well as with the SCF^{TIR} complex to mediate auxin responses. Thus, light signalling has an effect on a wide range of hormonal and other responses (Schwechheimer et al. 2001).

3.3 Interactions Between Light and Hormonal Effects on Hypocotyl Growth

In wild-type plants, all light signals act together in the regulation of plant development. An example of integration of multiple light and hormonal signals can be found in the shade-avoidance response (Vandenbussche et al. 2004). Canopies selectively remove red (R) and blue light, while far-red (FR) light is still transmitted. High R : FR ratios severely inhibit hypocotyl elongation, and conversely, low R : FR ratios promote elongation. In low R : FR conditions, the PHYB-mediated low fluence response (LFR) and the PHYA-mediated high irradiance response (HIR) counteract one another to regulate plant growth. In addition, PHOT1 detects asymmetric light on the hypocotyl caused by near neighbours or other obstacles and likely alters the distribution of auxin in the hypocotyl, resulting in phototropic bending (Christie et al. 1998; Friml et al. 2002). The induction of several auxin-responsive genes by low R : FR demonstrates a central role for auxin in shade avoidance (Morelli and Ruberti 2002; Devlin et al. 2003; Folta et al. 2003). Ethylene also appears to be involved in this process, as supported by the exaggerated shade-avoidance response of an ethylene-overproducing mutant and the increased ethylene production in low light intensities and low R : FR ratios for wild-type *Arabidopsis* (Vandenbussche et al. 2003).

Interaction between light and hormones at the transcriptome level was revealed by a microarray study of the *cry1* mutant (Folta et al. 2003). Several AUX/IAA genes and GA biosynthesis genes were up-regulated in *cry1* as compared to wild-type. Moreover, treatment of the wild-type with a combination of IAA and GA_4 induced a long hypocotyl phenotype characteristic of *cry1*, whilst the application of either one of these hormones was not sufficient to phenocopy the mutant. Together these data indicate that in blue light, suppression of hypocotyl growth through CRY1 is at least partially mediated by repressing GA and auxin levels and/or sensitivity (Folta et al. 2003). The inferred higher levels of (or sensitivity to) auxin in *cry1* seedlings are also consistent with previous work that linked blue light to changes in auxin transport or sensitivity (Stowe-Evans et al. 2001). Interaction with auxin was confirmed by the response of *nph4/msg1/tir5/ARF7* to unilateral blue light (Sect. 2.1), suggesting that NPH4/ARF7 is not only an auxin-responsive transcription factor, but is also active late in the phototropic signal-response pathway (Stowe-Evans et al. 1998; Harper et al. 2000).

Microarray research further indicated that COP1 is involved in ethylene and BR synthesis (Ma et al. 2002). Additional evidence for the interaction between BRs and light signalling was found in the *bas1-D* mutant, which is BR-hyperresponsive in a light-dependent manner. This mutant suppresses the long hypocotyl phenotype of the *phyB-4* mutant (Neff et al. 1999), suggesting that BAS1 positively modulates hypocotyl photomorphogenesis by inactivating the negative regulation through BR action (Turk et al. 2005).

Direct evidence of crosstalk between auxin- and phytochrome signalling was provided by the interaction of PHYA and phosphorylated Aux/IAA protein in vitro, probably enhancing its stability and/or activity (Colón-Carmona et al. 2000). Since the *phyB* mutant appears to have wild-type levels of endogenous GAs (GA_1, GA_8, GA_{12}, GA_{19}, GA_{20} and GA_{53}) and shows enhanced hypocotyl elongation in response to exogenous GAs, PHYB appears to control GA-dependent hypocotyl elongation by decreasing GA sensitivity rather than by regulating GA biosynthesis (Reed et al. 1996). Additional support for this hypothesis is the fact that *GA4* expression was not altered in the *phyB-1* mutant (Yamaguchi et al. 1996).

4 Circadian Rhythms

Defects in the circadian rhythm can also result in elongated hypocotyls (Dowson-Day and Millar 1999). The circadian clock is an endogenous biological timer that controls a wide range of rhythmic processes, all of which maintain rhythmic periods close to 24 h under constant environmental conditions (Lumsden and Millar 1998; Sweeney 1987). Circadian rhythms in *Arabidopsis* include rhythmic leaf movements (Engelmann et al. 1992), rhythmic opening of stomata (Somers et al. 1998; Webb 1998) and the transcription of a number of genes, e.g. *CAB* (reviewed by Fejes and Nagy 1999). In addition, circadian gating has been shown to control hypocotyl elongation and hook-opening responses, since mutants in components of the central circadian oscillator (*LHY*- and *CCA1*-overexpressor lines) have long hypocotyls (Horwitz and Epel 1978; King et al. 1982; Schaffer et al. 1998; Wang and Tobin 1998). Furthermore, Dowson-Day and Millar (1999) demonstrated that the circadian clock controls the elongation of *Arabidopsis* hypocotyls immediately upon germination. The pattern of hypocotyl elongation in constant light includes a daily growth arrest spanning subjective dawn and an interval of rapid growth at subjective dusk. Maximal hypocotyl growth coincides with the phase during which the cotyledons are raised, in a rhythm similar to that of cotyledon movement. Since the phase of inhibited hypocotyl elongation overlaps with the phase of *CAB* gene activation (Millar and Kay 1996), it has been suggested that at least one component (*TOC1*) is shared with the system controlling *CAB* expression (Dowson-Day and Millar 1999). This component is repressed by the Myb-like transcription factors *CCA1* and *LHY* (Schaffer et al. 1998). Interestingly, the nuclear localization of PHYB appears to follow a circadian fluctuation even after plants are shifted to complete darkness or continuous light, showing the involvement of red light signalling in the control of circadian rhythms (Nagy 2001).

Microarray studies such as the one by Harmer et al. (2005) can further improve our knowledge about the circadian regulation. In this study both activators and repressors of gene expression were found to act through "morning" or "evening" elements, motifs that are over-represented in the promoters of dawn- and evening-phased genes, respectively. The experiments also confirmed that both CCA1 and LHY are acting as repressors through interaction on the evening elements (Harmer et al. 2000, 2005; Alabadí et al. 2001).

Future research needs to clarify the specific role of plant hormones in determining the circadian rate of growth. The best candidates are GAs and auxins. Evidence for the role of GA can be found in the *spy* mutant, which has a long hypocotyl and is defective in input components of the circadian clock (Jacobsen and Olszewski 1993). Auxin levels in *Arabidopsis* stems also show circadian rhythmicity. When plants are transferred to continuous light, auxins peak at subjective dusk, as does the maximal velocity of hypocotyl extension (Downson-Day and Millar 1999; Jouve et al. 1999). Interestingly, transcript levels of genes involved in auxin transport and cell wall loosening culminate at the same moment (Harmer et al. 2000). Since the maximum ethylene production is out of phase with the peak of hypocotyl elongation, ethylene signalling is presumably not affecting the rhythmicity of elongation and does not provide significant input signals to the circadian clock controlling gene expression (Thain et al. 2004). Although no direct effect of cytokinins on the circadian timing could be noticed, mutants in cytokinin signalling (ARR3 or ARR4) had a long-period phenotype, as noticed in *phyB* mutants. Moreover, a direct interaction between ARR4 and PHYB has been demonstrated (Sweere et al. 2001), although there are several indications that this interaction is not the only explanation for the long-period phenotype (Salomé et al. 2006).

5 Other Factors

5.1 Environmental Influences

Since plants are sessile organisms, they must have mechanisms to adapt to several external influences which can cause stress. One such factor is heat. Gray et al. (1998) proved that dramatic hypocotyl elongation occurred when *Arabidopsis* seedlings were grown on a rich medium, in white light at high temperatures (29 °C) compared to plants grown at 20 °C. Dark-grown seedlings were indistinguishable at all tested temperatures. The temperature dependence of hypocotyls in white light was not influenced by mutations in the gibberellin, ethylene or abscisic acid pathways. However, hypocotyl elongation was reduced by mutations in the auxin response or transport pathways and in seedlings containing reduced levels of free IAA. This confirms the contention that endogenous auxin is capable of promoting cell elongation in intact plants with a suboptimal auxin level (Sect. 2.2). Consistent with this, Collett et al. (2000) observed no additional stimulus of exogenous auxin at a relatively high temperature (26 °C) indicating that auxin levels are at their optimum under these conditions. This hormonal involvement was also observed in temperature-induced stem and leaf elongation, which was due to an elevated level of auxins, while the endogenous gibberellin content remained unaltered (Thingnaes et al. 2003).

On the other hand, hypocotyl elongation of 2.5-day-old dark-grown seedlings grown for 60 min at 45 °C was used to screen for altered thermotolerance. The *hot1* mutant, which showed reduced acquired thermotolerance, was identified in this way (Hong and Vierling 2000). Although mutants from the ethylene and abscisic acid signalling pathways showed altered thermotolerance, none of them showed a difference in hypocotyl elongation under these stress situations (Larkindale et al. 2005).

A second stress factor is touch. In general, plants that grow in windy environments or that are exposed to repetitive touch stimulation are shorter, stockier, and often have altered flexibility to enhance their resistance. In contrast, Johnson et al. (1998) showed that vibration can cause hypocotyl elongation in *Arabidopsis* seedlings. Although the mechanostimulus signal transduction pathway is not yet well defined, cytoplasmic calcium seems to play a major role as secondary messenger, transducing the stimulus into an intracellular signal (Braam and Davis 1990; Knight et al. 1991; Haley et al. 1995). While touch stimulation results in an increase in ACC synthase activity and in a rapid increase in ethylene evolution (Goeschl et al. 1966; Poovaiah 1974; Takahashi and Jaffe 1984), touch-induced hypocotyl elongation appears not to be regulated through ethylene (Johnson et al. 1998). Yet, it remains still possible that other hormones are involved in this process.

5.2 Nutrients

The influence of sucrose is demonstrated by the phenotype of transgenic lines with reduced sucrose content in the seeds (antisense *PCK1* (phosphoenolpyruvate carboxykinase) lines), displaying a short hypocotyl in the dark when germinated on media without a carbohydrate supplement (Rylott et al. 2001, 2003). The same is seen for germinating wild-type embryos of which the endosperm, the storage place for sucrose and lipids, has been removed. Both phenotypes can be rescued by addition of exogenously supplied sucrose (Penfield et al. 2004), showing that the reduction in hypocotyl length is a result of the lack of sucrose arriving in the embryo from the endosperm. Moreover, *PCK1*, an enzyme catalysing a fundamental step of the gluconeogenesis pathway in the early post-germinative seedling growth, is also expressed in elongating hypocotyl cells in the dark. Hence, carbon exported from the endosperm contributes significantly to the elongating hypocotyl in the dark and gluconeogenesis is required for skotomorphogenesis, with *PCK1* as a major factor in this process. From these results, it can be concluded that sucrose preserves the hypocotyl elongation until photosynthesis takes over (Penfield et al. 2004).

6 Conclusions and Further Perspectives

Although many aspects remain to be uncovered, it is clear that hypocotyl elongation is controlled by the interaction of external and internal signals, with light and hormones playing a predominant role. Future research on hormonal interactions and cellular expansion using integrative approaches, including transcriptome and proteome analysis, will expand our knowledge of the general regulation of this process. Besides new hormonal interactions, there are certainly additional signalling crossroads to be discovered, which would improve our understanding of cellular elongation in general and hypocotyl extension in particular.

References

Achard P, Vriezen WH, Van Der Straeten D, Harberd NP (2003) Ethylene regulates *Arabidopsis* development via the modulation of DELLA protein growth repressor function. Plant Cell 15:2816–2825

Aharoni N, Anderson JD, Lieberman M (1979) Production and action of ethylene in senescing leaf discs. Effect of indoleacetic acid, kinetin, silver ion, and carbon dioxide. Plant Physiol 64:805–809

Alabadí D, Gil J, Blázquez MA, Garcia-Martínez JL (2004) Gibberellins repress photomorphogenesis in darkness. Plant Physiol 134:1050–1057

Alonso JM, Hirayama T, Roman G, Nourizadey S, Ecker JR (1999) EIN2, a bifunctional transducer of ethylene and stress responses in *Arabidopsis*. Science 284:2148–2152

Ang LH, Chattopadhyay S, Wei N, Oyama T, Okada K, Batschauer A, Deng XW (1998) Molecular interaction between COP1 and HY5 defines a regulatory switch for light control of *Arabidopsis* development. Mol Cell 1:213–222

Asami T, Oh K, Jikumaru Y, Kaneko I, Nakano T, Takatsuto S, Fujika S, Yoshida S (2004) A mammalian steroid action inhibitor spironolactone retards plant growth by inhibition of brassinosteroid action and induces light-induced gene expression in the dark. J Steroid Biochem Mol Biol 91:41–47

Benkova E, Michniewicz M, Sauer M, Teichmann T, Seifertova D, Jurgens G, Friml J (2003) Local, efflux-dependent auxin gradients as a common module for plant organ formation. Cell 115:591–602

Berger F, Linstead P, Dolan L, Haseloff J (1998) Stomata patterning on the hypocotyl of *Arabidopsis thaliana* is controlled by genes involved in the control of root epidermis patterning. Dev Biol 194:226–234

Blakeslee J, Bandyopadhyay A, Peer W, Makam S, Murphy A (2004) Relocalization of the PIN1 auxin facilitator plays a role in phototropic responses. Plant Physiol 134:28–31

Bleecker A, Estelle M, Somerville C, Kende H (1988) Insensitivity to ethylene conferred by a dominant mutation in *Arabidopsis thaliana*. Science 241:1086–1089

Boerjan W, Cervera M, Delarue M, Beeckman T, Dewitte W, Belline C, Caboch M, Van Onckelen H, Van Montagu M, Inzé D (1996) *Superroot*, a recessive mutation in *Arabidopsis*, confers auxin overproduction. Plant Cell 7:1405–1419

Bouquin T, Meier C, Foster R, Nielsen ME, Mundy J (2001) Control of specific gene expression by gibberellin and brassinosteroid. Plant Physiol 127:450–458

Braam J, Davis RW (1990) Rain-, wind-, and touch-induced expression of calmodulin and calmodulin-related genes in *Arabidopsis*. Cell 60:357–364

Cary AJ, Liu W, Howell SH (1995) Cytokinin action is coupled to ethylene in its effects on the inhibition of root and hypocotyl elongation in *Arabidopsis thaliana* seedlings. Plant Physiol 107:1075–1082

Celenza JL, Grisafi PL, Fink GR (1995) A pathway for lateral root formation in *Arabidopsis thaliana*. Gen Dev 9:2131–2142

Chae HS, Faure F, Kieber JJ (2003) The eto1, eto2, and eto3 mutations and cytokinin treatment increase ethylene biosynthesis in *Arabidopsis* by increasing the stability of ACS protein. Plant Cell 15(2):545–559

Chattopadhyay S, Puente P, Deng XW, Wei N (1998) Combinatorial interaction of light-responsive elements play a critical role in determining the response characteristics of light-regulated promoters in *Arabidopsis*. Plant J 15:69–77

Chaudhury AM, Letham S, Craig S, Dennis ES (1993) *amp1*—a mutant with high cytokinin levels and altered embryonic pattern, faster vegetative growth, constitutive photomorphogenesis and precocious flowering. Plant J 4:907–916

Chory J, Nagpal P, Peto CA (1991) Phenotypic and genetic analysis of *det2*, a new mutant that affects light-regulated seedling development in *Arabidopsis*. Plant Cell 3:445–459

Chory J, Reinecke D, Sim S, Washburn T, Brenner M (1994) A role for cytokinins in de-etiolation in *Arabidopsis*. Plant Physiol 104:339–347

Christie JM, Reymond P, Powell GK, Bernasconi P, Raibekas AA, Liscum E, Briggs WR (1998) *Arabidopsis* PNH1: a flavoprotein with the properties of a photoreceptor for phototropism. Science 282:1698–1701

Clark KL, Larsen PB, Wang XX, Chang C (1998) Association of the *Arabidopsis* CTR1 Raf-like kinase with the ETR1 and ERS ethylene receptors. Proc Natl Acad Sci USA 95:5401–5406

Collett CE, Harberd NP, Leyser O (2000) Hormonal interactions in the control of *Arabidopsis* hypocotyl elongation. Plant Physiol 124:553–561

Colón-Carmona A, Chen DL, Yeh KC, Abel S (2000) Aux/IAA proteins are phosphorylated by phytochromes in vitro. Plant Physiol 124:1728–1738

Cosgrove DJ (1994) Photomodulation of growth. In: Kendrick RE, Kronenberg GHM (eds) Photomorphogenesis in plants. Kluwer, Dordrecht, pp 631–658

Cowling RJ, Harberd NP (1999) Gibberellins control *Arabidopsis* hypocotyls growth via regulation of cellular elongation. J Exp Bot 50:1351–1357

Creelman RA, Mullet JE (1995) Jasmonic acid distribution and action in plants: regulation during development and responses to biotic and abiotic stress. Proc Natl Acad Sci USA 92:4114–4119

De Grauwe L, Vandenbussche F, Tietz O, Palme K, Van Der Straeten D (2005) Auxin, ethylene and brassinosteroids: tripartite control of growth in the *Arabidopsis* hypocotyl. Plant Cell Physiol 46:827–836

Deshaies RJ (1999) SCF and Cullin/Ring H2-based ubiquitin ligases. Annu Rev Cell Dev Biol 15:435–467

Devlin PF, Yanovsky MJ, Kay SA (2003) A genomic analysis of the shade avoidance response in *Arabidopsis*. Plant Physiol 133:1617–1629

Dowson-Day MJ, Millar SA (1999) Circadian dysfunction causes aberrant hypocotyl elongation patterns in *Arabidopsis*. Plant J 17:63–71

Ecker JR (1995) The ethylene signal transduction pathway in plants. Science 268:667–675

Engelmann W, Simon K, Phen CH (1992) Leaf movement rhythm in *Arabidopsis thaliana*. Z Naturforsch 47c:925–928

Ephritikhine G, Fellner M, Vannini C, Lapous D, Barbier-Brygoo H (1999) The *sax1* dwarf mutant of *Arabidopsis thaliana* shows altered sensitivity of growth responses to abscisic acid, auxin, gibberellins and ethylene and is partially rescued by exogenous brassinosteroids. Plant J 18:303–314

Fejes E, Nagy F (1998) Molecular analysis of circadian clock-regulated gene expression in plants: features of the "output" pathways. In: Lumsden PJ, Millar AJ (eds) Biological rhythms and photoperiodism in plants. BIOS Scientific, Oxford, pp 99–118

Folta KM (2004) Green light stimulates early stem elongation, antagonizing light-mediated growth inhibition. Plant Physiol 135:1407–1416

Folta KM, Pontin MA, Karlin-Neumann G, Bottini R, Spalding EP (2003) Genomic and physiological studies of early cryptochrome 1 action demonstrates roles for auxin and gibberellin in the control of hypocotyl growth by blue light. Plant J 36:203–214

Frankhauser C, Chory J (1997) Light control of plant development. Annu Rev Cell Dev Biol 13:203–229

Friml J, Wisniewska J, Benkova E, Mendgen K, Palme K (2002) Lateral relocation of auxin efflux regulator PIN3 mediates tropism in *Arabidopsis*. Nature 415:806–809

Fu X, Harberd NP (2003) Auxin promotes *Arabidopsis* root by modulating gibberellin response. Nature 421:740–743

Fu X, Richards DE, Ait-ali T, Hynes LW, Ougham H, Peng J, Harberd NP (2002) Gibberellin-mediated proteasome-dependent degradation of the barley DELLA protein SLN1 repressor. Plant Cell 14:3191–3200

Fuchs Y, Lieberman M (1968) Effects of kinetin, IAA and gibberellin on ethylene production, and their interactions in growth of seedlings. Plant Physiol 43:2029–2036

Geldner N, Friml J, Stierhof Y, Jürgens G, Palme K (2001) Auxin transport inhibitors block PIN1 cycling and vesicle trafficking. Nature 413:425–428

Gendreau E, Traas J, Desnos T, Grandjean O, Caboche M, Höfte H (1997) Cellular basis of hypocotyl growth in *Arabidopsis thaliana*. Plant Physiol 114:295–305

Goeschl JD, Rappaport L, Pratt HK (1966) Ethylene as a factor regulating the growth of pea epicotyls subjected to physical stress. Plant Physiol 41:877–884

Gray WM, Östin A, Sandberg G, Romano CP, Estelle M (1998) High temperature promotes auxin-mediated hypocotyl elongation in *Arabidopsis*. Proc Natl Acad Sci USA 95:7197–7202

Guzman P, Ecker JR (1990) Exploiting the triple response of *Arabidopsis* to identify ethylene-related mutants. Plant Cell 2:513–523

Haley A, Russell AJ, Wood N, Allan AC, Knight M, Campbell AK, Trewavas AJ (1995) Effects of mechanical signalling on plant cell cytosolic calcium. Proc Natl Acad Sci USA 92:4124–4128

Harmer SL, Kay SA (2005) Positive and negative factors confer phase-specific circadian regulation of transcription in *Arabidopsis*. Plant Cell 17:1926–1940

Harmer SL, Hogenesch JB, Straume M, Chang HS, Han B, Zhu T, Wang X, Kreps JA, Kay SA (2000) Orchestrated transcription of key pathways in *Arabidopsis* by the circadian clock. Science 290:2110–2113

Harper RM, Stowe-Evans EL, Luesse DR, Muto H, Tatematsu K, Watahiki MK, Yamamoto K, Liscum E (2000) The *NPH4* locus encodes the auxin response factor ARF7, a conditional regulator of differential growth in aerial *Arabidopsis* tissue. Plant Cell 12:757–770

Helliwell CA, Chin-Atkins AN, Wilson IW, Chapple R, Dennis ES, Chaudhury A (2001) The *Arabidopsis AMP1* gene encodes a putative glutamate carboxypeptidase. Plant Cell 13:2115–2125

Holm M, Ma LG, Qu LJ, Deng XW (2002) Two interacting bZIP proteins are direct targets of COP1-mediated control of light-dependent gene expression in *Arabidopsis*. Gen Dev 16:1247–1259

Hong SW, Vierling E (2000) Mutants of *Arabidopsis thaliana* defective in the acquisition of tolerance to high-temperature stress. Proc Natl Acad Sci USA 97:4392–4397

Horwitz BA, Epel BL (1978) Circadian changes in activity of the far-red form of phytochrome: physiological and in vivo spectrophotometric studies. Plant Sci Lett 13:9–14

Jacobsen SE, Olszewski NE (1993) Mutations in the SPINDLY locus of *Arabidopsis* alter gibberellin signal transduction. Plant Cell 5:887–896

Jensen PJ, Hangarter RP, Estelle M (1998) Auxin transport is required for hypocotyl elongation in light-grown but not dark-grown *Arabidopsis*. Plant Physiol 116:455–462

Johnson KA, Sistrunk ML, Polisensky DH, Braam J (1998) *Arabidopsis thaliana* responses to mechanical stimulation do not require ETR1 and EIN2. Plant Physiol 116:643–649

Jouve L, Gaspar T, Kevers C, Greppin H, Degli Agosti R (1999) Involvement of indole-3-acetic acid in the circadian growth of the first internode of *Arabidopsis*. Planta 209:136–142

Kauschmann A, Jessop A, Koncz C, Szekeres M, Willmitzer L, Altmann T (1996) Genetic evidence for an essential role of brassinosteroids in plant development. Plant J 9:701–713

King JJ, Stimart D, Fisher RH, Bleecker AB (1995) A mutation altering auxin homeostasis and plant morphology in *Arabidopsis*. Plant Cell 7:2023–2037

King RW, Schafer E, Thomas B, Vince-Prue D (1982) Photoperiodism and rhythmic response to light. Plant Cell Environ 5:395–404

Knight MR, Campbell AK, Smith SM, Trewavas AJ (1991) Transgenic plant aequorin reports the effects of touch and cold-shock and elicitors on cytoplasmic calcium. Nature 352:524–526

Koornneef M, Rolff E, Spruitt CJP (1980) Genetic control of light-inhibited hypocotyl elongation in *Arabidopsis thaliana* L. Heynh. Z Pflanzenphysiol 100:147–160

Larkindale J, Hall JD, Knight MR, Vierling E (2005) Heat stress phenotypes of *Arabidopsis* mutants implicate multiple signaling pathways in the acquisition of thermotolerance. Plant Physiol 138:882–897

Le J, Vandenbussche F, De Cnodder T, Van Der Straeten D, Verbelen J-P (2005) Cell elongation and microtubule behavior in the *Arabidopsis* hypocotyl: responses to ethylene and auxin. J Plant Growth Regul 24:166–178

Lehman A, Black R, Ecker JR (1996) *HOOKLESS1*, an ethylene response gene, is required for differential cell elongation in the *Arabidopsis* hypocotyl. Cell 85:183–184

Li H, Johnson P, Stepanova A, Alonso JM, Ecker JR (2004) Convergence of signalling pathways in the control of differential cell growth in *Arabidopsis*. Dev Cell 7:193–204

Lin C (2000) Plant blue-light receptors. Trends Plant Sci 5:337–342

Lin C, Yang H, Guo H, Mockler T, Chen J, Cashmore AR (1998) Enhancement of blue-light sensitivity of *Arabidopsis* seedlings by a blue light receptor cryptochrome 2. Proc Natl Acad Sci USA 95:2686–2690

Lincoln C, Britton JH, Estelle M (1990) Growth and development of the *axr1* mutants in *Arabidopsis*. Plant Cell 2:1071–1080

Liscum E, Hangarter R (1993) Light-stimulated apical hook opening in wild-type *Arabidopsis thaliana* seedlings. Plant Physiol 101:567–572

Lumsden PJ, Millar AJ (eds) (1998) Biological rhythms and photoperiodism in plants. BIOS Scientific, Oxford

Lyapina S, Cope G, Shevchenko A, Serino G, Tsuge T, Zhou C, Wolf DA, Wei N, Shevchenko A, Deshaies RJ (2001) Promotion of NEDD8-CUL1 conjugate cleavage by COP9 signalosome. Science 292:1382–1385

Ma L, Gao Y, Qu L, Chen Z, Li J, Zhao H, Deng XW (2002) Genomic evidence for COP1 as a repressor of light-regulated gene expression and development in *Arabidopsis*. Plant Cell 14:2383–2398

Mattoo AK, White WB (1991) Regulation of ethylene biosynthesis. In: Mattoo AK, Suttle JC (eds) The plant hormone ethylene. CRC, Boca Raton, pp 21–42

Millar AJ, Kay SA (1996) Integration of circadian and phototransduction pathways in the network controlling *CAB* gene transcription in *Arabidopsis*. Proc Natl Acad Sci USA 93:15491–15496

Morelli G, Ruberti I (2002) Light and shade in the photocontrol of *Arabidopsis* growth. Trends Plant Sci 7:399–404

Nagatani A, Reed JW, Chory J (1993) Isolation and initial characterisation of *Arabidopsis* mutants that are deficient in functional phytochrome A. Plant Physiol 102:269–277

Nagy F, Schafer E (2002) Phytochromes control photomorphogenesis by differentially regulated, interacting signalling pathways in higher plants. Annu Rev Plant Physiol Plant Mol Biol 53:329–355

Neff MM, Nguyen SM, Malancharuvil EJ, Fujioka S, Noguchi T, Seto H, Tsubuki M, Honda T, Takatsuto S, Chory J (1999) *BAS1*: a gene regulating brassinosteroid levels and light responsiveness in *Arabidopsis*. Proc Natl Acad Sci USA 96:15316–15323

Nicolas IL, Echeverria MA, Sanchez-Bravo L (2001) Influences of ethylene and Ag^+ on hypocotyl growth in etiolated lupin seedlings. Effects on cell growth and division. Plant Growth Regul 33:95–105

Okushima Y, Overvoorde PJ, Arima K, Alonso JM, Chan A, Chang C, Ecker JR, Hughes B, Lui A, Nguyen D, Onodera C, Quach H, Smith A, Yu G, Theologis A (2005) Functional genomic analysis of the *AUXIN RESPONSE FACTOR* gene family members in *Arabidopsis thaliana*: unique and overlapping functions of *ARF7* and *ARF19*. Plant Cell 17:444–463

Osterlund MT, Hardtke CS, Wei N, Deng XW (2000) Targeted destabilization of HY5 during light regulated development of *Arabidopsis*. Nature 405:462–466

Parks BM, Quail PH (1993) *hy8*, a new class of *Arabidopsis* long hypocotyl mutants deficient in functional phytochrome A. Plant Cell 3:39–48

Peng ZH, Serino G, Deng XW (2001) A role of *Arabidopsis* COP9 signalosome in multifaceted developmental processes revealed by the characterization of its subunit 3. Development 128:4277–4288

Penfield S, Rylott EL, Gilday AD, Graham S, Larson TR, Graham IA (2004) Reserve mobilization in the *Arabidopsis* endosperm fuels hypocotyl elongation in the dark, is independent of abscisic acid, and requires *PHOSPHOENOLPYRUVATE CARBOXYKINASE1*. Plant Cell 16:2705–2718

Poovaiah BW (1974) Promotion of radial growth by 2-chloroethylphosphonic acid in bean. Bot Gaz 135:289–292

Quail PH (2002) Photosensory perception and signalling in plant cells: new paradigms? Curr Opin Plant Biol 14:180–188

Quail PH, Boylan MT, Parks BM, Short TW, Xu Y, Wagner D (1995) Phytochromes: photosensory perception and signal transduction. Science 268:675–680

Raskin I, Kende H (1984) Role of gibberellin in the growth response of submerged deepwater rice. Plant Physiol 76:947–950

Raz V, Ecker J (1999) Regulation of differential growth in the apical hook in *Arabidopsis*. Development 126:3661–3665

Raz V, Koornneef M (2001) Cell division activity during apical hook development. Plant Physiol 125(1):219–26

Reed JW, Nagatani A, Elich T, Fagan M, Chory J (1994) Phytochrome A and phytochrome B have overlapping but distinct functions in *Arabidopsis* development. Plant Physiol 104:1139–1149

Reed JW, Foster KR, Morgan PW, Chory J (1996) Phytochrome B affects responsiveness to gibberellins in *Arabidopsis*. Plant Physiol 112:337–342

Refrégier G, Pellentier S, Jaillard D, Höfte H (2004) Interaction between wall deposition and cell elongation in dark-grown hypocotyl cells in *Arabidopis*. Plant Physiol 135:959–968

Rylott EL, Hooks MA, Graham IA (2001) Co-ordinate regulation of genes involved in storage lipid mobilization in *Arabidopsis thaliana*. Biochem Soc Trans 29:283–287

Rylott EL, Gilday AD, Graham IA (2003) The gluconeogenic enzyme phosphoenolpyruvate carboxykinase in *Arabidopsis* is essential for seedling establishment. Plant Physiol 131:1834–1842

Saibo NJM, Vriezen WH, Beemster GTS, Van Der Straeten D (2003) Growth and stomata development of *Arabidopsis* hypocotyls are controlled by gibberellins and modulated by ethylene and auxins. Plant J 33:989–1000

Salomé PA, To JPC, Kieber JJ, Robertson McClung C (2006) *Arabidopsis* response regulators ARR3 and ARR4 play cytokinin-independent roles in the control of circadian period. Plant Cell 18:55–69

Schaffer R, Ramsay N, Samach A, Corden S, Putterill J, Carre IA, Coupland G (1998) The late elongated hypocotyl mutation of *Arabidopsis* disrupts circadian rhythms and the photoperiodic control of flowering. Cell 93:1219–1229

Scheres B, Wolkenfelt H, Willemsen V, Terlouw M, Lawson E, Dean C, Weisbeek P (1994) Embryonic origin of the *Arabidopsis* primary root and root meristem initials. Development 120:2475–2487

Schwechheimer C, Deng XW (2001) COP9 signalosome revisited: a novel mediator of protein degradation. Mol Cell Biol 18:576–589

Smalle J, Haegman M, Kurepa J, Van Montagu M, Van Der Straeten D (1997) Ethylene can stimulate *Arabidopsis* hypocotyl elongation in the light. Proc Natl Acad Sci USA 94:2756–2761

Smets R, Le J, Prinsen E, Verbelen JP, Van Onckelen HA (2005) Cytokinin-induced hypocotyl elongation in light-grown *Arabidopsis* plants with inhibited ethylene action or indole-3-acetic acid transport. Planta 221:39–47

Somers DE, Webb AAR, Pearson M, Kay SA (1998) The short-period mutant, *toc1-1*, alters circadian clock regulation of multiple outputs throughout development in *Arabidopsis thaliana*. Development 125:485–494

Stowe-Evans EL, Harper RM, Motchoulski AV, Liscum E (1998) NPH4, a conditional modulator of auxin-dependent differential growth responses in *Arabidopsis*. Plant Physiol 118:1265–1275

Stowe-Evans EL, Luesse DR, Liscum E (2001) The enhancement of phototropin-induced phototropic curvature in *Arabidopsis* occurs via a photoreversible phytochrome A-dependent modulation of auxin responsiveness. Plant Physiol 26:826–834

Sweeney BM (1987) Rhythmic phenomena in plants. Academic, San Diego

Sweere U, Eichenberg K, Lohrmann J, Mira-Rodado V, Baurie I, Kudia J, Nagy F, Schafer E, Harter K (2001) Interaction of the response regulator ARR4 with phytochrome B in modulating red light signalling. Science 294:1108–1110

Su W, Howell SH (1995) The effects of cytokinin and light on hypocotyl elongation in *Arabidopsis* seedlings are independent and additive. Plant Physiol 108:1423–1430

Suzuki G, Yanagawa Y, Kwok SF, Matsui M, Deng XW (2002) *Arabidopsis* COP10 is a ubiquitin-conjugating enzyme variant that acts together with COP1 and the COP9 signalosome in repressing photomorphogenesis. Gene Dev 16:554–559

Szekeres M, Nemeth K, Koncz-Kalman Z, Mathur J, Kauschmann A, Altmann T, Redei GP, Nagy F, Schell J, Koncz C (1996) Brassinosteroids rescue the deficiency of CYP90, a cytochrome P450, controlling cell elongation and de-etiolation in *Arabidopsis*. Cell 85:171–182

Takahashi H, Jaffe MJ (1984) Thigmomorphogenesis: the relationship of mechanical perturbation to elicitor-like activity and ethylene production. Physiol Plant 61:405–411

Terzaghi WB, Cashmore AR (1995) Light-regulated transcription. Annu Rev Plant Physiol Plant Mol Biol 46:445–474

Thain SC, Vandenbussche F, Laarhoven LJ, Dowson-Day MJ, Wang ZY, Tobin EM, Harren FJ, Millar AJ, Van Der Straeten D (2004) Circadian rhythms of ethylene emission in *Arabidopsis*. Plant Physiol 136:3751–3761

Thingnaes E, Torre S, Ernstsen A, Moe R (2003) Day and night temperature responses in *Arabidopsis*: effects on gibberellin and auxin content, cell size, morphology and flowering time. Ann Bot 92:601–612

Tobin EM, Kehoe DM (1994) Phytochrome regulated gene expression. Semin Cell Biol 5:335–346

Turk EM, Fujioka S, Seto H, Shimada Y, Takatsuto S, Yoshida S, Wang H, Torres QI, Ward JM, Murthy G, Zhang J, Walker JC, Neff MM (2005) *BAS1* and *SOB7* act redundantly to modulate *Arabidopsis* photomorphogenesis via unique brassinosteroid inactivation mechanisms. Plant J 42:23–34

Vandenbussche F, Vriezen WH, Smalle J, Laarhoven LJ, Harren FJ, Van Der Straeten D (2003a) Ethylene and auxin control the *Arabidopsis* response to decreased light intensity. Plant Physiol 133:517–527
Vandenbussche F, Smalle J, Le J, Saibo NJM, De Paepe A, Chaerle L, Tietz O, Smets R, Laarhoven LJJ, Harren FJM, Van Onckelen H, Palme K, Verbelen JP, Van Der Straeten D (2003b) The *Arabidopsis* mutant *alh1* illustrates a cross talk between ethylene and auxin. Plant Physiol 131:1228–1238
Vandenbussche F, Pierik R, Millenaar FF, Voesenek LA, Van Der Straeten D (2004) Reaching out of the shade. Curr Opin Plant Biol 8:462–468
Vierstra RD (1994) Phytochrome degradation. In: Kendrick RD, Kronenberg GHM (eds) Photomorphogenesis in plants. Kluwer, Dordrecht, pp 141–162
Vogel JP, Woeste KE, Theologis A, Kieber JJ (1998) Recessive and dominant mutations in the ethylene biosynthetic gene ACS5 of *Arabidopsis* confer cytokinin insensitivity and ethylene overproduction, respectively. Proc Natl Acad Sci USA 95:4766–4771
von Arnim A, Deng XW (1996) Light control of seedling development. Annu Rev Plant Physiol Plant Mol Biol 47:215–243
Vriezen WH, Achard P, Harberd N, Van Der Straeten D (2004) Ethylene-mediated enhancement of apical hook formation in etiolated *Arabidopsis thaliana* seedlings is gibberellin dependent. Plant J 37:505–516
Wang ZY, Tobin EM (1998) Constitutive expression of the CIRCADIAN CLOCK ASSOCIATED 1 (CCA1) gene disrupts circadian rhythms and suppresses its own expression. Cell 93:1207–1217
Webb AAR (1998) Stomatal rhythms. In: Lumsden PJ, Millar AJ (eds) Biological rhythms and photoperiodism in plants. BIOS Scientific, Oxford, pp 69–80
Wei N, Deng XW (1999) Making sense of the COP9 signalosome. A regulatory protein complex conserved from *Arabidopsis* to human. Trends Genet 15:98–103
Weissman AM (2001) Themes and variations on ubiquitylation. Nat Rev Mol Cell Biol 2:169–178
Whitelam GC, Smith H (1993) Retention of phytochrome-mediated shade avoidance responses in phytochrome-deficient mutants of *Arabidopsis*, cucumber and tomato. J Plant Physiol 139:119–125
Yamagami T, Tsuchisaka A, Yamada K, Haddon WF, Harden LA, Theologis A (2003) Biochemical diversity among the 1-amino-cyclopropane-1-carboxylate synthase isozymes encoded by the *Arabidopsis* gene family. J Biol Chem 278:49102–49112
Yamaguchi S, Smith MW, Brown RG, Kamiya Y, Sun T (1996) Phytochrome regulation and differential expression of gibberellin 3beta-hydroxylase genes in germinating *Arabidopsis* seeds. Plant Cell 10:2115–2126
Yanagawa Y, Sullivan JA, Komatsu S, Gusmaroli G, Suzuki G, Yin J, Ishibashi T, Saijo Y, Rubio V, Kimura S, Wang J, Deng XW (2004) *Arabidopsis* COP10 forms a complex with DBB1 and DET1 in vivo and enhances the activity of ubiquitin conjugating enzymes. Gen Dev 18:2173–2181
Young JC, Liscum E, Hangarter RP (1992) Spectral dependence of light-inhibited hypocotyl elongation in photomorphogenic mutants of *Arabidopsis*: evidence for a UV-A photosensor. Planta 188:106–114

Subject Index

MIX
Papier aus verantwortungsvollen Quellen
Paper from responsible sources
FSC® C105338

If you have any concerns about our products,
you can contact us on
ProductSafety@springernature.com

In case Publisher is established outside the EU,
the EU authorized representative is:
Springer Nature Customer Service Center GmbH
Europaplatz 3, 69115 Heidelberg, Germany

Printed by Libri Plureos GmbH
in Hamburg, Germany